HEALTH AND DEVELOPMENT

CONCEPTION TO BIRTH

This course is dedicated to the memory of Dr Richard Holmes, a Senior Lecturer in Biology (1971–1993) and former Pro-vice-chancellor for Student Affairs at the Open University. Richard's vision was an inspiration in laying the foundations for *Human Biology and Health*.

SK220 book 1
Science: a second level course

HEALTH AND DEVELOPMENT

CONCEPTION TO BIRTH

HUMAN BIOLOGY AND HEALTH

BOOK 1

Edited by Brian Goodwin

The SK220 Course Team

Course Team Chair
Michael Stewart

Course Manager
Verena Forster

Course Team Secretary
Dawn Partner

Academic Editors
Brian Goodwin (Book 1)
Michael Stewart (Books 2 and 3)
Jill Saffrey (Book 3)
Frederick Toates (Book 4)
Heather McLannahan (Book 5)

Authors
Janet Bunker (Books 1, 2 and 3)
Melanie Clements (Book 3)
Basiro Davey (Books 1 and 2)
Brian Goodwin (Book 1)
Linda Jones (Book 1)
Jeanne Katz (Book 5)
Heather McLannahan (Book 5)
Hilary MacQueen (Books 1 and 4)
Jill Saffrey (Book 3)
Moyra Sidell (Book 5)
Michael Stewart (Book 2)
Margaret Swithenby (Book 1)
Frederick Toates (Books 2, 3 and 4)

Editors
Andrew Bury
Sheila Dunleavy
Sue Glover
Gillian Riley
Margaret Swithenby

Design Group
Sarah Hofton (Designer)
Steve Best (Graphic Artist)
Andrew Whitehead (Graphic Artist)

BBC
Sandra Budin
Rissa de la Paz
Phil Gauron
Ian Thomas
Nick Watson

OU Course Consultant
Chris Inman

External Course Consultant
Bill Tuxworth (University of Birmingham)

External Course Assessor
Professor Jennifer Boore (University of Ulster)

First published 1997. Reprinted 2002

Copyright © 1997, 2000 The Open University.

All rights reserved. No part of this publication may be reproduced, stored in a retrieval system or transmitted in any form or by any means, without written permission from the publisher or a licence from the Copyright Licensing Agency Limited. Details of such licences (for reprographic reproduction) may be obtained from the Copyright Licensing Agency Ltd, 90 Tottenham Court Road, London, W1P 9HE.

Edited, designed and typeset in the United Kingdom by the Open University.

Printed and Bound in Singapore under the supervision of MRM Graphics Ltd, Winslow, Bucks.

ISBN 0 7492 81529

This text forms part of an Open University Second Level Course. If you would like a copy of *Studying with The Open University*, please write to the Course Reservations and Sales Centre, PO Box 724, The Open University, Walton Hall, Milton Keynes, MK7 6ZS. If you have not enrolled on the Course and would like to buy this or other Open University material, please write to Open University Educational Enterprises Ltd, 12 Cofferidge Close, Stony Stratford, Milton Keynes, MK11 1BY, United Kingdom.

1.3

CONTENTS

CHAPTER 1 HUMAN BIOLOGY AND HEALTH: AN INTRODUCTION

During your study of this chapter (at the beginning of Section 1.7) you will be asked to listen to Audio sequence 1, *Holism: the whole truth?*
TV programme 1 is relevant to Section 1.8 and you are advised to view it at the end of that section.

1.1 The philosophy of SK220

This short chapter has a difficult task to fulfil. Its first aim is to put you in a particular frame of mind, which will enable you to get the most out of the course of study you have just embarked upon. It is important that you begin with a clear understanding of what this course is going to be about and the philosophy that has informed its making. A secondary aim is to introduce and explain the meaning of some of the most fundamental terms and concepts that will recur many times in this and subsequent books.

The problem is how to achieve these aims without the benefit of having much content to use as illustrations. Some students of this course will bring to it a sophisticated knowledge of human biology and health, but we cannot assume at this point that all readers have studied any relevant subjects since school, which may have been many years ago. Theory without content can be pretty dry stuff, so we have had to be inventive in seeking ways to illustrate the central themes of the course before it has properly begun. So what *are* those themes? They can be simply stated.

First, this is a course about human *health* rather than human *disease*.

Our emphasis will be on how health is maintained and restored, rather than on how we become ill when things go wrong. Inevitably, there will be many references to disease states scattered throughout the books, but we have set out quite firmly to avoid the traditional syllabus of medical biology, with its focus on disease. We recognize health as more than the absence of disease, but this presents us with the challenge of developing our own definition of health.

Second, we define health as a dynamic *process* not a fixed state.

It is impossible to characterize health in absolute terms, even in the relatively precise realm of biology, because the living world is changing from moment to moment. All living organisms are in a continual internal flux. Biological health is the sum total of a vast and complex number of simultaneous chemical reactions and physiological processes, interacting with a huge range of stimuli from the outside world. Add to this the equally important contributions of human psychology and social relationships, which are also part of what generates human health, and you have some idea of why health must be seen as a dynamic process.

Moreover, it is dynamic in another sense: what health means to you may not be quite the same as it means to someone else, even of a similar age and experience living in the same culture. The concept of health is fluid and culture-bound: in other times and places it can have very different meanings. For example, in contemporary Western medicine the tendency has been to define health as the absence of diagnosable disease, whereas in traditional Chinese medicine health is seen as the balancing of opposing energies in the body. Moreover, your understanding of what it means to be healthy is not the same now as it was when you were a child, and it can be expected to undergo change as each of us grows up and grows old.

Third, we believe that health emerges from the *interaction* of human biology, psychology and social relationships, in a given environment.

The emphasis on interaction can be illustrated by our definition of **environment**. In this course, the term environment must be understood in its broadest sense: not simply the geography, climate and physical surroundings in which a certain human life takes place, but also the social circumstances of that life, within the wider framework of culture and politics, legislation and technology, religion, science, the media and so on. We shall clarify still further what we mean by 'environment' and 'environmental influences' on health later in this chapter.

The interaction of human biology, psychology and social relationships does not simply happen in the here and now – it also has a history which affects health in the present and in the future. The historical dimension that concerns us most is that of **biological evolution**, the gradual changes to the structures and functions of living things which have been occurring from the moment that the earliest life forms appeared on Earth perhaps 3 500–4 000 million years ago. But we also recognize the importance of **cultural evolution**, the gradual changes to the customs, beliefs, values, knowledge and actions of human societies, which have been occuring for about the last million years.

In a course of this limited size, we cannot hope to consider such a huge agenda in detail. As the title of this course clearly advertises, our main focus will be on the *biological* dimensions of health. But it is our aim to teach you about human biology within a psychological and sociological context, which emphasizes the interactive nature of the forces that shape human health.

1.2 Some provisional definitions

We recognize that few students will have had contact with all three of the major disciplines featured in this course, so we had better start by offering definitions of their principal domains of interest. A word of caution: the definitions that follow are necessarily a cryptic 'shorthand' which cannot do justice to the subtleties and richness that professional practitioners of these disciplines understand – but they will have to do as a starting point. By the time this course is over, you will have acquired a much more detailed understanding of their unique characters and the ways in which they overlap and inform each other.

Biology – the discipline we have drawn on most heavily in writing this course – is the study of living organisms, their body structures and functions and their interrelationships. It is the most diverse branch of science, with many subdisciplines, each focusing on a particular aspect either of the inner workings of the bodies of living things, or on their interactions with other organisms and the environment. The biological material in SK220 has come primarily from the subdisciplines of human *physiology* – the study of the structure and functions of body systems, and *biochemistry* – the study of the chemical interactions that occur within the cells and fluids of the body. Biologists often study closely related organisms and make careful extrapolations between them; knowledge of human biology has often been informed by studying other *mammals* such as chimpanzees, monkeys, rats, mice, dogs, cats, etc. which (like us) are warm-blooded 'furry' animals that give birth to their offspring and feed them with their mother's milk.

Sociology is the study of human interactions and social relationships, their organization, functions, development and significance. Sociologists investigate what aspects of the organization and dynamics of human societies and social groups affect human behaviour, and with what outcomes. Sociology is also concerned with the influences on personal and cultural *meaning*, i.e. what beliefs, values, customs, etc. seem to mean to the individuals concerned and more generally within a culture. For example, sociologists have studied belief systems about when it is considered appropriate to consult a doctor, and revealed differences between men and women, between individuals of different ages and social classes, and between different cultures. This course has drawn its sociological material primarily from the subdiscipline called the *sociology of health and illness*, which focuses on social interactions in the context of health care, by both professional health workers and lay people. Sociologists study only humans and do not attempt to extrapolate from observations of other animals.

Psychology occupies something of a middle ground between sociology and biology. Psychologists study the behaviour of humans and their mental states. Humans are observed in either artificial (e.g. laboratory) or natural conditions and their behaviour is then recorded. The condition of a person's mind can either be inferred from his or her behaviour, or by asking people to describe their thoughts and feelings. Psychology has its own theories and language in explaining behaviour, but also calls upon knowledge from the biological and social sciences. For example, in studying feeding, psychologists would consider such things as attitudes to food, cultural practices and early experiences, but would also look to biological science for an understanding of such things as the hormones and brain regions involved in generating hunger. Psychologists commonly also study the behaviour of other animals, principally rats, and attempt to make careful extrapolation to human behaviour and mental processes.

In attempting provisional definitions of these three disciplines, we must take care to reiterate that we cannot expect to do equal justice to them in the space available. Our focus is primarily on the *biological* contribution to human health. But the books and other course materials are unusual in the extent to which they stress the interaction of human biology with psychological and sociological influences. We can expand on this point, briefly, by means of the following exercise.

❑ Suppose a person wakes up bounding with health – fit and well, full of energy yet mentally relaxed, confident about the day ahead, content with the world. The first thought that comes to mind is 'I must be feeling so healthy this morning because ...'. Suggest a number of very different ways that the sentence might end.

■ The range of possible answers is immense, but here are a few ideas. (Feel free to add to the list.)

'I got 8 hours' sleep at last – the baby didn't wake up once.'

'I threw up after drinking all that alcohol!'

'I'm on holiday for two weeks and the sun is shining.'

'Yesterday's pay-rise means I'll soon be out of debt.'

'I always feel good around the middle of my menstrual cycle.'

'Jogging to work each day has done wonders for my energy.'

'I'm so relieved that the blood-test results were normal.'

'I'm in love!'

❑ Now try to decide which (if any) of these explanations for feeling healthy is primarily a biological one, which is psychological, and which sociological.

■ We hope you found this an impossible task. Although some of the explanations include aspects that affect or concern the biological functioning of the body (sleep, alcohol, menstrual cycle, blood-test), there are equally obvious psychological aspects to all of these explanations, and if you think harder you can probably find sociological dimensions too (e.g. factors in society influence social drinking and the ways that women feel about menstruation). Explanations in terms of the holiday and the pay rise might be classified as mainly psychological with sociological implications (because having a holiday implies having a job, financial security, etc., and the debt suggests a problem in the person's social world). But if someone 'feels good' for whatever reason, this must also be manifesting itself as biological changes in the body. Jogging to work conjures up aspects of all three domains, and how would you classify 'being in love'?

We hope this brief exercise has convinced you that all three domains of influence – biological, psychological, sociological – are acting simultaneously in producing the dynamic, continuously changing property we refer to as health. Trying to divide them neatly is a fruitless task. But their interactive nature also poses us with a problem in writing this course.

We cannot describe and discuss all these influences on health simultaneously because a fully integrated interdisciplinary 'language' has not yet been developed. Research on the *interactions* within what is sometimes called the **biopsychosocial world** is in its infancy. This conglomerate term seeks to encompass all the influences on an individual from his or her biological and psychological states, within the context of the prevailing social world. Collaborative work to bridge the gap between biology and psychology on the one hand, and between psychology and sociology on the other, has been growing since the 1950s, but few attempts have been made to keep all three plates in the air at once. This course is an attempt to increase the fluency with which we handle interdisciplinary concepts in the area of human health. However, substantial sections of each of the five books have inevitably been written from within one of the major disciplinary frameworks. The majority of the chapters concentrate on biological aspects of human health, with several on psychological aspects and fewer still with a primarily sociological focus, but wherever possible the connections between the three disciplines will be made transparent.

1.3 Levels of explanation

Attempts at interdisciplinary analysis of complex human phenomena such as 'health' generally have to start with a review of what each of the relevant disciplines can contribute from within its field of expertise. In order to signify that several, equally valuable, disciplinary perspectives coexist, they are often referred to as occupying different **levels of explanation**. In this course, we are focusing on three levels of explanation for health: the biological, psychological and sociological levels. The term is useful because it signals that other levels of explanation exist and merit attention even if, for the moment, the discussion is located entirely from within a single disciplinary level. For example, a complex health problem such as stress is studied within each of the three disciplines represented in this course. If concepts, hypotheses and research findings are presented as emanating from within (say) the sociological level of explanation, this indicates to the reader that other disciplines also have a contribution to make in attempting a complete explanation of the phenomenon.

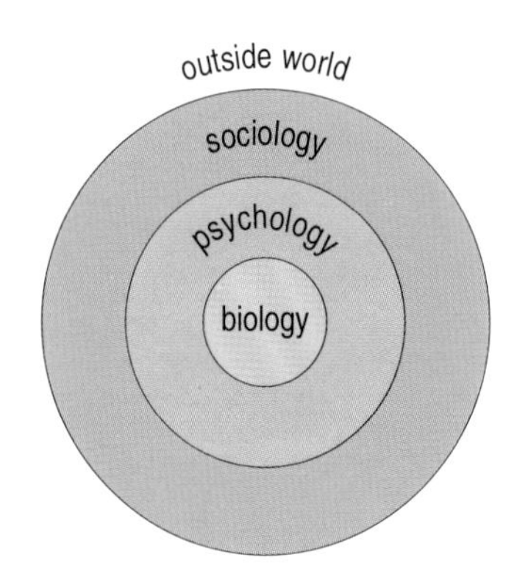

Figure 1.1 Influences on human health can be organized into three major levels of explanation, arranged here in terms of their relative emphasis on factors arising from outside the body.

However, one of the drawbacks of the term 'levels of explanation' is that it can misleadingly suggest a sort of hierarchy, in which the explanations offered by one discipline are superior or inferior to those of others. This false impression is sometimes reinforced by those increasingly rare academics or practitioners who indeed assert that 'their' discipline generates all the really important knowledge.

With this caveat in mind, and without implying any hierarchical relationship, we think it is useful to draw your attention to one aspect that differs between the biological, psychological and sociological levels of explanation for health phenomena: the extent to which they focus on the world outside the physical body. Figure 1.1 represents this schematically in a simple diagram. The biological level of explanation is depicted as the inner circle in Figure 1.1 to signify that, of the three disciplines, biology pays the least attention to influences on health arising from outside the body; the sociological level of explanation occupies the outer ring because sociology places the most emphasis on the external world. This claim demands some justification.

Think first about the biological level of explanation. Some aspects of health are determined by biological factors such as anatomy, the chemical reactions inside cells, the genes a person inherits, etc. on which the outside world can have only a limited impact. For example, your diet, education, income or the security of your childhood upbringing cannot change the colour of your eyes, or enable you to develop an extra pair of hands. The human body has a relatively stable biological structure and organization which is fundamentally similar in all members of the human species. Certainly, there is considerable variation in the details between one individual and the next, but these are variations on a common biological theme which has changed only very slowly over millions of years of evolution.

Even when we consider a single individual as a biological entity, rather than the human species as a whole, a person's individual genetic make-up imposes certain limits on health. (We shall explain what genes are in some detail in Chapter 3, but for now even a vague idea is enough.) The genes a person inherits cannot be exchanged for others – at least not yet! And although their activity may be influenced by stimuli from the outside world – sometimes profoundly – genes impose fundamental biological limits on each individual's capacity to grow and develop, and to respond to disturbances that affect health. This is most obvious when a person inherits genes with altered structures that interfere with normal biological functioning, but it is equally true (though often forgotten) that the state of 'bounding health' we referred to in the exercise earlier also has inherited and relatively stable biological components.

Although we are arguing that the biological level of explanation is less concerned with the external world than is the case for the psychological and sociological levels, the growth and development of the human body and its ability to function, in biological terms, can be profoundly affected by external influences.

❑ Can you suggest some examples?

■ You may have thought of the need to shield pregnant women from X-rays to avoid causing harm to developing babies; the damaging effects of infectious organisms are another example, and the quality and quantity of nutrition has many effects on health.

However, the body appears a relatively stable entity by comparison with the huge capacity for flexibility, variation and change in interactions with the outside world, which is apparent in the sociological level of explanation for health. This is essentially concerned with the impact on health of factors emanating *entirely* from outside the person. Little attention is paid by sociologists to the inner psychological world and none at all to the biological dimension. Humans, as sociological entities, show huge variations across time and distance. Whereas a biological description of a 14th-century English peasant would be much the same as that of an Egyptian pharaoh or an Ethiopian lawyer, their sociological description would be vastly different.

❑ Suggest some examples of external factors arising from social interactions, which seem to you to be important aspects of sociological explanations for health.

■ You may have suggested: relationships within families and other social groups which may influence health; the social customs of the surrounding culture and the ways in which these promote or inhibit health-related behaviours (e.g. peer pressure to smoke); the prevailing economic and political climate and its impact on health factors such as housing, employment and nutrition; and constraints on access to education, health and welfare services.

The psychological level of explanation can be seen as falling somewhere in between the other two, in terms of its emphasis on external factors. The explanations for health that it offers clearly encompass the flexible, changeable mental processes of the individual. Most psychologists would accept that the functioning of the brain is involved in these processes and hence they are subject to certain biological limitations. At the same time, psychological explanations for health acknowledge that mental functioning affects human behaviour, and both are subject to significant influences from the outside world, including those emanating from social relationships and social structures.

1.3.1 Moving between levels

In practice it is rarely (if ever) the case that any investigation or discussion of a health issue encorporates all three levels shown in Figure 1.1. Attempts to do so are often termed **holistic** explanations, because they strive to keep the 'whole' in view, with all its many interacting possibilities. A truly holistic explanation for health would incorporate other levels not included in Figure 1.1, which are beyond the scope of this course.

❐ Can you suggest other levels of explanation that might be considered if a 'complete' holistic account of a health issue was being sought?

■ An important additional level of explanation is concerned with the molecules and atoms from which matter is constructed; yet another would focus on spiritual or religious explanations for health.

Keeping all the possible levels of explanation in view would be a hugely difficult task, requiring sophisticated and specialized knowledge of several fundamentally different areas of study. Not surprisingly, it is usual to find that a given health issue (like issues in many other fields) is investigated and discussed primarily in terms of one level of explanation – for example, the biological – with little more than lip service paid to the others. But even an acknowledgement that other levels exist and have a value is to take an holistic *perspective*.

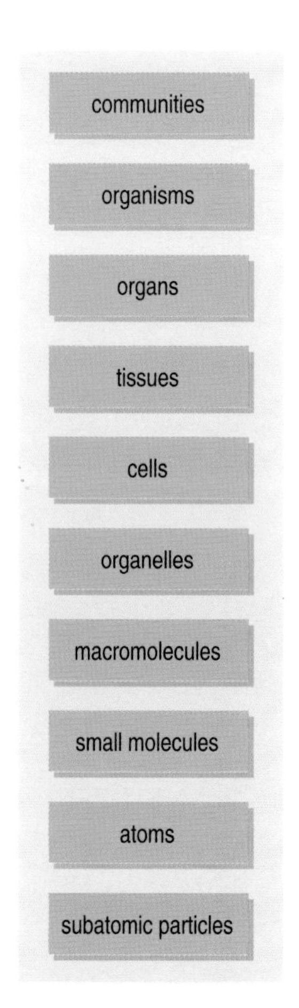

Figure 1.2 Levels of biological organization.

In contrast, an account that denies the value of all but one level of explanation and promotes that level as having all the answers is termed **reductionist**. In effect, reductionist explanations reduce the possible interpretations of a complex phenomenon, such as a health issue, to the domain covered by a single field of knowledge or belief. So, for example, an explanation for health and disease as acts of divine providence is reductionist in the sense that a complete explanation is offered from within a single level – in this case, the religious one. All other apparent levels of explanation are either discarded as irrelevant or reinterpreted as subsidiary ramifications of the 'true' level. In the example we are pursuing here, if evidence of a certain health phenomenon is found at (say) the biological level, then this can be reinterpreted as evidence that divine power controls all the biological manifestations of life itself. This example is worth emphasizing because the term reductionist sounds as though it means 'interprets complex phenomena in terms of their smallest possible constituents' and, indeed, it is often misused in this way; but in reality it means 'interprets complex phenomena in terms of a single, all-embracing level of explanation' which may be as vast as a divine creator.

We shall discuss holistic and reductionist explanations again, later in this chapter. For the moment, you should be clear about the philosophy of this course and its authors, that there is no generally preferred level of explanation. Our task is to identify the most *appropriate* level at which to investigate or discuss a particular phenomenon, given the purpose of that enquiry. So the fact that the greatest number of pages in this and subsequent books contains biological information and debate, should not be misunderstood as an attempt to promote biology as a superior level of explanation to psychology or sociology. Our intention – wherever possible – is to consider all three levels in discussing issues of central importance to human health. Our perspective is holistic, even if we cannot live up to the ideal of providing holistic accounts of all the health phenomena we discuss.

Within each of the major levels of explanation that concern us in SK220, further subdivisions can be distinguished. We shall skim through them quickly to give you some idea of what they involve.

1.4 Levels of biological explanation

The biological level of explanation is traditionally subdivided on the basis of the size and organization of the physical components of living things. The emphasis on physical size as a distinguishing feature means that the levels are usually arranged in a sort of hierarchy, with the largest components at the top and the smallest at the bottom, as in Figure 1.2. All but the topmost level in this hierarchy will be described and discussed in some detail in Chapter 3 of this book, so we will do no more than make a general introduction here.

At the lowest level of this biological hierarchy of explanations, the state of the human body can be described in terms of the activity of its smallest components – the *subatomic particles*, which are combined to form *atoms*. There are many different types of atom, of which the most abundant in living things are oxygen, carbon and hydrogen (together they comprise over 90% of human body weight). Atoms are combined in larger assemblies known as *molecules*, one of which – water – is considered to be a basic ingredient for the evolution of life forms. The discovery of traces of water on distant planets has increased scientific confidence that life may have evolved elsewhere in the Universe.

Some biological molecules contain hundreds of thousands of atoms of several types in complex assemblies. They are termed *macromolecules*; common examples in living things are the proteins, fats and carbohydrates, which are familiar components of foodstuffs. The very largest macromolecules are just about visible with the most powerful microscopes, for example DNA (Figure 1.3). This huge macromolecule contains in its structure, in the form of *genes*, the coded information necessary to direct the assembly of other molecules, and hence it exerts a basic influence on all higher levels of organization in the hierarchy. A particular gene contains the coded instructions to make a specific *protein* chain. You will learn how this occurs in Chapter 3.

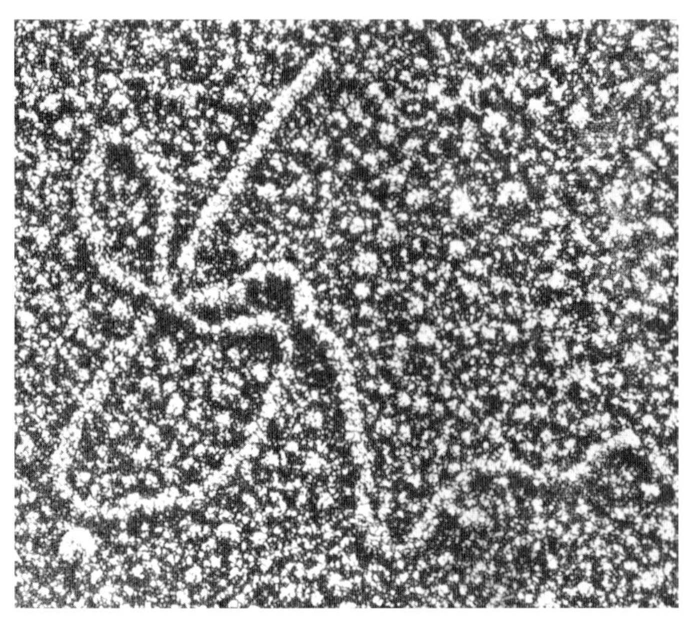

Figure 1.3 A strand of DNA, magnified about 200 000 times.

Cells are the lowest level of biological organization capable of performing all the activities necessary to sustain independent life. Some *organisms* stop here, at a single cell; for example, bacteria are single-celled organisms, and so is the tiny parasite that causes malaria. Each cell has an outer membrane, within which is a complex, organized internal environment consisting of fluid packed with dissolved molecules and subcellular structures called *organelles*, such as the nucleus and the mitochondria (described in Chapter 3). In humans, only the very largest cells are just visible with the naked eye: most are so small that a new-born baby's finger contains more cells (about 10 billion) than there are people in the world. Many different types of cell, each with a distinctive appearance and specialized functions, occur in the bodies of multicellular animals such as ourselves. This complexity and specialization has enabled humans to exploit a far greater range of environments than any other type of multicellular organism.

The bodies of plants and animals are composed of *tissues*: groups of cells with a shared structure and function, associated together as a recognizable sheet, block, matrix or fluid. Examples include the muscular tissue from which our muscles are constructed, the connective tissue which binds everything together, and blood – a fluid tissue. *Organs* such as the brain and heart are specialized structures with distinct boundaries, engaged in particular functions within the body. They are generally composed of several different tissues, organized into very precise structures. Assemble all these components in a certain relationship to one another and the result is a multicellular *organism* – a complex living entity which is capable of feeding, growing, reproducing, responding and so on. Organisms do not live isolated lives: we all inhabit a shared Earth, so we must also consider the interaction of organisms of the same kind (species) living in *communities*, and their interactions with the members of other species in the same environment.

All of these levels are studied by systematic observation and measurement, using specialized technologies such as microscopy to reveal aspects of structure and function that are too small to be seen with the naked eye. Experiments are conducted at all levels to learn more about how the components interact. The experimental biologist intervenes in the system in carefully controlled ways, such that the outcome can be reliably attributed to the intervention.

Even when the organism under investigation is human, biologists generally ignore the psychological and social worlds absolutely. The aim is to understand the underlying fundamental 'laws' which govern either the organization of matter into life, or the organization of living things into communities, and the ways in which these undergo biological evolution. We shall discuss evolution at the end of this chapter, but first we return to the subject of reductionism, which has been a particularly influential force in biology.

1.4.1 Biological determinism and the nature–nurture debate

You will meet examples in this course of biological explanations for health which are framed primarily in terms of one or other of the levels in Figure 1.2, with some reference to those above or below it in the hierarchy. Holistic explanations even within biology are rare, and there has been a long tradition of reductionism, in which explanations at one level (for example, the cellular) are considered less powerful than those at lower levels (biologists tend to lose interest below the molecular level and leave the rest to physicists!). The reason for this is generally rather simple: the scope of biological knowledge is so great that very few biologists consider themselves competent to give a detailed analysis outside the confines of their home discipline. So biochemists focus on interactions at the molecular and macromolecular levels, geneticists are concerned principally with one particular macromolecule (DNA) and the proteins whose manufacture it instructs, physiologists know about what goes on in tissues and organs, and so on.

In itself, this partiality is understandable and need not cause problems, as long as the act of focusing on one level is a *provisional* one, taken in full understanding that all the other levels are equally valid and important. If they are set aside, it is simply because there is just too much material for a single mind to handle all at the same time. The difficulty has been that the reductionist tradition in biology has also led to some abuses of power.

The belief that every aspect of human life – including attributes such as intelligence, personality and artistic talent – can ultimately be explained in terms of human biology, is known as **biological determinism**. Advocates of this view claim that even though the organism expresses its characteristics in its behaviour and in the structure and activity of its cells, tissues and organs, the ultimate explanation for all its characteristics lies in the organism's *genes*. The argument, at its most extreme, states that genes direct a person's outward appearance, inner biological functioning, thoughts, feelings and behaviours as surely as if a puppet-master were pulling the strings.

In the first half of the 20th century, the *eugenics* movement gained supporters not only in Nazi Germany, but in many other parts of the Western world as well. Advocates of eugenics believed that there were genetic differences between races and between individuals, which gave inherent advantages and disadvantages in terms of intellect and other abilities. It was therefore morally right, they claimed, to prevent those with low inherent abilities from passing on their defective genes to future generations, and equally defensible for the master races to assume authority over the rest.

In the 1990s, we can hear the echoes of this philosophy in the claim that by the next century each of us will be carrying a 'smart card' imprinted with our personal genetic details. By implication, those found to have the 'healthiest' collection of genes could be awarded advantages in employment, insurance and possibly marriage or parenthood, which would be denied to, or at least made more expensive for, the rest. You will have already encountered examples of biological determinism in the media, as scientists announce the discovery of genes which they claim (variously) to be responsible for criminal behaviour, or homosexuality, or mathematical excellence, to name but a few. No doubt, there must be a genetic contribution to all of these attributes, but this is a far cry from asserting that genes alone are *responsible* for them.

Reducing human health to the activity of human genes ignores the fact that the surrounding environment in which those genes are operating has a profound influence on the ways in which their effects are manifested. For example, *resistance* to many complex diseases has a genetic component which acts in concert with environmental factors to promote health. This is simply the corollory of the more frequently stated argument that *susceptibility* to many complex diseases has a genetic component which acts in concert with environmental factors to promote disease. The focus on genes has been at the expense of attempts to study the whole organism, undoubtedly in part because this is intrinsically difficult, but also because some biologists have argued that it was simply not a worth-while enterprise. For example, Jim Watson – one of the scientists who worked out the structure of DNA in the early 1950s – is reported to have said that 'everything else is merely social work'. As a consequence of this single-minded attention to one level of explanation (the genetic), influences on complex phenomena falling outside that field of study are ignored and interactions between different influences remain invisible. Moreover, attacks on the reputation of other disciplines can damage their ability to compete for research funds and students.

The interaction of genes and environment is at the heart of the so-called **nature–nurture debate**, an often badly formulated and thereby largely fruitless argument about 'how much' of a particular human characteristic is due to genetic inheritance (nature) and 'how much' to the influence of environmental factors (nurture). For example, how much of a person's intelligence is due to his or her genes and how much to nutrition, health and education during childhood? However, there are serious questions about the interaction of genes and the environment on growth and development, which have been investigated principally by studying identical twins who were separated at birth and brought up by different families: their genes are identical but they are raised in different environments. If they turn out to have different characteristics, then (in theory) these are likely to be due to differences in their environments.

There may be some useful purpose in determining the relative contributions of genes and environment to human growth and development, but the nature–nurture debate has been conducted mainly on ideological rather than practical grounds. The discovery of a number of diseases (e.g. cystic fibrosis) in which the inheritance of a *single* gene can profoundly damage a person's health, has lent support to arguments that *every* aspect of biological functioning will also turn out to be determined by genes, with little input from the environment. The opposing view has also been fiercely argued: that humans are products of their upbringing, social circumstances, education and access to resources, rather than the slaves of their genes.

Most biologists would now agree that nature and nurture interact. An often-quoted example demonstrates this point. Manipulation of the environment can prevent disease from developing in some individuals who have inherited a *lethal gene defect* – a faulty gene which causes fatal damage if its effects are not counteracted. Babies born with a faulty gene that prevents them from breaking down a common constituent of the diet (called phenylalanine), suffer brain damage as this substance builds up in the bloodstream (a condition known as phenylketonuria, or PKU). However, if the diet is controlled to exclude excess phenylalanine, then the damage cannot occur: changing the environment counteracts the effects of the gene.

The example of PKU is instructive in another sense. The very power of reductionist science is its ability to home in on the critical details of molecules, cells, tissues and organs, and to reveal their inner workings – sometimes with breathtaking precision. The fact that PKU can be successfully treated by dietary manipulation rests on knowledge of a single gene among 100 000 others in human DNA, and its effect on a single molecule, phenylalanine, among the multitude of different kinds of molecule in the human body. This is reductionism in all its glory. By setting aside all distractions and focusing research on the smallest components of life, a wealth of information about human biology has been gathered – much of which has been put to good use. Reductionism in itself is not a bad thing. The problems that flow from it are the consequence of elevating biological science to the status of supreme truth.

1.5 Levels of psychological explanation

Psychology does not represent a unified body of knowledge to which all psychologists can subscribe. Rather, psychology has been torn by factional in-fighting and disputes concerning the very nature of what a science of psychology should be all about, or even whether psychology should consider itself a science at all. Disputes often arise because psychologists cannot agree on what is the nature of a human, and therefore what is the appropriate method to employ to study humans. The tension between reductionism and holism is central to this debate.

Consider, for example, the perspective of a school of thought termed *humanistic* psychology. The central assumption of this branch of psychology is that humans are free agents who can exert choice and agency in the world. The possession of consciousness – the capacity to self-reflect upon their own conscious minds – means that there is something fundamental about human beings that cannot be reduced to anything simpler. Only by treating humans as responsible agents with a unique insight into their own behaviour and thought processes can we hope to gain some understanding of them. To humanistic psychologists, to treat humans as machines or as something like rather clever rats running in urban mazes, is to lose the essence of what it means to be human. Therefore, although the humanistic psychologist would not deny that biology can give useful insights into how brains work, he or she would insist that we never lose sight of the fundamental nature of humans who are essentially holistic and cannot be reduced to the sum of any of their component parts.

Other psychologists of various types adopt a variety of reductionist approaches and thereby would attach their flags to the mast of conventional science. For example, *behaviourist* psychologists see humans as the product of their environment. They are shaped by it rather as the potter shapes a vase. Behaviour is the sum total of a history of environmental influences impressed upon the individual in one way or another. To this school of thought, to understand how rats come to learn their way through mazes is to give insight into how a human negotiates his or her decisions of life. Behaviourist psychologists believe that humans can be reduced to the sum total of their bits and understood mechanistically by analogy with rats. They play down the role of consciousness.

Whatever one might think of the more extreme claims of behaviourism, there is no doubt that it has a useful application to health as understood in a broad context. A careful look at the behaviour of people can often reveal that maladaptive behaviour is maintained by its consequences. As an example to illustrate the principle, children who are disruptive in class might be deriving some kind of behaviour-strengthening effect from the attention that comes to them from teacher and classmates. A behaviourist psychologist would look carefully at the classroom situation and would suggest modifications. For example, it could be recommended that the child might be given attention for good behaviour whereas bad behaviour should be ignored. In some cases, schizophrenic patients have been persuaded to come out of their isolation by a careful application of reward to any behaviour that shows signs of social responsiveness. The advocates of applying such behavioural principles would suggest that they can be used to encourage people to change to a healthier lifestyle, e.g. by participating in exercise programmes or by giving up risky behaviours such as smoking.

Psychologists of even more reductionist persuasions see biology as the answer to understanding their subject matter. The psychologist of this orientation is one who is looking over his or her shoulder to see what is happening in brain research and then using that biological knowledge to gain insight into human psychology. Some psychologists believe that humans are *nothing but* the collection of cells that make up their bodies, albeit connected together in interesting ways. Therefore, they would assert, the more understanding we get of the activity of cells in brains and other nervous tissue, the nearer we are to our goal of describing and explaining human behaviour.

Most psychologists do not take such an extreme reductionist approach to biology but rather look to biological science for insights that can be applied in a psychological context. So, for example, a researcher looking into the psychology of eating might wish to know about the chemicals in the brain that have a particular role in stimulating or inhibiting appetite. However, the psychologist would interpret this in the broad context of other factors that also play a role in appetite such as the availability of attractive foods, the person's self image and the comfort value of eating.

Sigmund Freud has had another, quite different influence on psychology. He emphasized the power of the unconscious mind as a determinant of our behaviour. Thus, in a sense, the *Freudian* psychologists are reductionists. They tend to see our behaviour as the product of deep-seated unconscious desires; the role of the psychoanalyst is that of trying to gain insight into these unconscious processes. Our early childhood is a determining experience for our subsequent life. The analyst is the expert who will interpret things. Even an apparently harmless slip of the tongue is labelled a 'Freudian slip' and is the kind of stock-in-trade of the analyst who probes deep-seated motivations and unconscious wishes.

❐ Contrast the humanistic psychologist with the others you have met so far.

■ The humanistic psychologist sees the whole individual as unique and with insight into his or her own world and problems. People cannot be reduced to a collection of desires, slips and inhibitions, nor to a collection of nerve cells, nor to a series of responses like a rat running a maze. To the humanist, the unique individual is the expert on his or her own existence. To the others, the scientist or therapist is the expert who is using a 'tool-kit' of specialist techniques in order to try to understand the subject matter of the person. These techniques inevitably mean simplifying the complex whole.

Some psychologists concentrate on individuals, and others – the *social* psychologists – look at humans in groups. The social psychologist would emphasize that when we look at humans in groups, new properties appear which cannot be reduced to events within a particular individual. For example, individuals will either comply with, or rebel against, what they perceive to be group norms. A crowd can become violent as in soccer hooliganism, but it is not necessarily the case that a bit of that hooliganism resides in the head of each hooligan. Properties *emerge* as a function of being in a group and new intellectual tools are needed for understanding groups, over and above the techniques required to investigate individual humans.

1.6 Levels of sociological explanation

Sociology began as an academic discipline in the 19th century with its roots firmly in the scientific tradition of observation and experiment (its founder, August Comte, originally called it 'social physics'). It was based on the premise that certain universal principles were driving human social organization – an analogy with the presumption in biology that fundamentally stable processes are directing the organization of cells into organisms. If human behaviour and interactions were observed systematically, then it was believed that the 'invariable laws' governing the social world could be derived and understood. Such knowledge would enable the creation of a rational society, based on cooperation. This tradition within sociology is known as *positivism,* and although its adherents have long since disappeared, its echoes linger on in sociological schools of thought that pay little or no attention to the inner meaning of social relationships and interactions. The argument in its purest form is that since we cannot directly observe or measure the inner meanings of human actions, any more than we can do so for non-human animals, then we should simply ignore them.

❐ Does positivism seem closest to a holistic or a reductionist way of thinking about human social interactions?

■ It is essentially a reductionist approach to explanation, since only the directly observable level of human interactions is considered important.

Health as an area of social behaviour has been of increasing interest to sociologists since about the middle of the 20th century. Sociologists want to know why people respond in certain ways to health issues and what factors in the social world influence their responses. The 19th-century emphasis on systematic observation remains central, but most sociologists are now very interested in what social interactions *mean* to the people who participate in them.

One branch of sociology concerned with investigating health matters is known as *ethnography*. It derives its methods as well as its name from the study of primitive tribal societies (*ethno* refers to groups of people), whose actions were observed and written down (*graphy*), before being interpreted to reveal their underlying meaning and significance for the individuals concerned. When this technique was applied to the observation of interactions in health-care settings, such as hospitals and consulting rooms, the ethnographers struck a rich vein of material. In order to get a more subtle and complete understanding of what is going on in such interactions, sociologists in this tradition sometimes become participants in the interaction themselves (participant observers).

Ethnographic research has also been used to gain insight into *lay health beliefs* – the inner world of meaning and action in relation to health which lay people reveal. These beliefs are often contrasted to those of health professionals and others with an expert interest, such as biologists and psychologists. But sociologists are not particularly interested in generating explanations at the level of the *individual*, except insofar as each person's experience sheds light on a more fundamental knowledge of human social behaviour. For example, in a study of the interactions and speech of 50 mothers with new babies meeting their health visitors for the first time, the aim is usually to elucidate meanings from this sample which can lead to generalizable insights into the mother–health visitor relationship.

Observation of naturally occurring interactions is often supplemented by carefully constructed interviews with the participants, usually recorded on audiotape, sometimes on videotape, for later analysis. The interview is structured in such a way as to allow comparisons between respondents to be drawn afterwards. The interviewees can be encouraged to reflect on what was happening in the social interaction, how it felt, what it meant, and so forth. This technique is called *qualitative* interviewing, partly because it generates information about the qualities of the relationship under observation, and partly to distinguish it from *quantitative* sociological methods, which generate numerical, statistical data.

Qualitative sociologists focus primarily on understanding the inner meaning of interactions, thoughts, feelings and experiences, within a certain social setting. The age, social status, education and ethnic group of the participants in any interaction would usually be considered as important influences, whereas the psychological and biological dimensions are generally ignored. Qualitative social research has sometimes been criticized for paying too little attention to the wider political and economic aspects of social interactions. For example, the relationship between new mothers and health visitors may be influenced by the organization of local services and the budgets allocated to health visiting. Observing and interpreting human social interactions is time-consuming work and is therefore usually based on relatively small samples of intensively studied individuals.

By contrast, quantitative sociologists have focused on much larger, often representative, samples of people drawn at random from defined populations and groups. Quantitative sociologists use tools such as postal questionnaires and formal interviews which follow a predetermined series of questions with relatively constrained choices of answer, to allow easy conversion of the data into numerical tables. This type of sociological investigation provides explanations at the level of the group, and makes estimates of how prevalent a particular viewpoint, behaviour or experience is likely to be in the population from which the group was selected. For example, it might ask 'What do lay people of different ages, gender, social circumstances or ethnic group, believe keeps them healthy? What actions do they take to promote health, or restore it when they feel ill?' Age, gender, social circumstance and ethnicity are the four traditional categorizations singled-out by sociologists when seeking to understand the underlying meaning of social interactions. Interpretations of the social world frequently make reference to the importance of these factors.

There is another school of thought in sociology – termed *social constructionism* – which has been extremely influential in the area of health and illness. It is based on the premise that we cannot understand any phenomenon except from within the culture in which we exist. Since that culture is shaping our experience of that phenomenon, we cannot stand back and observe and interpret what is going on in any objective sense. To go back to the example of lay health beliefs, many sociologists would say that they have been 'socially constructed' and so too have our interpretations of what they mean.

The social constructionists argue that we cannot view the body and its internal workings except from within the perspective imposed on us by our culture. Modern Western culture has so trained us to view the body as a machine formed from molecules, cells, tissues and organs, that we believe this to be an objectively true description in some absolute sense. Yet if we had lived in another culture, medieval Britain for example, we would have a very different understanding of the body: a vessel in which the four humours (blood, phlegm, yellow and black bile) intermingle under the influence of intrinsic qualities (hot, cold, wet, dry) and elements (earth, air, fire and water). The meanings we would have attributed to bodily experiences in the Middle Ages would be consistent with that vision. To a social constructionist, to say that other visions, other ways of 'constructing' the body, are *wrong* is to commit the commonest sin of reductionism – placing one level of explanation in supreme authority over all others.

The social constructionists have had a key role in promoting a more holistic framework within sociology, in which all levels of explanation are (in theory at least) supposed to be accorded equal importance. A powerful advantage of holistic approaches to any complex phenomenon is that knowledge generated by one discipline can inform and enhance the investigations carried out in another field of study, and reveal the interactions between them. Curiously enough, one of the leading figures in social constructionism – the English sociologist, David Armstrong – has also been a strong critic of the extent to which holism as a philosophy has begun to claim the moral high ground. Armstrong (1986) argues that it has become almost heretical to point out the limitations of holistic approaches to understanding human societies and social relations, particularly in the area of health and health care. Holistic approaches in any area of study have several weaknesses: for example, it can be hard to get anything done if all possible levels of investigation have to be included; potentially valuable single-discipline research can be underrated; and, since no-one has full command of all the knowledge and debates within every discipline, apparently holistic explanations often turn out to be superficial. If the holistic view of human health assumes the status of 'supreme truth' it is in danger of becoming reductionist!

1.7 Holism and reductionism in health care

The strengths and weaknesses of holistic and reductionist approaches to explanation can be illustrated by reference to health care.

1.7.1 Holistic approaches to health care

If we were able to give a fully integrated account of the many and varied biological, psychological and sociological influences on human health, it could properly be termed holistic.

What does the term holism mean to you? Jot down a few words or phrases that come to mind. Then listen to Audio sequence 1, *Holism: the whole truth?*

As the tape clearly illustrates, definitions of holism are as numerous and diverse as the people who gave them. However, what they have in common could be rather crudely summed up in the phrase 'the whole is more than the sum of its parts'.

The words holism and health both have their roots in the concept of wholeness: health comes from the Anglo-Saxon and holism from the Greek words for whole. The two concepts were indivisible in medieval European culture, where sickness was understood to be a disorder of the whole person in disharmony with inner and outer forces, and the return to health relied on restoring the balance between opposites. The rise of scientific medicine from the 18th century onwards drove out this view, replacing it with a model of health as the 'absence of disease'. Holistic concepts of health did not resurface in Western culture until about the

1950s, when the term 'biopsychosocial' began to be used in the context of the mental-health movement. Psychiatry was criticized for its focus on disease and reliance on drugs and other medical interventions, while ignoring the social, psychological and behavioural dimensions of the patient's experience and rehabilitation. In 1958, the World Health Organization (WHO) developed an holistic definition of health, as 'a state of complete physical, mental and social wellbeing' (WHO, 1958).

In Chapter 2, you will learn about other definitions of health, including the 'health field' concept promoted in Canada in the 1970s by Marc Lalonde, then Minister for Health and Welfare. This was holistic in its philosophy and sought to promote health by achieving balanced interventions in four areas: human biology, the environment, personal lifestyles, and health care organization.

The idea of considering the 'whole person' began to appear in nursing models of care far sooner than it did in medicine, with its emphasis on the 'doctor-centred' model of treating the disease rather than the patient. Nurses are now routinely trained to take account of the patient's psychological and social worlds, as well as their biological state. The concept of nurses aiding patients in caring for themselves (self-care) is playing an increasingly important part in how nurses see their role. Empowering the patient to self-care can be seen as liberating the patient from professional dominance. Alternatively, cynics claim that it is a way of running a health service more cheaply and with fewer trained staff, expecting patients and their friends and family to undertake most of the caring.

The holistic approach has been further strengthened in nursing by the inclusion of some alternative or complementary therapies in the range of available treatment options. These therapies are 'alternative' to orthodox medicine because they respond to the *person* who is seeking help to recover or maintain health, and engage that person as an active agent in the process. However, there are legitimate doubts about the extent to which these therapies are effective in treating or preventing illness and promoting well-being. This is at least partly because the reductionist scientific methods used to evaluate conventional therapies are not well suited to investigate the 'personalized' outcomes that alternative therapies claim to achieve. In recent years, collaborative research between alternative and conventional practitioners has been steadily increasing, as Chapter 3 briefly discusses. A drawback for some users of alternative therapies is that taking personal responsibility for becoming ill is part and parcel of taking responsibility for getting well again. Holistic approaches to cancer treatment have been highly praised by many patients, but some have suffered the added distress of self-doubt and personal failure if they could not 'conquer' the disease.

1.7.2 Reductionist approaches to health care

Reductionism has been the dominant influence on Western medical thought for the past three centuries, and it has become fashionable for advocates of holistic health care to view this as malign. Certainly it has led to the person being 'reduced to' the disease, with no greater responsibility than to transport the symptoms to a doctor for diagnosis and treatment, and then to comply faithfully with instructions. People who delay seeking medical help tend to be chastized, as though the disease 'belongs' to the professionals. Patients have sometimes complained that they were expected to be passive recipients of medical care and accept that 'doctor knows best', but it is worth noting that confidence in orthodox medicine is justifiably widespread. If its successes have often been overplayed, they are nonetheless considerable.

But the point of this chapter is not to rehearse the arguments about whether, and if so to what extent, Western medicine has been a success. The aim here is to reveal its reductionist credentials and point to the consequences. The basis of modern medicine is the scientific study of the human body, so disease has come to be understood as disordered functioning of specific molecules, cells, tissues or organs. The medical task is to identify the malfunction and put it right. The most disputed disease categories are those (like schizophrenia and post-viral fatigue syndrome) where no *consistent* physical damage or malfunction has been detected with the technology currently available, that reliably distinguishes these conditions from other related disorders. If the disease cannot be reduced to a precise description of this or that cellular or molecular derangement, then uncertainty about its legitimacy cannot be banished. This philosophy has led to considerable frustration and misery for people who feel ill but cannot 'prove' their illness exists, according to the criteria set by the reductionist nature of medical science.

Reductionism in medicine has also meant that the outcomes of orthodox medical interventions are categorized as either 'central' effects or 'side' effects – a view that tends to undervalue the importance of side-effects for *patients* as long as the central effect is to counteract the *disease*. For example, the surgical treatment of breast cancer by mastectomy was conducted for many years without much consideration given to its psychological side-effects on women. However, as we pointed out earlier, the power of reductionist science has enabled a huge range of useful drugs, vaccines and surgical interventions to be developed, and there is little serious argument that they should be abandoned in favour of alternative therapies which claim to be holistic. The movement in modern health care seems to be towards a greater partnership between reductionist and holistic approaches.

1.7.3 Developing a fruitful dialogue

We should, by now, have dispelled the false notion that holism is 'good' and reductionism is 'bad' by noting the strengths and weaknesses of each approach. It is our intention in this book – and in the course as a whole – to foster a dialogue between these two approaches, while using each of them to their best advantage. One of our aims is to encourage you to travel more easily between these often falsely opposed poles.

The discussion of holism in health care in late 20th-century industrialized countries can be seen as a rediscovery of the *variety* of ways of looking at, and caring for, health. In place of the medical model, a collaborative multidisciplinary team approach is increasingly being advocated, in which the patient is a key team member along with a range of professionals. However, the coordination and resources necessary to bring all these agencies together and the educational task in teaching them to speak a common language presents a huge challenge. But by studying human health as a multidimensional and dynamic phenomenon, it is hoped that new understanding will emerge and new methods of tackling health problems will be developed. It is our aim in this course to foster that development.

1.8 Evolution and human health

It may seem as though we are now heading off into a completely new topic, but there are two reasons for bringing evolution into the discussion at this point. First, it is an area of undisputed importance for human health, which has nonetheless provoked bitter controversy between holistic and reductionist interpretations of evolution that continue to the present day. So in one sense it is an instructive 'case study' of what we have been discussing in previous sections. Second, it is also essential for you to understand something of the processes of evolution, so it is useful to get an overview of evolutionary change before you meet some of the details in later chapters, and in later books.

1.8.1 Biological and cultural evolution

At the beginning of this chapter we clarified the distinction between *biological evolution* and *cultural evolution*.

❐ How do these two processes differ in what they transmit to future generations and in the mechanisms they use?

■ Biologicial evolution involves the transmission of genes, the basic units of inheritance, which are passed on from parents to offspring in their eggs and sperm (a process which is discussed in detail later in this book). Cultural evolution involves the handing down of social customs and values, and knowledge passed on from teacher to pupil; it is also transmitted in artefacts such as books, technologies of many kinds, and works of art and architecture.

In both cases, what is passed down to one generation differs in some respects from the inheritance of previous generations, so the forms and structures of biology and culture both 'evolve' – they change over time. Moreover, they *interact* and this has great importance for human health.

Fossil evidence tells us that the organisms we refer to as modern humans emerged at least one million years ago, and that aspects of human culture have been influencing human biology, and vice versa, ever since. For example, cultural changes in diet and food preparation have had an impact on human biology in terms of the structure of our digestive tracts, bones and teeth. The anatomical features required to grind down raw foods have gradually changed over the last million years, as the cultural 'breakthrough' of harnessing fire for cooking enabled us to exploit a huge new range of foodstuffs. As new foods and hence new habitats became accessible, communities of humans spread into them, and new cultures evolved. Biological changes occur much more slowly than cultural change: the human appendix may be one vestige of a portion of the digestive tract that was useful in the very distant past, but which now has no known digestive function and may gradually disappear in the far distant future.

Despite their different time-scales, it is important to recognize that the influences between human biology and human culture are two-way: they interpenetrate. This process is demonstrated if we reconsider the *environment* in which both biological and cultural evolution take place.

❐ What does the term 'environment' mean to you?

■ In everyday language, the environment is often taken to mean the physical world around us – land masses, rivers, mountains, etc. and their vegetation, together with the built environment of cities and roads. A wider definition includes the climate and weather, the quality and availability of food, water and air, and the amounts of natural and industrial radiation. Biologists add to this all the other organisms sharing the same habitat, including microscopic ones such as bacteria. Psychologists would use the term to include any feature of the external world that influences human behaviours or mental states. Social scientists and historians would include all the aspects of culture and technology that characterize human societies.

In the discussion that follows, we are using 'environment' in this all-inclusive sense, which encompasses the physical, biological, psychological and social worlds in continuous interaction.

1.8.2 Evolution and natural selection

If biological evolution is the history of biological change, what are the forces that prompt change to occur? This is an area of considerable agreement between biologists – even those who sit in opposing camps when it comes to discussing the relative contributions of genes (nature) and environment (nurture) to the outcome.

The theory of biological evolution is associated with the British naturalist Charles Darwin (1809–1882), who noted the range of *variation* in appearance and behaviour even among the members of a single **species** – that is, a population of similar organisms that usually interbreed in their natural habitat and produce fertile offspring. This variation is extremely important because it provides the potential for the species to survive environmental change. If the conditions in a certain environment gradually alter – for example, because the average temperature rises, or the environment becomes overcrowded – some members of the species will find it easier to survive than others, because tiny differences in their body structure, or function, or behaviour, give them a small advantage. For example, if the average temperature rises, then individuals who are better able to lose excess heat (perhaps because they have less body hair, or produce more sweat) will be at a slight advantage.

The characteristics of body structure, or function, or behaviour, which give some members of a species a survival advantage are termed **adaptive characteristics**. If the advantage is sufficiently great, then the proportion of better-adapted individuals who survive long enough to reproduce and leave surviving offspring, will exceed the proportion of less well adapted survivors. This is the basis of **natural selection**, the driving force of evolution as described by Darwin. The theory relies on the fact that a population produces more offspring than can survive to parent the next generation, because the resources available in terms of food, shelter and so on are insufficient for all. In the competition for resources, a natural process of 'selection' occurs in that individuals with the most adaptive characteristics in the context of *that* environment at *that* time are the most likely to survive and reproduce. This principle is often translated as 'survival of the fittest' – a phrase that needs to be used with care, because it has the power to mislead in two ways.

First, we have to distinguish between the meaning of 'fittest' in everyday language and its meaning in evolutionary biology. Depending on the environmental conditions, the 'fittest' members of a species could be the slowest, sleepiest and most obese – if these characteristics give them a survival advantage over the rest. Second, the characteristics that denote the 'fittest' at any given time and place, are not 'the best' characteristics in any absolute sense.

❐ Will the same individuals still be the fittest members of a species if the environmental conditions change, even by a little?

■ No; the accolade of 'fittest' will pass to other individuals with slightly different characteristics, who now find themselves better able to survive and reproduce in the new conditions than they were before.

To a biologist, then, **fitness** means *lifetime reproductive success* (i.e. the total number of surviving offspring) relative to the number produced by other members of the same species. But why is there this emphasis on reproductive success? A characteristic is only considered to be adaptive in an evolutionary sense if it can be passed on from parent to offspring. Thus, the survival advantage conferred by that characteristic on the parent is also enjoyed by the offspring. Over many generations, if the characteristic continues to give an advantage, then the proportion of individuals in the species who share that characteristic can be expected to increase.

❐ Can you explain why?

■ More offspring of the better-adapted individuals will survive to reproduce in their turn; more of *their* offspring will survive to reproduce, and so on. At the same time, fewer offspring of the less well adapted individuals will survive, because they lose out in the competition for resources. So the proportion of individuals in the species who share the adaptive characteristic rises.

1.8.3 Variation and change

If environments stayed the same, the members of a species would (in theory) become more and more adapted to survive and reproduce successfully in that environment from one generation to the next. But in reality, *environments never stay the same.* The natural world is in a continual state of change, so evolution by natural selection cannot produce 'perfection'. It is a common misconception to think of evolution as a process of 'onward and upward', when a better analogy might refer to a process of 'continually moving the goalposts'! The variation within a species is its strength: the species *as a whole* has a chance of surviving continual environmental change if its members have sufficient variation between individuals, such that *some of them* will survive when conditions change.

One of the beneficial outcomes of understanding evolutionary theory in this way is that it enables us to view human variation in a rather different light. Humans display immense variation – each face, each set of fingerprints is unique for six billion individuals, and that's just counting the ones alive today. The ability to colonize almost every habitat on Earth, from the deserts to the ice-caps and everything in between, is a testament to human evolution by natural selection which has resulted in so many variations on a basic theme. This variability is our 'hedge' against extinction. Species like the dodo and the dinosaurs paid the ultimate price when the environment changed around them: the variations between individuals were insufficiently great and none could survive and reproduce in the new conditions. So, from a biological standpoint alone, the intolerance that human societies often display towards individuals who look different and behave differently from the norm is illogical.

1.8.4 Nature and nurture revisited

This is the point at which we head into more troubled waters and revisit the earlier debates about nature and nurture in the context of evolution. Just as there has been argument about the *sufficiency* of genes to explain everything about the structure and function of the individuals we see around us today, the role of genes in the vast timescale of biological evolution is also open to debate.

All biologists agree that characteristics are passed on from one generation to the next in the form of genes, donated from parents to offspring in sexually reproducing species such as ourselves, in the eggs and sperm. Organisms survive or die out in a given environment because variations between individuals confer advantages on those with the best-adapted characteristics. Since many of these variations are a consequence of variations in the genes each individual has inherited, so natural selection can be said to be 'selecting' organisms with the best-adapted genes in that environmental context. The problem lies in deciding whether anything *in addition to* genetic inheritance should also be taken into account when considering evolutionary change.

There is at one extreme, a gene-centred theory of evolution popularized by the biologist Richard Dawkins in his book *The Selfish Gene* (1976), which views organisms as nothing more than vehicles for ensuring the survival of genes. In this highly reductionist account, all the characteristics of an organism (behavioural, psychological and physical) can be traced to the activity of its genes. This theory has been enormously influential in popular culture as a result of its promotion in documentaries, best-selling books and newspaper articles. It leads to a view of the body as a collection of interacting and more-or-less independent parts – analogous to a construction model kit in which parts can be removed, added or modified independently of each other without fundamentally changing the nature of the model itself. In terms of evolutionary change, if a 'part' is useful to the survival of the organism as a whole, then this theory predicts it will be found increasingly in future generations; if not, then it will gradually fade out of the species. Since the 'parts' themselves are programmed to have certain characteristics by the organism's genes, then it can be said that the individual genes – rather than the whole organism – are subject to natural selection.

Critics of this view of evolution argue that it presents organisms as opportunistic 'survival machines', with no other function than to pass on their genes to as many offspring as possible. In sharp contrast, the holistic view of evolutionary change sees organisms as more than the 'sum of their parts'; this theory argues that evolution has produced organisms which are complex integrated wholes. To return to the construction kit analogy, parts within the model are not independently variable and dispensable; parts could not change independently of changes to the model as a whole. This approach to an understanding of evolution asserts, just as Darwin did, that whole organisms are subject to natural selection.

Supporters of this holistic view of evolutionary change offer three arguments in making their case. First, humans and other organisms don't have perfectly adapted bodies so – despite natural selection – there are some genes preserved in the species which confer less-than-optimum characteristics (for example, the genes that influence the development of the appendix in humans). According to the gene-centred theory of evolutionary change, these genes ought to 'die out', unless you construct an argument that says 'well, there must be other genes that ensure the preservation of these less-than-optimum genes'. The logic then begins to get circular! Health depends upon a combination of heredity and habit, of nature and nurture. We don't have perfect bodies – over time there have been compromises. After all, our ancestors had tails and we have lost them, except for a remnant that is quite prominent in the early human embryo, as you will see in Chapter 5. All we can say is that some organs appear to be more difficult to modify or eliminate than others, but this is not an explanation. Rather, it points to an interesting problem and to the second argument that counters the gene-centred theory of evolution.

The second argument is that it appears that there may be constraints on the construction of body structures. Consider this example. In every four-limbed animal that exists or (on fossil evidence) has ever existed, the limb starts at the shoulder or pelvis with a single long bone, which then makes a joint with two bones (elbow or knee), and these then articulate with a complex of bones ending in a number of digits. This structure is known as the *tetrapod limb.* No-one disputes that variations on this theme (e.g. compare the limbs of a gazelle, a frog and a human) are adaptations to life in different environments, but no matter how useful it might be to have a different *basic* limb structure, evolutionary processes have not produced an alternative to the tetrapod limb in millions of years. You might think that in some species it would be more useful to have two bones to start with – for example, birds need a strong, light, flat structure for the wing, and two struts together can be lighter and stronger than one. However, since no such structure has emerged, it could be argued that there may be *intrinsic* constraints on how the body is constructed – no matter which genes you inherit. Constraints such as this may account for many aspects of our body structures which are certainly not optimal, but work well enough.

Third, the gene-centred theory ignores the *interpenetration* of organism and environment. Since all species modify their environments, they transmit not only their genes but an altered environment to successive generations as a result of their behaviour. Humans do this more than any other species: we are undeniably transmitting an altered environment to our progeny which is favourable to their survival in some respects and unfavourable in others. The open question is the extent to which the environment we inherit is acting on our genes, at the same time as our genes are acting on the environment!

To conclude, there is a consensus in biology that genes set limits to the *range* of behaviour, body structures and functions that any individual can express, while the environment or culture adds a further component of specificity in determining *which* of the possibilities an individual actually expresses. Health depends on a combination of genetic inheritance and environmental factors, of nature and nurture. To attempt to disentangle and measure their *relative* influence is fraught with difficulty. However, the reductionist tradition in biology sees the task as a useful pursuit of knowledge, while the holistic tradition decries it as an attempt to rip apart an indivisible whole.

We, like all the other species on our planet, are involved in a continuously creative process that results from, and depends upon, diversity – of individuals, species, environments and cultures. Health can then be seen as a creative response to whatever circumstances are encountered, so it involves a different strategy for each person in realizing a creative life. Health is an expression of this integrated whole, which involves living with compromises and imperfections.

TV programme 1 looks at the interaction between nature and nurture in the development of a variety of characteristics. You should watch this programme as soon after reading the above section as you can.

1.9 The scope of this book

With all the themes, definitions and debates introduced in previous sections in mind, we conclude by casting a glance forward to the rest of this book. It will already be obvious to you from its title that its principal territory is health and development in the period from conception to birth. You may already have been surprised by the wide-ranging content of the material presented in this first chapter, so be prepared for diversity in subsequent chapters. This is the way of multidisciplinary teaching!

In Chapter 2, we develop the first theme outlined at the very start of this chapter, by discussing a range of approaches to defining and measuring health. The variation is striking, ranging from subjective accounts provided by lay people, sociological concepts of health as the fulfilment of potential, medical definitions based on the absence of disease, and measurements of biological parameters such as blood pressure and birth weight. This leads to an analysis of the most influential explanations for variations in health experience. To what extent are we personally responsible for our health, given the importance of biological, psychological and sociological factors, many of which are beyond our direct control? The chapter illustrates this question by discussing the long-term health consequences of biological, psychological and sociological influences on the developing embryo.

Chapter 3 begins in earnest the biological study of human life, from the chemical constituents of cells to their organization into tissues and organs.

We discuss the mechanisms by which genes exert control over the molecular interactions within a single cell – the basic biological unit of life. Coordination is vital for sustaining life at the biological level, and this is illustrated by the interactions of cells and molecules in wound healing – a process which has much in common with the interactions seen in developing embryos.

Chapters 4–6 demonstrate the indivisibility of biological, psychological and sociological influences on conception and contraception, fertility and infertility, which lead to the development of a new-born baby. We trace this process from a single fertilized egg, through all the complex and delicate movements of cells into organized and specialized structures, which gradually become recognizably human. As the tissues and organs grow, their contacts with each other are rearranged; the senses develop and control mechanisms become more sophisticated, until at last the process is complete. Along the way, there are many opportunities for 'errors' of development to occur, but the organism has in-built methods of detecting and correcting many of them. If all goes well, pregnancy culminates in labour and the delivery of a healthy full-term baby – the climax of Chapter 6, the last chapter of this book.

As you study the subsequent chapters, bear in mind the major themes we outlined at the beginning of this chapter and look for evidence of them: first, the emphasis on health rather than disease; second, a view of health as a dynamic process rather than a fixed state; and third, the emergence of health from the interaction of human biology, psychology and social relationships in a given environment.

Objectives for Chapter 1

After completing this chapter you should be able to:

1.1 Define and use, or recognize definitions and applications of, each of the terms printed in **bold** in the text.

1.2 Use examples to demonstrate your understanding of health as a dynamic process, influenced by the interaction of biological, psychological and sociological factors. (*Question 1.1*)

1.3 Discuss the strengths and weaknesses of holistic and reductionist approaches to health and health care, using examples from biology, psychology and sociology. (*Question 1.2*)

1.4 Explain what is meant by biological determinism and the nature–nurture debate. (*Question 1.3*)

1.5 State the basic principles of Darwin's theory of biological evolution and natural selection. (*Question 1.4*)

Questions for Chapter 1

Question 1.1 (*Objective 1.2*)

Note down a few of the possible ways in which biological, psychological and sociological factors might interact in contributing to the good health of a person in their 90s. At this early stage in the course, you will have to rely partly on your general knowledge of any of the three disciplines that you have not previously studied. See how far you can get in attempting an interdisciplinary account, then read our answer; you may well have thought of valid points that we have not included.

Question 1.2 (*Objective 1.3*)

Sum up the main strengths and weaknesses of reductionist and holistic approaches to researching the influences on health.

Question 1.3 (*Objective 1.4*)

Identical twins have identical genes. What would a person who believes in biological determinism predict about the growth and development of identical twins who were separated at birth and brought up by different families?

Question 1.4 (*Objective 1.5*)

The Inuit people of the Arctic circle have small pads of fat around their eye sockets, which push the eyelids forward, so that they cover more of the eyeball than is the case in people who live in more temperate climates. It has been argued that these pads help to protect the eyes from the damaging brilliance of a snow-bound landscape.

How would the evolution of the fat pads be explained by an evolutionary biologist? Include the following terms in your answer: natural selection; adaptive characteristic; survival advantage; fitness; evolution.

References

Armstrong, D. (1986) The problem of the whole person in holistic medicine, *Holistic Medicine*, **1**, pp. 27–36.

(An edited version appears in Davey, B., Gray, A. and Seale, C. (eds) (1995) *Health and Disease: A Reader*, 2nd edn, Open University Press, Buckingham.)

Dawkins, R. (1976) *The Selfish Gene*, Oxford University Press, Oxford.

WHO (1958) *Constitution of the World Health Organization*, Annex 1, Geneva.

CHAPTER 2
HEALTH: LIFE PROCESS AND LIFESTYLE

2.1 Introduction

This chapter examines different ways of defining health and how these relate to the various aspects of a person's life. The differing sociological, psychological and biological emphases of these definitions will be explored, in order to identify the assumptions on which they are based and to examine their adequacy in providing a concept of health that is consistent with the multi-level approach described in Chapter 1. We end this chapter with an example that introduces the biological level, the detailed examination of which begins in Chapter 3.

2.2 Defining health

What is 'health'? What do we mean when we say that someone is 'healthy'? At first glance these seem easy questions to answer. We could say that health is about not feeling ill or not having a disease. Or we could define it in a more positive way by saying that health means being well and fit. We might argue that someone is healthy if they are capable of keeping going all day at a strenuous job or running a certain distance without getting exhausted.

Being healthy is often linked in our minds with physical fitness. We suggest that people live a healthy life if they take lots of exercise and call them unhealthy 'couch potatoes' if they do not. But for many people health is also about how they feel mentally. If you feel good about yourself – content, happy, at peace – then you might also say that you feel healthy. If you feel bad about yourself you might not define yourself as healthy, even if you were fit, took lots of exercise and generally had a healthy 'lifestyle'. On the other hand, for some people – such as athletes – feeling mentally and physically healthy might involve being in peak physical condition. For others, managing to cope unaided and lead an independent life might be the most important aspect of health.

This highlights the fact that views about health are not static but are influenced by people's life situation. In other words, health is a state of being which is subject to wide individual, social and cultural interpretation. It is produced by the interplay of individual perceptions and social influences. All of us, whether we are lay people or professional health workers, create and re-create meanings of health (and illness) through our lived experience.

❑ Why might people's views of health differ and change?

■ Age, gender, culture, social background, state of mind and upbringing may all influence people's views about health. The definition you offered at the age of 20 (if you thought about it then at all) will probably have changed considerably by the time you reach retirement age. It would be surprising if a child – likely to be fit, active and with no serious impairment – viewed health in exactly the same way as an older person who may be coping with a chronic ailment, limited mobility as well as several other moderate impairments.

A study of beliefs about 'what health is', called the Health and Lifestyle Survey, was undertaken on a representative sample of 9 000 adults in the UK in the 1980s and updated in the early 1990s (Cox *et al.*, 1987, 1993). The results of this survey suggest that people's perception of their own state of health influences how they define health. Older people are more likely to view health in terms of function and coping: carrying out household tasks, managing to work, being able to get around. Older people, in particular, may conceptualize health in a way that includes illness or disability. Herzlich (1973) characterized this as 'health despite disease' and it is a reminder that physical frailty does not preclude positive feelings of health and well-being. Young people frequently define it in terms of fitness, energy, vitality and strength, emphasizing positive attainment and a healthy lifestyle, as shown in Table 2.1 (Cox *et al.*, 1987).

Table 2.1 Descriptions of health: (a) concepts of health used for describing someone else; (b) concepts of health used for describing what it is to be healthy oneself. Results are expressed are percentages and the values in parentheses are the numbers of individuals in each group (e.g. there were 1 668 males aged 18–39).

Part (a)

	Males			Females		
Age	18–39	40–59	60+	18–39	40–59	60+
Unable to answer, can't think of anyone, don't know why I call them healthy	10	14	23	13	12	21
Never ill, no disease	25	37	35	43	49	34
Physical fitness	44	26	10	28	20	10
Functionally able to do a lot	12	15	19	13	17	18
Psychologically fit	8	9	12	10	8	5
Leads a healthy life	22	16	11	24	16	12
In good health for age (applied to an older person)	2	8	12	3	8	15
Mean no. of concepts used by each individual offering any reply	1.3	1.3	1.2	1.4	1.4	1.2
(100%)	(1 668)	(1 240)	(997)	(2 150)	(1 596)	(1 352)

Part (b)

	Males			Females		
Age	18–39	40–59	60+	18–39	40–59	60+
Unable to answer	16	12	10	11	7	8
Never ill, no disease	14	17	16	12	10	10
Physical fitness, energy	39	27	12	41	32	16
Functionally able to do a lot	22	26	43	22	36	34
Psychologically fit	31	40	36	48	52	44
Other	5	6	10	6	8	12
Mean no. of concepts used by each individual offering any reply	1.3	1.3	1.3	1.4	1.5	1.3
(100%)	(1 668)	(1 240)	(997)	(2 150)	(1 596)	(1 352)

At this point note down your own definition of health. Explain, if you can, why you have defined health in this way. Then compare your account with the following definition from Mrs Brown, an 80-year-old Yorkshire woman.

I've lived in the village all my life and there's rarely a day when I haven't been out. It doesn't feel right if we can't get out for a walk. You can see the hills from my window, we're the last house down the lane, you just step outside. Walking keeps you warm, getting around, that's what I call being well. I've never felt the cold except in the really bad weather, but now I can't walk like I used to, I need a stick. I always have the fire on and keep busy doing my jobs, tidy the house, cook the tea, but my legs go stiff and I have to stop, and then I do feel the cold. Rain's the worst, you can always be cheerful if the sun shines, can't you? (quoted in Jones, 1994)

How far did you share Mrs Brown's view of health as the ability to 'get around' and 'keep busy'? Did you find yourself equating health with physical fitness and energy, like the respondents in the Health and Lifestyle Survey? You may have pointed out – as Mrs Brown implied – the emotional dimension of health. Mrs Brown's sense of contentment is threatened by her increasing lack of mobility, and other factors – like the weather – may significantly influence people's perception of their health as well as their physical and mental health state.

A person's age is only one of many factors that may influence their definition of health. The Health and Lifestyle survey documented differences in responses between men and women. Younger women tended to link energy and vitality to undertaking household tasks, whereas younger men linked energy and fitness to participating in sports. Having a traditional female role as a housewife influenced Mrs Brown's perception of health, in the sense that she measured her health partly in terms of her ability to cope with what she saw as her 'jobs' – that is, the housework. You may find that your own work and role play some part in defining your view of health. You may also be realizing that a straightforward definition of health – even if we could frame it satisfactorily – would not help very much in uncovering the variety of different experiences, beliefs and assumptions that make up people's views of health.

2.2.1 Health as 'absence of disease'

Many seemingly simple definitions of health – 'not being ill', 'an absence of disease', 'the ability to function normally', 'a state of fitness' among them – contain within them complex ideas about what it is to be healthy, whose responsibility it is to maintain health, and how illness and disease should be interpreted (Jones, 1994). They may project officially sanctioned ways of viewing health which have passed into public circulation and become part of popular thinking. Consider 'absence of disease', which appears as a category in many research findings (Herzlich, 1974; Blaxter, 1990). It has been the most pervasive official definition of health in the Western world.

❑ What ideas and messages inform the definition of health as 'an absence of disease'?

■ The 'absence of disease' definition presents a rather negative view of health – as 'absence' rather than as a positive state. It is also a polarized view: you are either suffering from a disease or you are in good health. Quite apart from the problems surrounding diagnosis – are you diseased if neither you nor your doctor recognize it? – there are clearly powerful signals in this definition about what health is and is not. Health is not about feeling well, at ease, energetic, or even necessarily about not feeling ill; it is about not having a disease. And since diagnosing disease is the specialist activity of medical doctors it follows that everyone is healthy until diagnosed as diseased by their doctor! Indeed, in Britain and much of the rest of Europe a person must be officially diagnosed in order to qualify for sickness pay beyond the first few days; so a state of health is what the vast majority of us are otherwise assumed to enjoy.

'Absence of disease' derives from a medical concept of disease as a pathological state which can be diagnosed and categorized, or as deviation from measurable biological variables which represent 'normal' parameters in the 'healthy' body. This view of disease, which has come to dominate Western thinking about health during the past two centuries, is often termed the 'medical model' (Table 2.2). It is linked to the rise of clinical pathology and the scientific investigation of disease by a growing body of specialist doctors and researchers, and to the emergence of health work as a formal, professionalized area of expertise (Jones, 1994).

Table 2.2 The medical model.

- Health is predominantly viewed as the 'absence of disease' and as 'functional fitness'.
- Health services are geared mainly towards treating sick and disabled people.
- A high value is put on the provision of specialist medical services, in mainly institutional settings.
- Doctors and other qualified experts diagnose illness and disease and sanction and supervise the withdrawal of patients from productive labour.
- The main function of health services is remedial or curative – to get people back to productive labour.
- Disease and sickness are explained within a biological framework that emphasizes the physical nature of disease; i.e. it is biologically reductionist.
- It works with a pathogenic focus, emphasizing risk factors and establishing abnormality (and normality).
- A high value is put on using scientific methods of research (hypothetico-deductive method) and on scientific knowledge.
- Qualitative evidence (given by lay people or produced through academic research) generally has a lower status as knowledge than quantitative evidence.

Many people in surveys offer 'absence of disease' as one of their definitions of health, so there is presumably quite widespread acceptance of such a view. This at least enables individuals of all ages and backgrounds to see themselves as being 'in good health'. Some evidence of the extent to which health is viewed as 'an absence of disease' is provided by the Health and Lifestyle survey, in which 11% of women and 16% of men interviewed (averaged over the three age groups) offered a definition of their own health as 'never ill' or 'no disease' (Table 2.1). Moreover, the popularity of 'absence of disease' as a definition offered by ordinary members of the public (not only professional health workers), suggests that people's definitions of health arise from the sifting of broader ideas and theories as well as from 'personal' beliefs and experiences.

As noted above, in the Western world health has been defined most often in terms of absence of disease and death. Improved health is measured in terms of life expectancy, calculated from **standardized mortality** (death) **rates**, i.e. deaths per 1 000 of the population (calculated by applying observed age-specific rates to a standard population). In recent years, health standards have been measured in terms of potential years of life lost and changes in **life expectancy** (Figure 2.1). Life expectancy is the average number of years which a new-born baby could be expected to live if its rates of mortality at each age were those that occur in that calendar year.

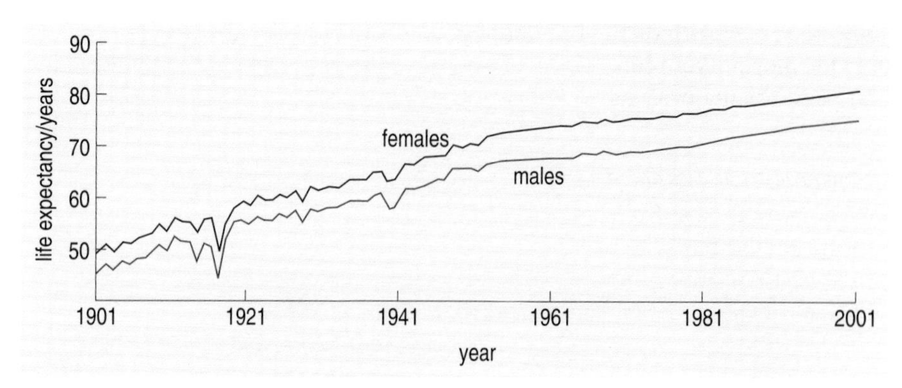

Figure 2.1 Changes and projected changes in life expectancy for males and females in the UK over the 20th century.

Regional and social differences in health are calculated using the **standardized mortality ratio** (SMR), which measures the chances of death at a stated age of people in particular groups or regions compared with the standardized mortality rate for the whole population. Mortality statistics began to be collected and published by governments in the 1840s, although official estimates of death rates go back long before this. All deaths (and births and marriages) had to be notified to the local registry office, so that the Registrar General's office in London was able to build up detailed information on death rates and causes of death. This work is still carried on by the Office for National Statistics today. (Until 1996 this was the Office for Population, Censuses and Surveys, OPCS.)

Mortality and **morbidity** (reported incidence of sickness per 1 000 of the population) are still widely used as measures of health, and lowering morbidity is equated with improving the nation's health. For example, in 1992 the goverment published a strategy for the English health service – called 'The Health of the Nation' – which identified five key areas of objectives and targets for improving health (Table 2.3). All of them were conceptualized in terms of 'reducing' and 'preventing' disease and sickness. For example, coronary heart disease (CHD) and stroke was justified as a key area 'because of the scope for preventing illness and death from these conditions, and because reductions in risk factors ... would also help to prevent many other diseases'. Cancers were selected 'because of the toll that cancers take in ill-health and death' (Department of Health, 1992). There are obvious advantages in measuring health in terms of mortality and morbidity, because it is generally relatively easy to record reported sickness and actual death.

Table 2.3 Health of the Nation targets for improving health. (The baseline year for mortality targets represents an average of three years centred around that year; the majority have 1990 as the baseline.)

Coronary heart disease and stroke

To reduce death rates from both CHD and stroke in people under 65 by at least 40% by the year 2000 (baseline 1990).

To reduce the death rate from CHD in people aged 65–74 by at least 30% by the year 2000 (baseline 1990).

To reduce the death rate for stroke in people aged 65–74 by at least 40% by the year 2000 (baseline 1990).

Cancers

To reduce the death rate for breast cancer in the population invited for screening by at least 25% by the year 2000 (baseline 1990).

To reduce the incidence of invasive cervical cancer by at least 20% by the year 2000 (baseline 1986).

To reduce the death rate for lung cancer under the age of 75 by at least 30% in men and by at least 15% in women by 2010 (baseline 1990).

To halt the year-on-year increase in the incidence of skin cancer by 2005.

Mental illness

To improve significantly the health and social functioning of mentally ill people.

To reduce the overall suicide rate by at least 15% by the year 2000 (baseline 1990).

To reduce the suicide rate of severely mentally ill people by at least 33% by the year 2000 (baseline 1990).

HIV/AIDS and sexual health

To reduce the incidence of gonorrhoea by at least 20% by 1995 (baseline 1990), as an indicator of HIV/AIDS trends.

To reduce the rate of conceptions amongst girls aged under 16 by at least 50% by the year 2000 (baseline 1989).

Accidents

To reduce the death rate for accidents among children aged under 15 by at least 33% by 2005 (baseline 1990).

To reduce the death rate for accidents among young people aged 15–24 by at least 25% by 2005 (baseline 1990).

To reduce the death rate for accidents among people aged 65 and over by at least 33% by 2005 (baseline 1990).

The 'morbidity model' of health, as we might call it, has encouraged us to think of health as 'absence of disease'. There are many advantages in this: it has enabled us to take minor complaints in our stride; we can have aching feet, period pains, a bad cold, backache, and still consider ourselves healthy. On the other hand, it has tended to label people with disabilities and chronic conditions of various kinds as inevitably 'sick' or even 'diseased', even if they are otherwise healthy. This kind of thinking, together with the measurement of health in terms of avoided or delayed death, has left lay people and professional health workers alike with a rather negative view of health which is only slowly being overcome.

2.2.2 Health as a positive state

From a simple definition of health as 'absence of disease' we have begun to tease out a series of explanations which expose it as complex and multi-dimensional, highlighting certain issues and pushing others to the margins. Now let's look at some other definitions which conceptualize health as a *positive* state.

Consider another 'official' definition of health. What ideas and messages inform this definition?

> *Health is a state of complete physical, mental and social well-being, not merely the absence of disease. (World Health Organization, 1958)*

If you are a health student you are probably already familiar with this World Health Organization (WHO) definition, which seems to equate health with all-round well-being. You might have been influenced – as many have been – by its idealism and its positive message. Although almost all of us, in these terms, could be labelled 'unhealthy' – in a curious reversal of the 'absence of disease' scenario – this phrase has become a standard definition of health today. The WHO definition has been widely used within contemporary health care, particularly in community settings where professionals want to enhance people's health rather than merely to treat established disease. It highlights health as a positive goal rather than just a neutral state of 'no disease', and indicates that this is to be achieved by personal and social change as well as by medical advance.

As a definition, it contains almost as many new problems as it tries to solve. Its idealistic, even utopian nature has been commented upon by critics. How is anyone to achieve a state of complete health? What would it feel like if we reached it? How can well-being be measured and what kinds of individual and societal changes are required to achieve it? In a similar way to the 'absence of disease' approach, the apparent simplicity of the WHO definition conceals a range of assumptions about what health should be. Thus freedom from disease is not health; real health is viewed as the transformation of 'no disease'-type health into all-round well-being. Health becomes a personal struggle and a goal to be worked towards on a community, national and global level (Jones, 1994).

This sense of health as action and adaptation is captured in the WHO Working Group Report (1984) on health promotion, which conceptualized health as:

> *... the extent to which an individual or group is able, on the one hand, to realize aspirations and satisfy needs and on the other hand, to change or cope with the environment. Health is therefore seen as a resource for everyday life, not the objective of living: it is a positive concept emphasizing social and personal resources as well as physical capabilities. (WHO, 1984)*

❑ Reflect on this comment about health. Do you think it has any advantages over the two earlier definitions?

■ This later WHO comment emphasizes that health is embedded in the processes and actions of everyday life. It relates health to our ability to cope and adapt within a particular environment. This deliberately avoids objectifying health; instead, health is viewed as 'a resource for everyday life'. It also identifies health as a multidimensional and shifting concept which can't be easily analysed or measured.

2.2.3 Health as the fulfilment of potential

A positive view of health in which satisfying needs and realizing aspirations are seen as central is obviously quite far removed from one in which 'absence of disease' is the main objective. Whereas absence of disease is at best a neutral state, a positive view of health opens up the territory of 'well-being' and 'quality of life'. It becomes important to think about personal growth, health enhancement and fulfilment of potential. This might be termed a 'process model of health', a model in which health is not a state or a fixed point but a process of creative fulfilment of potential or self-actualization throughout life. Figure 2.2 illustrates one way in which this might be represented.

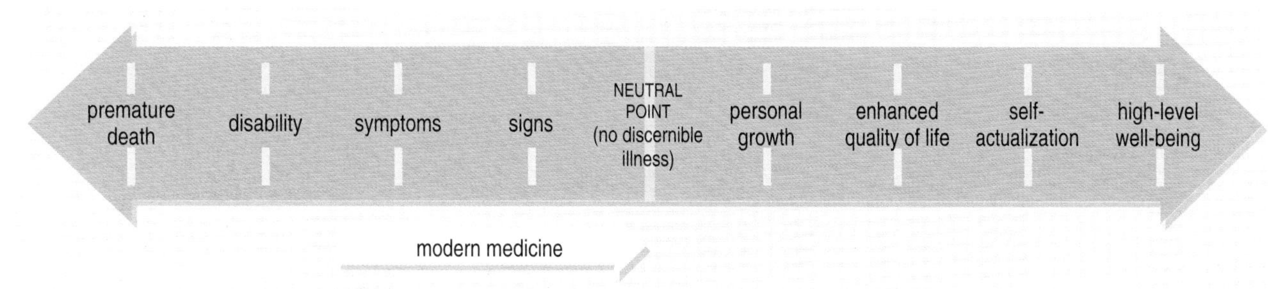

Figure 2.2 A process model of health.

The centre of the scale – the 'neutral point' – shows the absence of ill health. Moving from this neutral point leftwards shows a progressively worsening state of health, whereas moving right along the scale indicates an increasingly positive state of health, personal growth and fulfilment of personal potential (self-actualization). Traditionally, modern medicine has sought to cure disease but that only brings the individual back to the neutral point on the scale (white arrow), whereas a process definition focuses on growth and enhancement leading to high-level well-being.

From this viewpoint, a person who may not have any manifest physical illness still has considerable scope to enhance his or her health. Being bored, depressed, tense, anxious or generally dissatisfied with life can undermine health, so taking steps to solve one's problems and modify one's lifestyle may be important. Some of the respondents in the follow-up Health and Lifestyle Survey (Cox *et al.*, 1993) had tried to cut down smoking, take more exercise, eat more sensibly and drink less alcohol, for example.

- ❑ What might be the advantages and drawbacks of this way of representing illness and well-being?

- ■ The advantage of this approach is that it does not polarize health and ill health – people with disabilities and disease can still develop towards high-level well-being. But the shift towards high-level well-being is conceptualized largely in personal terms, whereas it might be greatly influenced by socio-economic circumstances.

As we noted earlier, the concept of 'health despite disease' figures as quite significant in research findings. Older people such as Mrs Brown might have increasing mobility problems but still regard themselves as well and able to fulfil their potential. Potential, after all, is not an absolute concept but closely linked to circumstances and opportunities.

2.2.4 An experiment in enhancing health: the Peckham Health Centre

An early experiment in enhancing health and developing people's potential for high-level 'wellness' can be seen in the work of the Peckham Health Centre in South London. The idea for the centre was developed by Dr G. Scott Williamson and Dr Innes Pearse. Pearse was a pioneer in the infant welfare centre movement of the 1920s and Williamson was a pathologist with a keen interest in the links between state of health and susceptibility or resistance to disease. Pearse and Williamson proposed that a centre dedicated to the maintenance and enhancement of people's health would form the ideal testing ground for theories about natural immunity (Pearse and Crocker, 1943). In other words, they hoped to test this idea that humans are endowed with a biological organization that naturally resists disease and restores a condition of working balance.

This pioneer centre was built as a community resource and was open to any families living within one mile of it – 'pram walking distance'. It was geared towards enhancing the health of basically healthy adults and children, rather than treating sickness. Families were offered periodic health checks, a chance to share in the results of scientific studies of themselves and access to all the facilities of the centre. The Peckham centre offered a wide range of services designed to enhance health, such as antenatal clinics, vocational guidance, swimming (it had its own pool), sex education, music, debates, drama, keep fit and youth clubs. Participation was encouraged, rather than high achievement, and the families organized and ran the centre themselves.

There was a weekly subscription fee of one shilling per family, and extra charges to adults for activities, which in effect excluded the poorest families in the area. It was mainly the skilled working class who became members and by 1939, 875 families had joined. However, after 1945 numbers declined and Williamson and Pearse sold the centre to the local authority. It seems that the creation of the National Health Service, with its curative and disease focus, was inimicable to the spirit of Peckham and the experiment foundered (Ashton and Seymour, 1988). However the Peckham vision of health as an expression of human potential, enhanced by locally controlled community structures that integrate a diversity of activities, remains a relevant model in which health is seen as a normal life process which includes disability and disease.

2.2.5 Measuring wellness?

There have been other attempts to define wellness. For example, some of the categories in Figure 2.2 have been used in Abraham Maslow's conceptualization of a 'hierarchy of needs', where the satisfaction of basic needs, such as the need for shelter, food and heating, enabled the realization of higher-level needs. Maslow (1954) envisaged a state of self-actualization in which the individual realizes his or her full potential as a human being. The Wellness Index questionnaire (Table 2.4) is one of several questionnaires designed to find out 'well' people's state of emotional and physical health.

Table 2.4 The Wellness Index questionnaire. The subject is asked to circle the category that most closely answers the question. (R = rarely, S = sometimes, VO = very often.)

Question no.	Question	Response		
1	I am conscious of the ingredients of the food I eat and their effect on me.	R	S	VO
2	I avoid overeating and abusing alcohol, caffeine, nicotine and other drugs.	R	S	VO
3	I minimize my intake of refined carbohydrates and fats.	R	S	VO
4	My diet contains adequate amounts of vitamins, minerals and fibre.	R	S	VO
5	I am free from physical symptoms.	R	S	VO
6	I get aerobic cardiovascular exercise. (VO is at least 12–20 minutes, five times per week, vigorously running, swimming or bike riding.)	R	S	VO
7	I practice some form of limbering/stretching exercise.	R	S	VO
8	I nurture myself. (Nurturing means pleasuring and taking care of oneself.)	R	S	VO
9	I pay attention to changes occurring in my life and am aware of them as stress factors.	R	S	VO
10	I practice regular relaxation.	R	S	VO
11	I am without excess muscle tension.	R	S	VO
12	My hands are warm and dry.	R	S	VO
13	I am both productive and happy.	R	S	VO
14	I constructively express my emotions and creativity.	R	S	VO
15	I feel a sense of purpose in life and my life has meaning and direction.	R	S	VO
16	I believe I am fully responsible for my wellness or illness.	R	S	VO

Using your answers to the questionnaire you can create a graphic representation of your wellness, your Wellness Index, by mapping the results onto the circle in Figure 2.3. Each numbered sector of the circle corresponds to the same numbered question in the questionnaire. The circle is divided into quarters representing the four major dimensions of wellness. To construct your personal Wellness Index, colour in an amount of each sector corresponding to your response to the question with the same number: the inner broken circle corresponds to 'rarely', the middle one to 'sometimes', and the outer one to 'very often'. After you have done this, see what shape your index has. Is it lopsided or balanced? You may wish to design your own definition of health and an appropriate questionnaire to evaluate it.

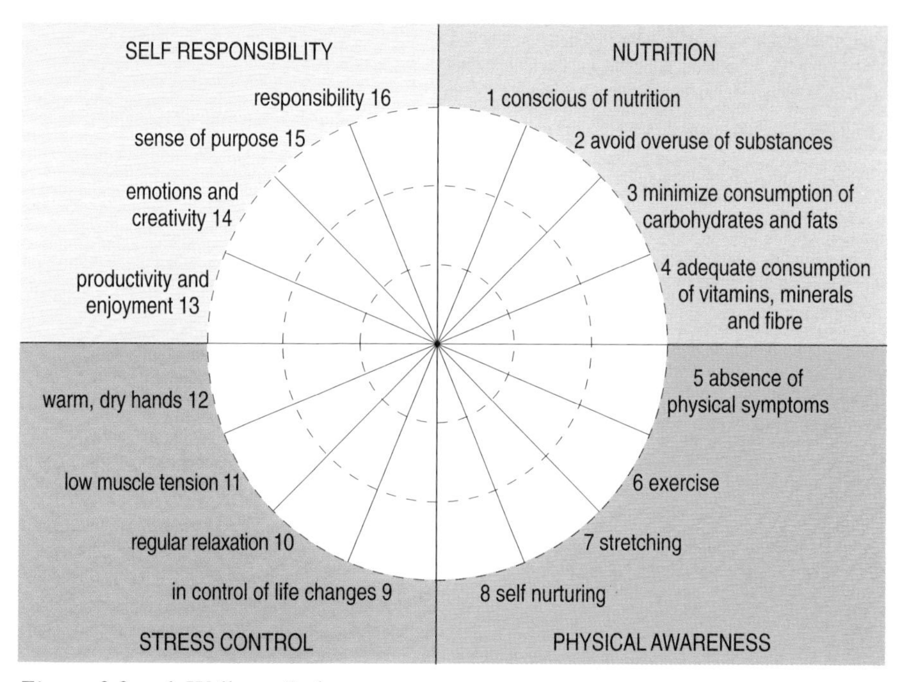

Figure 2.3 A Wellness Index.

❑ What is your response to being asked to complete this type of questionnaire?

■ You may have found the questionnaire fun to do and it may have made you stop and think about aspects of your life that you took for granted. You may often think about your diet, for example, because nutrition is always in the news, but you may rarely think about your emotional health or make time for relaxation. However, there are inbuilt assumptions in the questionnaire and you may have spotted these and been irritated by them. For example, there is a whole set of assumptions about what is healthy and what is not, which you may want to challenge. People who are really healthy, the questionnaire suggests, check their diet carefully, exercise thoroughly, relax, express their emotions and nurture themselves. The implication is that if you don't do these things you can't expect to be really healthy.

The clear message underlying the questionnaire is that people should take responsibility for their own health and if they don't do so, that is their choice. This may be a message with which you have some sympathy, especially if you work in a clinical arena where part of your role is to give patients health education advice – which they then ignore even though their own health is at stake. On the other hand, it is self-evident that not everyone can control their health in the sense suggested by the questionnaire. For example, some people work in dead-end jobs which are unlikely to give a 'sense of purpose' or 'meaning and direction' to their lives or to feel 'productive'. Many people are poorly paid and more likely to think about getting enough to eat than about checking the ingredients of their food or avoiding excess refined carbohydrates and fats. People who work in heavy physical jobs may not respond well to being asked about aerobic exercise. Even if people know they are stressed they can't always escape from the stressful situation.

Summary of Section 2.2

Individuals' perceptions and definitions of health change with age and are influenced by class, sex, job and personality. A dominant Western definition of health has been 'absence of disease', which is linked to the medical model that focuses on the scientific investigation of disease and has no concept of positive health. By contrast, other approaches seek to characterize well-being as a state that goes beyond absence of disease and relates to fulfilment of individual potential in life. This often emphasizes individual responsibility for health.

2.3 Individual and social responsibilities for health

The WHO (1984a) definition of health acknowledges individual and social responsibilities for health by emphasizing that health grows out of social resources, as well as personal resources and physical capabilities, and that its realization depends on how far individuals and groups are able to 'change or cope with the environment'. In other words, it is recognized that the realization of 'wellness' depends not just on personal attitudes and lifestyles but on material and other resources. For all people, adequate housing, food and heating are essential prerequisites to the achievement of good health. A worthwhile job with a decent level of pay is important as well. Beyond this, community support in the form of adequate shared facilities – parks, transport, health care and so on – is also needed. Clearly, looking after yourself and taking some responsibility for having a healthy lifestyle contributes to the enhancement of individual health. But it must be set against evidence which highlights the importance of poverty and unemployment as detrimental to physical and emotional health.

2.3.1 The lifestyles approach

Lifestyle is a term widely used in contemporary health work and is now a central focus for health education and health promotion. An interest in modifying and changing individual lifestyle has long been a minor feature of a medical approach to health but it has recently emerged as a focus for health action, as part of a wider attempt to improve people's health.

❑ How would you define the term lifestyle?

■ Lifestyle is a word currently used to describe a bundle of linked attitudes and behaviour patterns. For health workers, the most prominent aspect of lifestyles is the focus on 'risk' behaviour such as smoking, heavy drinking, physical inactivity, 'unsafe' sex and drug abuse.

These lifestyle factors are highlighted in the 1992 Health of the Nation Strategy for England and in the other national health strategies in the UK. The consequence is that smoking behaviour, eating habits and exercise routines, which were hitherto thought of as private concerns, are now public issues and the focus for professional health action. However, there is considerable debate between those who would target individual lifestyle 'deficiencies' and those who focus on broader public action designed to make 'healthy choices easier choices' (WHO, 1986).

An influential figure was Marc Lalonde who, as Canada's Minister of Health and Welfare, endorsed the 'health field' concept (Figure 2.4). Lalonde and his team argued that since prospects for improving health, and limitations upon it, arose in all the four 'fields' shown in Figure 2.4, interventions to promote health must do likewise. An overwhelming focus on health care organization – and by implication on a 'medical model' of health – had obscured the equally important environmental, behavioural and biological determinants of health (Lalonde, 1974). Lifestyle, with its focus on helping people to change unhealthy habits, was seen by many health workers as offering the best prospects for enhancing people's health.

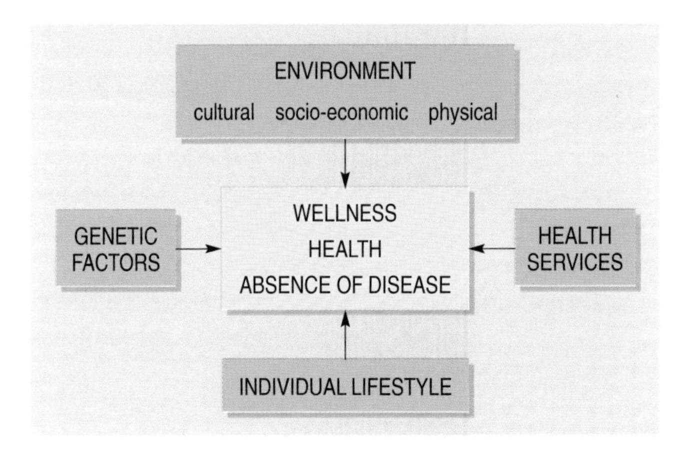

Figure 2.4 The health field concept. Health is seen as influenced by four 'fields': environment; genetic factors; health services; individual lifestyle.

Although Lalonde and others highlighted social and environmental influences on health – nutrition, public health measures, government legislation, falling family size – it is lifestyle aspects that have received most attention from governments in recent decades. The British government published a White Paper on Promoting Better Health in 1987, in which behavioural change was highlighted:

> *Much distress and suffering could be avoided if more members of the public took greater responsibility for looking after their own health... family doctors and primary health care teams should increase their contribution to the promotion of good health... [they] are very well placed to persuade individuals of the importance of protecting their health; of the simple steps needed to do so; and of accepting that prevention is better than cure. (DHSS,* Promoting Better Health, *HMSO, 1987, p. 3.)*

In this respect, the government could claim to be reflecting public opinion. In spite of considerable research (Townsend *et al.*, 1987, 1988), which highlighted the relationship between poverty and ill health, there is widespread support from surveys for the view that personal lifestyle is the main influence upon health. Pill and Stott (1982) reported that women blamed themselves for their inability to modify their lifestyle – for example to eat a more 'healthy' diet – even though their failure was largely the result of material factors such as lack of money, time pressures and so on. Cornwell (1984) noted that although illness tended to be seen by people in East London as a separate entity which 'happened' to them and for which they were not responsible, they believed that health (as an ability to function) was maintained by having 'the right attitude'. 'The moral prescription for a healthy life is in fact a kind of cheerful stoicism, evident in the refusal to worry, or to complain, or to be morbid.'

❑ Suggest why people might be likely to blame themselves for their 'deficient lifestyles'?

■ The most obvious answer is that we are most likely to offer explanations couched in personal terms. After all, we 'choose' our own diet, decide whether or not to smoke and set our pattern of exercise.

While a number of studies have indicated that these patterns of living are bound up in social relationships and are influenced by economic and cultural factors, there is nevertheless strong support for the view that personal behaviour is a major influence. Respondents in the Health and Lifestyle survey (Cox *et al.*, 1987) echoed this view and in the 1993 follow-up this belief had become even stronger.

However, there is another point worth considering. It is difficult to tell how far people's responses to researchers represent what they feel they ought to say or what they really believe. How far are people offering a public, officially sanctioned view of health and keeping their real opinions to themselves? Mildred Blaxter (1990) commented that 'there is a high level of agreement within the population that health is, to a considerable extent, dependent on behaviour and in one's own hands' but she went on to say that 'at least it is recognized that these are the 'correct' and 'expected' answers to give'. Smaller-scale, more intensive, contextual studies (such as Cornwell's study of the perceptions and lifestyles of 17 families in East London) have provided stronger evidence of people's belief that environmental factors also influence their health (Cornwell, 1984).

Lifestyle approaches have received support in the 1985 WHO 'Health for All' targets for Europe (see Table 2.5) although here they are set within a broader framework of public policy changes.

Table 2.5 Some of the WHO health targets for Europe (WHO, 1985). This is the sublist entitled 'Health for all in Europe by the year 2000'.

Target	Description
1 Reducing the differences	The actual differences in health status between countries and between groups within countries should be reduced by at least 25%, by improving the level of health of disadvantaged nations and groups.
2 Developing health potential	People should have the basic opportunity to develop and use their health potential to live socially and economically fulfilling lives.
3 Better opportunities for the disabled	Disabled persons should have the physical, social and economic opportunities that allow at least for a socially and economically fulfilling and mentally creative life.
4 Reducing disease and disability	The average number of years that people live free from major disease and disability should be increased by at least 10%.
5 Elimination of specific diseases	There should be no indigenous measles, poliomyelitis, tetanus in the new-born, congenital rubella, diphtheria, congenital syphilis or indigenous malaria.
6 Life expectancy at birth	Life expectancy should be at least 75 years.
7 Infant mortality	Infant mortality should be less than 20 per 1 000 live births.
8 Maternal mortality	Maternal mortality should be less than 15 per 1 000 000 live births.
9 Diseases of the circulation	Mortality from diseases of the circulatory system in people under 65 should be reduced by at least 15%.
10 Cancer	Mortality from cancer in people under 65 should be reduced by at least 15%.
11 Accidents	Deaths from accidents should be reduced by at least 25% through an intensified effort to reduce traffic, home and occupational accidents.
12 Suicide	The current rising trends in suicides and attempted suicides should be reversed.

A rather more restricted and individualist approach to lifestyle has also been endorsed by recent UK government reports. The Health of the Nation (1992), as we have seen, identified several aspects of individual behaviour as risk factors, and set 'risk factor targets' for the reduction of unwanted types of behaviour and consumption (see Table 2.3). Many health professionals have welcomed the idea of a strategy and the setting of health targets. Individual behaviour plays some part in many diseases; the connection between lung cancer and smoking is well established, for example, and dietary factors are now thought to be significant in relation to circulatory disease and some cancers. So persuading people to change risky aspects of their lifestyle seems highly appropriate in improving their **health status**, defined as a measure of a person's overall health experience. If clear targets and priorities are set, progress can be closely monitored and strategies adjusted if the hoped-for changes fail to materialize. The strategy also urges central and local government, individuals, statutory and voluntary agencies, and health professionals to work together to meet these targets.

Yet some health professionals have practical and ethical doubts about an approach that mainly targets personal lifestyles (Thomas, 1993). First, how far can health really be improved through targeting personal behaviour? Recent health campaigns warning young people to modify their sexual behaviour in the light of the HIV/AIDS danger seem to have had some immediate but far fewer lasting effects (Rhodes and Hartnell, 1996). Making changes in personal behaviour is very difficult.

A five-year study of working-class mothers in South Wales indicated that although half the women had made a change in their eating and smoking behaviour at some point in the study, most fairly quickly relapsed (Pill and Stott, 1982, 1985, 1986). Although the women tended to blame themselves for their relapses, many of the reasons for failure were actually to do with pressure from domestic or work circumstances and from partners and children. This raises a second point, about the ethical basis of a lifestyles approach. If many of the impediments to change are related to social circumstances, to the constraints imposed by income, housing, child care demands or social pressure from partners or peer group, how justifiable is it to target personal behaviour? Targeting individual lifestyle can come close to 'blaming the victims' if people are in a position where change is very difficult or even impossible, and yet are made to feel guilty for not making a change. Pill (1990) commented that 'some were only too ready to blame themselves' and their 'lack of willpower' and for a few 'this led to loss of self-esteem and strong guilt feelings'.

2.3.2 Social responsibility for health

The WHO definition connects to a 'social model' of health, which emphasizes the environmental causes of health and disease, in particular the dynamic interaction between individuals and their environment. Health is seen as being produced not just by individual lifestyle or medical intervention, but by conditions in the wider natural, social, economic and political environment, and by individual interaction with that environment.

When asked to describe what it is to be healthy, many people do respond with evidence about their life situation – having a healthy diet, taking enough exercise, participating in leisure activities, having reasonable working hours and conditions, having a decent roof over your head, getting through the day without feeling tired or depressed. Comments like these from health surveys remind us that health is contingent, that to realize good health depends not just upon genetic inheritance and personal behaviour but upon how 'lifestyle' is influenced by social and economic factors.

❑ Look back at your own definition of health. How far does it depend for its realization on broader social and economic factors?

■ It is likely that your definition of health acknowledged some of the underpinnings of health: the need for an adequate income level to provide decent housing, nutrition and heating, for example, or the need for social support. People's health chances – the likelihood that they will enjoy better or worse health – depend on a wide range of factors. Although 'having a healthy diet' and 'taking enough exercise' can be partly realized through personal decision-making, other decisions – about food pricing, marketing, transport provision and access, pedestrian safety, provision of open spaces, and so on – are made at an institutional level, largely beyond the control of individuals. And over the past 20 years considerable evidence has accumulated of the extent of health inequalities in the UK and of the influence of poverty on health.

Research into health inequalities has a long and distinguished record linking back to investigations of poverty at the end of the 19th century and to the work of the Medical Officers of Health, health visitors, midwives and district nurses. With the advent of the Welfare State in Britain in the 1940s, and the right to free, needs-based medical treatment in the new National Health Service, it was thought that health status would gradually be equalized. In the late 1960s, however, research by Townsend (1979) and others indicated that there was still a widespread incidence of poverty in prosperous post-war Britain. Then came the Black Report (Townsend *et al*, 1988) which indicated, at the extreme, that the death rate for adult men in occupational class V (unskilled workers) was nearly twice that of adult men in occupational class I (professional workers) (see Figure 2.5d).

The Black Report pointed to the existence of a growing occupational class gradient in all the major diseases. In particular it reported that between the 1950s and the 1970s the mortality rates for both men and women aged 35 and over in occupational classes I and II had steadily declined, whereas those in classes IV and V were the same or had marginally increased. In addition, there was some evidence that black and minority ethnic groups suffered differentially high rates of heart disease, evidence of regional variations, evidence of sex differences, and evidence linking disease risks to household type, with house-owners having the lowest risk of premature death.

Occupational class represented the major, though not the only, means by which health inequalities were measured in the Black Report. As the report commented, 'undoubtedly the clearest and most unequivocal – if only because there is more evidence to go on – is the relationship between occupational class and mortality' (Townsend *et al.*, 1988). The Registrar General classifies every male and every unmarried female worker 'in gainful employment' in one of six categories, ranging from higher professional (class I) through managerial (II), skilled non-manual and manual (IIIa and b) to semi-skilled (IV) and unskilled (class V). Although this classification tends to marginalize women's work, and creates various other distortions, it is a widely used and relatively reliable indicator not only of differences in income, occupational status, and living standards but also of relative levels of deprivation. Because researchers are able to draw on a long run of roughly comparable national data on mortality rates and causes of death, they can highlight the relative fortunes of the different classes over time; indeed, it was the discovery that class V mortality rates were not declining relative to those of the other classes that caused major concern in the report.

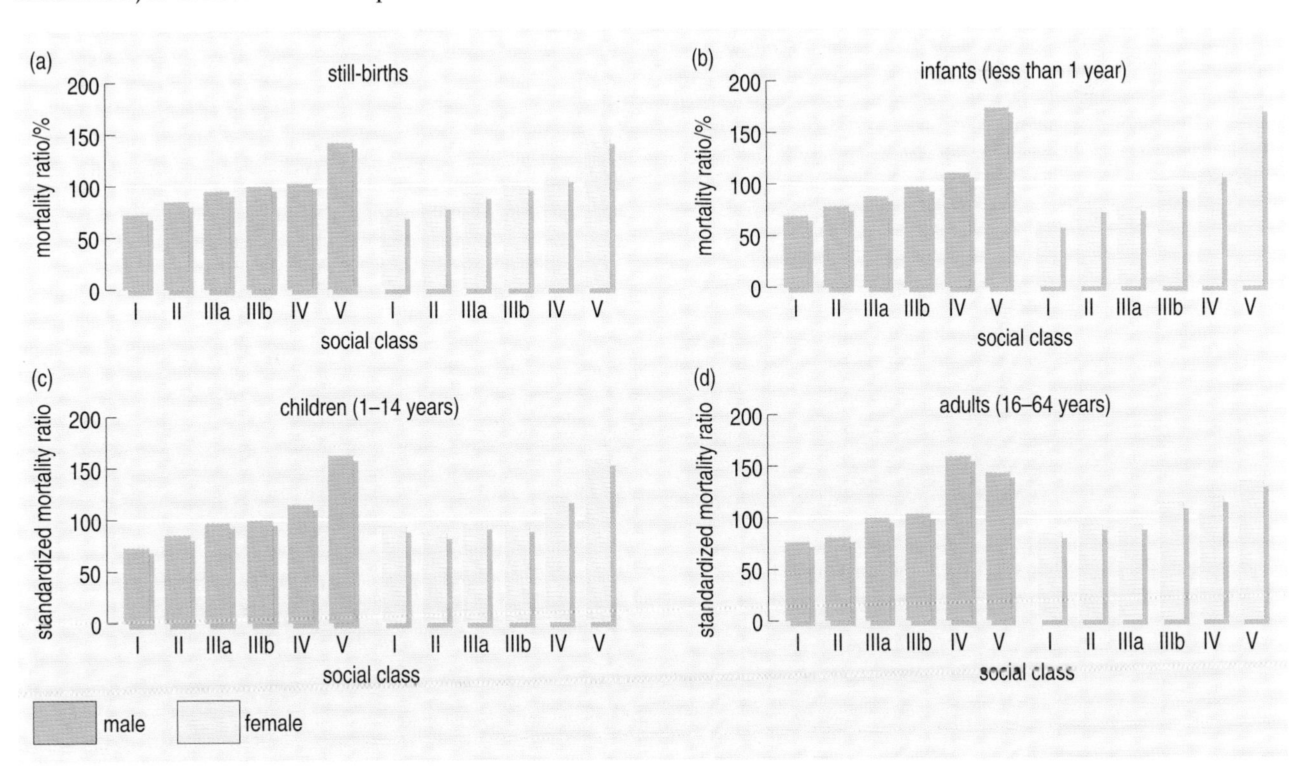

Figure 2.5 Mortality ratios by occupational class and age. (*Note*: it is not necessary to age-standardize the data for still-births and infant deaths, as both categories refer to small age ranges.)

A second survey of research carried out by Margaret Whitehead confirmed the Black Report's findings about 'class gradients' in health as well as documenting other differences in health chances and in access (Townsend *et al*, 1988). Having reviewed the findings made by researchers in the 1980s, Whitehead commented:

> *The results of these studies, taken together, give convincing evidence of a widening of health inequalities between social groups in recent decades, especially in adults. In general, death rates in adults of working ages have declined more rapidly in the higher than in the lower occupational classes, contributing to the widening gap. Indeed in some respects the health of the lower occupational groups has actually deteriorated against the whole background of a general improvement in the population as a whole. While death rates have been declining, rates of chronic illness seem to have been increasing, and the gap in the illness rates between manual and non-manual groups has been widening too, particularly in the over-65 age group. (Whitehead, in Townsend* et al., *1988,* Inequalities in Health: The Black Report and The Health Divide, *p. 266.)*

During the 1980s there was a considerable amount of new research which focused on health status and inequalities at a local and regional level, and which demonstrated that there were often wide variations within quite small areas (Townsend *et al.*, 1987). But at the national level the debate about health inequalities was stifled and it was claimed that poverty had disappeared (Moore, 1989). More recently this marginalization of environmental explanations of health status has been revised. Following the Health of the Nation strategy the Chief Medical Officer of Health established a group to review health variations. The report reasserted the need to systematically identify and tackle variations in health (Department of Health, 1995). It acknowledged that 'cumulative differential exposure to health damaging or health promoting physical and social environments is the main explanation for observed variations in health and life expectancy…'.

2.3.3 The interconnection of environment and lifestyles

Health research findings increasingly suggest that behavioural and cultural factors on the one hand and material, environmental, and structural factors (such as employment and income distribution) on the other are interrelated and interdependent. The Variations in Health Report (1995) noted that 'the expression of genetic or biological predispositions depends on behavioural or environmental factors, and behaviours may be closely related to available material resources, social roles, and living or working conditions'. If behaviour cannot easily be separated from its social context, then an attack targeted at how people behave – unless it also addresses the social, material and cultural environment in which that behaviour takes place – will have little chance of success. Hilary Graham's work on young working-class women who were regular smokers and had children under five, indicates that their smoking behaviour arose largely from their social circumstances (Graham, 1988). Smoking became the one activity that the women could choose to do 'for themselves'; it gave them a little time and personal space during a day filled with housework and responding to children's demands.

This suggests that wider material and structural changes – perhaps in child care provision, the availability of part-time work, educational provision – might be as useful, or more useful, in improving health as targeting individual behaviour. Another example is that of child accidents (Townsend *et al.*, 1988). The higher incidence of childhood accidents in lower social groups could be explained in terms of personal risk-taking and parental neglect, but it could equally well be seen as the result of unsafe local environments which create supervision problems for parents. 'In the latter view, the environment is dictating the behaviour of both mother and child' (Townsend *et al.*, 1988). It does not follow that, because individual and cultural factors influence health, lifestyle modification is necessarily the way forward.

Note down your own views about the relative importance of personal behaviour and environment in creating health status.

It is difficult to disentangle the relative importance of behaviour and environment in creating health chances. For example, there is evidence that middle-class households consume more of what are considered to be healthier foods than do working-class households (Townsend *et al.*, 1988). It may be argued that this is due to the greater willingness of middle-class people to modify their behaviour, to 'act responsibly' and rationally and to think more seriously about their health. In this view, it is the attitudes and behaviour of working-class people which need to be modified so that they too 'act responsibly'. On the other hand, it is clear that in some respects middle-class people, with higher living standards and greater material security, are able to make 'healthy choices' more easily than people from poor working-class backgrounds. A diet that meets national nutritional guidelines (NACNE, 1984) by including brown bread, lots of fresh vegetables and fruit, low-fat spreads and lean meat is more expensive to maintain than a nutritionally unsatisfactory diet of white bread, sugary and fatty foods. It may be that working-class women *are* acting responsibly and rationally – by buying the cheapest, most filling foods available to feed their families (Graham, 1984, 1993).

Claims about what is rational behaviour also seem to be class-biased: it is the lower social classes who are seen as the problem. But it is likely that middle-class people are just as liable to be influenced by 'peer group pressure' as others; witness the spectacular spread of jogging, aerobics, health clubs and fitness training with the attendant mass middle-class consumption of appropriate clothing and footwear. This may be a result of individuals making 'healthy choices', but it may also be the consequence of multi-million pound advertising campaigns and promotions which create fashions and persuade people that they must jog and work-out to keep up with their peers (Jones, 1994).

Summary of Section 2.3

Health depends upon material and social resources as well as individual action, implicating collective, political responsibility in reducing economic inequalities and adverse environmental conditions which can affect people's well-being. Too much emphasis on the lifestyle approach to health fails to acknowledge the constraints which restrict individual choice, associated with sex, class, employment, and geographical location.

2.4 The biological foundations of health

Personal lifestyle, physical environment and social influences all contribute to the circumstances which affect a person's realization of health, however this may be defined. But there is another dimension to health which is connected with biological processes. This is a primary focus of this course, and it is time now to locate this aspect of the life process within the concept of a health field that extends the perspective presented in Figure 2.3, in order to provide a holistic approach to health as described in Chapter 1. The following study introduces the biological level into the health field.

As described in Section 2.3.1, there has recently been a strong emphasis on the importance of lifestyle in connection with health. In particular, diet, exercise, and smoking have been strongly linked to one of the primary causes of death in industrialized nations, namely cardiovascular disease (diseases of the heart and blood vessels). The most familiar symptom is high blood pressure (hypertension), which is associated with strokes and cardiac damage leading to heart attack. A Medical Research Council Environmental Epidemiology Unit at Southampton University noticed a paradox: the lifestyle connection predicts that increasing affluence, resulting in heavier eating and less exercise, should increase the risk of cardiovascular disease. However, in the UK these diseases are commoner among poorer people. For instance, it is known that regions of the country that currently show high mortality from heart disease used to have high infant mortality, which is associated with poverty through inadequate nutrition, poor housing, stressful living and similar influences. The Southampton Group wondered if factors that adversely affect infant health might also cause disease in later life; that is, conditions to which the developing embryo or fetus in the womb and the early infant are exposed might have significant influences on health in the adult. To examine this possibility, the Group set up studies of individuals by using data more than 40 years old that gave relevant information about fetal development and early infancy in a group of people who had lived all their lives in a certain region.

2.4.1 Epidemiology

The type of study, in which relationships are sought between particular factors suspected of influencing contraction of disease and measures of health in communities and populations, goes under the name of **epidemiology**. This can take different forms, among which are *geographical studies* in which the incidences of particular diseases in different parts of the country are compared and related to other observations such as infant mortality, as mentioned above; and *longitudinal studies*, in which data on individuals are examined at different stages of life to see what factors may be influencing health over the life cycle. The work that will now be considered belongs to the second category. It depended upon finding accurate records on individuals that gave relevant information about the conditions they experienced as developing embryos or fetuses during pregnancy and as infants, and gathering data on the health of these individuals much later in their lives. Such data are not easy to come by, and you might think that the Southampton Group would have to set up a study with a cohort of pregnant women and wait for 40–50 years to examine the health of the individuals, then identifying whatever relationships emerged between early life and later health. However, that was hardly a practicable procedure since members of the group, and possibly the Medical Research Council itself, would not necessarily survive to complete the experiment.

❑ Can you think of a more scientific reason why setting up such a study now, to harvest data 50 years hence, is not a good procedure?

■ Because scientific understanding changes so rapidly, the questions being asked now are likely to be seen as obsolete or irrelevant in 50 years time, and the kinds of information which would be sought then will include many items not originally recorded.

❑ Is there an example within your own experience of a significant shift in attitude towards health or health therapies that has happened in the past 10 years or so?

■ You may have noticed changes in lifestyle reflected by more joggers on the road, more (or fewer) cyclists going to work, smoke-free zones in workplaces, restaurants, etc.; you may have encountered more people discussing and using complementary therapies such as homeopathy, aromatherapy, reflexology, acupuncture and spiritual healing, and these or others being offered at your local clinic.

The procedure followed by the Southampton Group was to search throughout the country for old maternity and infant welfare records that would provide relevant data on individuals now in their 40s or older. One such set of records was found at the Sharoe Green Hospital in Preston,

Lancashire (Barker *et al.*, 1990). Detailed information was available on infants born during the period 1935–43, including the baby's birth weight, placental weight, length from crown to heel, and head circumference. Those individuals who still lived in Lancashire in 1989, when the study was conducted (and so were in the age range 46–54), were traced and invited to participate in the investigation. This involved a visit by an investigator who recorded the following data: height, weight, blood pressure and pulse rate (subjects sitting down). The individuals were asked about their medical history, current medication, smoking habits, alcohol consumption and family history of cardiovascular disease. The father's occupation was used to define social class at birth, and current social class was derived from the subject's or her husband's occupation. After the interview blood pressure was measured again (subject seated throughout the interview) and the average of the two readings was used, to provide a more reliable measurement.

The investigator had not seen the birth data recorded for the subjects. This is known as working 'blind', often expressed as 'blind testing' or 'blind trials' in experiments. 'Double-blind trials' refer to experiments in which two individuals or groups carry out different parts of the study, each one ignorant of all previous history of the individuals or samples, and of each other's results. This is all designed to avoid unconscious observer bias.

Blood pressure measurements usually include two values: a higher pressure recorded just after the heart has contracted (**systolic pressure**) and a lower value recorded while the heart is relaxed and filling with blood (**diastolic pressure**). The pressure is registered as the height in millimetres (mm) of a column of mercury (Hg) which the pressure supports (1 millimetre = 0.001 metres). Weight was recorded in pounds. Table 2.6 presents some results of the analysis. On the left is the baby's birth weight, recorded as less than 5.5 lb (< 5.5), between 5.5 and 6.5 lb (5.5–6.5), between 6.5 and 7.5 lb (6.5–7.5) and greater than 7.5 lb (> 7.5). Placental weight appears along the top, divided into groups of less than 1.0 lb (< 1), between 1.0 and 1.25 lb (1.0–1.25), and so on. Systolic pressure is entered as the average for the individuals falling into different categories, the numbers of individuals in the category being recorded in parentheses beside the average systolic pressure. For example, 77 individuals had birth weights between 6.5 and 7.5 lbs, and placental weights between 1.0 and 1.25 lb, and their average systolic pressure was 148 mmHg. The average systolic pressure for all individuals within a birth weight range, irrespective of placental weight, is presented in the column headed 'all', with the number of individuals in the range shown in parentheses. The *row* marked 'all' shows the other average: the average systolic pressure for all individuals whose placental weight falls into a certain range, irrespective of birth weight. The lower part of the table gives the diastolic pressures for the same groups as the upper part.

Table 2.6 Average systolic and diastolic blood pressures of men and women aged 46–54, according to placental weight and birth weight.

	Placental weight/lb				
Birth weight/lb	< 1.0	1.0–1.25	1.25–1.5	> 1.5	all
	Systolic pressure/mmHg				
< 5.5	152 (26)	154 (13)	153 (5)	206 (1)	154 (45)
5.5–6.5	147 (16)	151 (54)	150 (28)	166 (8)	151 (106)
6.5–7.5	144 (20)	148 (77)	145 (45)	160 (27)	149 (169)
>7.5	133 (6)	148 (27)	147 (42)	154 (54)	149 (129)
all	147 (68)	149 (171)	147 (120)	157 (90)	150 (449)
	Diastolic pressure/mmHg				
< 5.5	84	87	87	97	86
5.5–6.5	84	88	85	93	87
6.5–7.5	84	84	84	90	85
> 7.5	78	85	85	88	86
all	84	86	85	89	86

A very rough indication of average systolic pressure for people older than 40 is given by the formula 100 + age.

❑ What range of systolic pressures would be expected for the sample of individuals in this investigation, using the above formula?

■ Since the individuals were in the age range 46–54, the average systolic pressures would be expected to fall in the range 146–154 mmHg.

❑ Do the figures in Table 2.6 conform to this expectation?

■ The overall average for the whole group is 150 mmHg, which is right in the middle of the expected range. The subgroups averaged for each of the four birth weight categories range from 149–154 mmHg, which is also within the expected range of 146–154 mmHg. But there are also groups whose average values fall outside the expected range, either below (133, 144 mmHg) or above (160, 166, 206 mmHg).

Using the table, answer the following questions.

❑ What is the relationship between birth weight and systolic pressure for individuals with placental weights less than 1 lb?

■ As birth weight increases, systolic pressure decreases. There is an inverse relationship.

❑ Is this inverse relationship followed consistently in each of the columns corresponding to different ranges of placental weight?

■ No. In the third column (placental weight between 1.25 and 1.5 lb), the lowest systolic pressure (145 mmHg) is for individuals with birth weights between 6.5 and 7.5 lb, while in those with weights greater than 7.5 lb, it rises to 147 mmHg. However, the overall trend in the data is the same.

❑ Is the same trend shown in diastolic pressures?

■ The trend is there, but it is very much weaker.

❑ What is the relationship between placental weight and systolic pressure for individuals with birth weights greater than 7.5 lb?

■ There is a tendency for systolic pressure to increase as placental weight increases.

❑ Is the same tendency shown in the other rows?

■ Yes.

The trends that occur in the data have to be checked for reliability by statistical methods, which allow one to determine the probability that such trends could appear just by chance, for the number of individuals measured. Clearly the more individuals, the more reliable is any trend revealed by the analysis. The numbers involved in this study allow for confidence in the conclusion that large placental weight and low body weight at birth are both correlated with higher blood pressure in later life. The Southampton Group analysed these data in many different ways to examine the relationships within them. In particular, they analysed the connection with hypertension, defined as a systolic pressure of greater than 160 mmHg. The strongest indicator of risk is given by the combination of placental weight and birth weight: highest blood pressures and risk of hypertension were among people who had been small babies with large placentas.

❑ What relationship would you expect to find between the weights of the adults in the Preston study and their blood pressures, and also between alcohol consumption and blood pressure, based on the connection between lifestyle and health (Sections 2.3.1 and 2.4)?

■ Both higher weight and higher alcohol consumption would be expected to correlate with higher blood pressure, and they did.

This is in agreement with the results of many other studies on the relationship between lifestyle and blood pressure. However, the relation of placental weight and birth weight to blood pressure was stronger, and it was independent of current body weight and alcohol consumption. That is to say, a large baby with a small placenta will tend to result in an adult with lower blood pressure, irrespective of lifestyle (as measured by the two indicators of alcohol consumption and weight), and conversely for a small baby with a large placenta. The two measurements taken at birth – weight of baby and weight of placenta – provide better predictors of blood pressure and hypertension than do lifestyle measures of the adult.

The study just described produced unexpected results and has provoked considerable comment and criticism. A single piece of work of this type could hardly stand on its own as a significant contribution to understanding the causes of cardiovascular disease. It has to be combined with many other studies that point in the same direction before one can be confident about the relationships that are indicated. The Southampton Group has produced an impressive amount of evidence from different analyses – both its own and those of other groups – supporting their hypothesis that there is a significant connection between the environment experienced by the developing fetus and the young infant, and later health (Barker, 1992). Lifestyle certainly influences health, but not as much as the earliest conditions of the developing individual. Measures of health used in these studies included not just blood pressure but proneness to respiratory infection, capacity to use glucose efficiently (related to diabetes), concentrations of fibrinogen in the blood (related to blood clot formation), among others. These all support the basic hypothesis about early environment and later disease.

However, there are many in the community of epidemiologists who are critical of the conclusions of the Southampton Group, and for good scientific reasons. They point to extensive studies carried out in other countries that failed to establish any connection between birth weight and blood pressure in late adolesence, for example, or which indicated the *opposite* relationship to the results obtained by the Southampton Group. Such disagreements are an intrinsic part of the scientific dialogue and have their origins in many different aspects of the research. Different studies are carried out in different ways, using different numbers of subjects, of different ages and in different countries. These introduce sources of variation that need to be investigated by yet more research to identify the factors responsible for the differences.

Another criticism is that the data might be explained by genetic mechanisms that determine both the blood pressure of the child *and* growth of the placenta. It is known that mothers' blood pressures are related to those of their children. So it is possible that placental weight is linked to adult blood pressure by a genetic mechanism that somehow determines both the blood pressure of the child and the growth of the placenta. To test this, the Southampton Group used data from a study by another investigator on 5 161 women undergoing a first pregnancy in Oxford during 1987–8 (Barker *et al.*, 1990). This used the blood pressures recorded at their first antenatal clinic attendance and examined relationships with birth weight of their babies, and placental weight. No correlations were found, so these results fail to support the hypothesis of a genetic link that could explain the observed connections.

There is no doubt that genes have a role to play in every aspect of an individual's development, and we shall be examining this in some detail in later chapters of this book. However, this does not mean that once all the genes of the human body have been determined, which is the goal of the Human Genome Project, we shall then understand how the human body is made and how defective genes cause genetic disease. This is because genes are only one set of contributory factors to the intricate process that is the life experience of an individual. The tangled web that links together gene activity, environmental influence and personal lifestyle is complex and subtle, and the different strands contribute in different ways at different times and circumstances. Each one has to be examined to find the dominant influences in particular contexts.

2.4.2 The programming hypothesis

The evidence that conditions experienced by the developing fetus in the womb can affect the health of the individual in later life has given rise to a hypothesis about how this may come about, called the **programming hypothesis**. It is known from studies on animals that undernourishment of pregnant rats results in stunted growth of the offspring, and that this cannot be reversed by feeding an optimum diet after birth. Even short-term exposure of fetuses or new-born animals to abnormal conditions can irreversibly affect the condition of the individual later on. For instance, a female rat injected with a small amount of the hormone testosterone during the first four days of life will develop perfectly normally until puberty. But then the animal fails to ovulate or to show normal female sexual behaviour. Some irreversible change has evidently occurred to the processes that are involved in the development of female reproductive activity. However, this happens only if the injection is given during the first four days following birth, after which it has no effect. For rats, then, the first four days of life is the *sensitive period* for sexual maturation.

The different organs and systems of the human body develop in the embryo at different times, and they each have their sensitive periods during which they are particularly prone to external influences. We shall be examining these in more detail in Chapter 5 of this book. The Southampton Group has used this knowledge to propose a programming hypothesis that could explain their observations. Because an embryo has sensitive periods in its development, environmental influences can affect its pathway of development and push the embryo into a permanently altered condition which has consequences for the future health of the individual. It can be said to have been reprogrammed relative to the developmental path of an embryo that did not experience the stressful environmental conditions. Thus we get the programming hypothesis. In relation to the results of the Preston study just described, it leads to the question: what influences on the embryo could induce a large placental to birth weight ratio and at the same time result in a baby that is prone to the development of hypertension more than 40 years later?

2.4.3 Seeing life whole

It is known from studies on animals that if there is a reduced amount of nutrients or oxygen in the mother's blood, so that the fetus experiences malnutrition or **hypoxia** (reduced oxygen), then there is a redistribution of the blood from the fetal heart such that proportionately more goes to the brain than to the body. This is a regulatory process that protects the brain from adverse conditions and is at the expense of the body (though the developing brain is relatively insensitive to changes of oxygen level until late in the pregnancy). The result is that the young have disproportionately large heads relative to the size of their bodies, though both tend to be smaller than in new-born animals that have experienced normal nutrient and oxygen supplies. It is also known that placentas tend to grow larger if the availability of oxygen or nutrients to the fetus is reduced. For example, mothers who are **anaemic** (deficient in haemoglobin, the oxygen-carrying protein in the blood) have a tendency to develop larger placentas. The same occurs with malnutrition: if the mother is undernourished, the placenta will tend to grow larger in an attempt, so to speak, to get more of the available nourishment for the fetus, though the fetus will still experience some malnutrition and so be on the small side, in general. In both instances, of hypoxia and malnutrition, the growth and development of the head is favoured over the body.

❑ What prediction could be made about the ratio of head to body size in babies with large placentas and small birth weights?

■ Since a large placental to birth weight ratio might result from either hypoxia or malnutrition (or both) in the mother, the regulatory mechanism that favours growth of the brain over that of the body under these circumstances would result in babies with small bodies relative to head size.

Since the Preston study (Section 2.4.1) included measurements of heel to crown length and head circumference of the new-born babies, this prediction could be checked. It was found that greater placental weight at any birth weight of the baby is associated with a decreased ratio of length to head circumference. So this is consistent with the possibility that a significant factor in the development of hypertension in adults is the experience of malnutrition and/or hypoxia by the fetus. The placenta would then overgrow in an attempt to compensate for these adverse conditions, but the result is likely to be babies with small bodies relative to head size. We now need to find a reason why such babies are at risk of developing hypertension as adults.

A possible link proposed by the Southampton Group involves the consequences of reduced blood flow to the body relative to the head in a fetus that is exposed to malnutrition or hypoxia. As the arteries form and develop in such a body, they would experience lower blood pressure. Both animal studies and observations on humans have shown that arteries that develop under conditions of normal blood pressure tend to be thick-walled and elastic, whereas arteries that have been exposed to low blood pressure during their development are thin-walled and inelastic (Berry, 1978). The processes involved in this response will be considered in more detail in Chapter 5. It is a general characteristic of tissues and organs that they respond to use and increased demand by becoming better able to resist and respond. Muscles become larger and stronger in response to work, smaller and weaker if they are not used; bones become stronger in response to increased load (within limits), and get thinner and weaker if they are not used.

❑ Astronauts on space missions experience no weight on their bones or normal tension on their limb muscles. How can they prevent their bones and muscles from becoming weaker during long space flights?

■ They must exercise in such a way that bones and muscles experience mechanical loads, such as by isometric exercises in which one muscle is pitted against another, which puts stress on the bones to which the muscles are attached.

Most of these responses of tissues and organs in the adult are reversible. However, during the development of the fetus and the early infant, developmental processes can be diverted along pathways that have irreversible consequences, as we have seen. The formation of thin-walled, inelastic arteries under conditions of low blood pressure is one of these irreversible developments. The arteries of the adult then have reduced elasticity and this can result in proneness to hypertension. So a possible link is revealed between the initially quite surprising relationship that the Southampton Group discovered between the ratio of placental to baby weight and the tendency for the adults to develop high blood pressure.

This points to health as a process which involves the conditions and experiences of an individual over the whole of a life-time, from conception to death. Responsibility for health is not in the hands of individuals alone. It also lies with the community and with society, in ensuring adequate nutrition of its members and an absence of severe environmental stress due to such factors as air and water pollution, homelessness, unemployment and personal isolation, that can act through pregnant mothers on the fetus. The fact that there has been a significant increase in inequality in the UK since 1979 means that the health of the nation will deteriorate simply because fundamentals are being ignored. (The Income and Wealth Inquiry conducted by the Rowntree Foundation in 1995 showed that 20–30% of the population suffer social hardship.)

Though well protected from a considerable range of influences, the developing person in the womb is prone to lasting damage from excessive stress, just as is an adult. The work of the Southampton Group reinforces what common sense suggests. It draws attention again to the formative period of human life, the relationship between the mother and the developing child in the womb, where the foundations of health are established. We shall return to a closer examination of the processes involved in the embryo and fetus in Chapters 4–6. But first, in Chapter 3, we introduce the basic chemical and biological background necessary to understand the development and functioning of the human body.

Summary of Section 2.4

Evidence for connections between environmental conditions experienced by individuals and their health can be obtained by epidemiological studies. One such study suggests that malnutrition or hypoxia experienced by a developing fetus due to an inadequate diet or smoking by the mother puts the child at risk of hypertension (high blood pressure) in later life. This emphasizes collective responsibility for the welfare of mothers to ensure the health of future generations.

Objectives for Chapter 2

After completing this chapter you should be able to:

2.1 Define and use, or recognize definitions and applications of, each of the terms printed in **bold** in the text.

2.2 Explain how and why health may be defined in different ways. (*Question 2.1*)

2.3 Describe and evaluate the contemporary 'lifestyles' approach to health. (*Question 2.2*)

2.4 Assess the significance of environmental explanations of health status. (*Question 2.3*)

2.5 Understand the interconnection of behavioural and environmental factors in determining health status. (*Question 2.4*)

2.6 Understand how epidemiological studies can be used to indicate possible causal factors in disease. (*Question 2.5*)

2.7 Explain the programming hypothesis and describe the proposed link between fetal environment and hypertension in later life. (*Question 2.6*)

Questions for Chapter 2

Question 2.1 (*Objective 2.2*)

The Peckham Health Centre pioneers, Scott Williamson and Innes Pearse, wrote 'by the study of health is implied study of the unfolding of the fullest human capacities' (Stallibrass, 1989). Explain what concept of health underpins this definition.

Question 2.2 (*Objective 2.3*)

A very overweight neighbour of yours with two small children and a husband who is unemployed is advised by her doctor to go on a diet. She stays on it for several weeks but relapses when her younger child falls ill. Talking to you about this, she obviously blames herself for her failure. How would you respond?

Question 2.3 (*Objective 2.4*)

How would you reply to a critic who claimed that health inequalities are 'an invention of medical researchers'?

Question 2.4 (*Objective 2.5*)

The incidence of accidents at work is highest in occupational class V males. How would you explain this?

Question 2.5 (*Objective 2.6*)

It could be claimed that epidemiological studies are of no value in the study of disease because they are unable to identify specific causes. What is your view on this?

Question 2.6 (*Objective 2.7*)

Suppose that a friend of yours who is pregnant and also a heavy smoker has heard that smoking tends to reduce the baby's weight thus making the delivery easier and safer. What would be your advice to her?

References

Ashton, J. and Seymour, H. (1988) *The New Public Health*, Open University Press, Buckingham.

Barker, D. J. P. (1992) *Fetal and Infant Origins of Adult Disease*, Tavistock Press.

Barker, D. J. P., Bull, A. R., Osmond, C. and Simmonds, S. J. (1990) Fetal and placental size and risk of hypertension in adult life, *British Medical Journal*, **301**, 259–62.

Berry, C. L (1978) Hypertension and arterial development: long-term considerations, *British Heart Journal*, **40**, 709–17.

Blaxter, M. (1990) *Health and Lifestyles*, Tavistock, London.

Cornwell (1984) *Hard-Earned Lives: Accounts of Health and Illness from East London*, Tavistock, London

Cox, C. B. *et al.* (1987) The *Health and Lifestyle Survey: Preliminary Report, The* Health Promotion Research Trust, London.

Cox, C. B. *et al.* (1993) *The Health and Lifestyle Survey: Seven Years On*, The Health Promotion Research Trust, London.

Department of Health (1992) *The Health of the Nation*, HMSO, London.

Department of Health (1995) *The Health of the Nation (Variations in Health Report): What Can the Department of Health and the NHS Do?*, DoH, London.

Department of Health and Social Security (1987), *Promoting Better Health*, HMSO, London

Graham, H. (1984) *Women, Health and the Family*, Wheatsheaf, Health Education Council, Chichester.

Graham, H. (1988) Women and smoking in the United Kingdom: implications for health promotion, *Health Promotion*, 3(4), pp. 371–82.

Graham, H. (1993) *Hardship and Health in Women's Lives*, Wheatsheaf, Health Education Council, Chichester.

Herzlich, C. (1974) *Health and Illness*, Academic Press, London.

Jones, L. J. (1994) *The Social Context of Health and Health Work*, Macmillan, Basingstoke.

Lalonde, M. (1974) *A New Perspective on the Health of Canadians*, Ministry of Supply and Services, Ottawa.

Maslow, A. (1954) *Motivation and Personality*, Harper and Row, New York.

Moore, J. (1989) The *End of the Line for Poverty*, speech delivered by the Secretary of State for Social security, 11 May, DHSS, London.

NACNE (National Advisory Committee for Nutritional Education) (1984) *Nutritional Guidelines for Health Education in Britain*, Health Education Council, London.

Pearse, I. H. and Crocker, L. H. (1943) *The Peckham Experiment: A Study of the Living Structure of Society*, Collins, London.

Pill, R. (1990) Change and stability in health behaviour: a five-year follow-up study of working-class mothers, in *Lifestyle, Health and Health Promotion*, Symposium Proceedings, Ch. 5, pp. 63–79, Cambridge Health Promotion Research Trust.

Pill, R. and Stott, N. C. H. (1982) Concepts of illness causation and responsibility: some preliminary data from a sample of working-class mothers, *Social Science and Medicine*, **16**(1), pp. 43–52.

PART I
SOME BIOLOGICAL BASICS

3.2 Constituents of matter

3.2.1 Atoms and elements

Everything around you, the walls of the room, the chair you are sitting in, the book in front of you, the air you breathe – in fact all matter – is made up of collections of individual particles called **atoms**. This concept has a long history and the word atom comes from a Greek word meaning indivisible. The Ancient Greeks believed that matter could not be subdivided indefinitely and they conceived the notion of the atom as a particle of matter that could not be broken down any further.

Until very recently, no one had seen individual atoms because they are so small; 20 million of even the largest atoms lined up would stretch only a centimetre. The existence of atoms was inferred from experimental observations and from theories about the nature and construction of matter. Now, however, we have pictures of real atoms, obtained using very powerful microscopes (Figure 3.1).

Figure 3.1 A thin layer of gold, in which the atoms appear as bright spots. The atoms are about 0.000 000 000 2 cm apart.

In photos such as Figure 3.1, the atoms have the appearance of solid spheres. However, the evidence from experiments that separate atoms into smaller components (i.e. doing what the Ancient Greeks thought impossible) has led to the theory that atoms are far from solid and in fact are made up of even smaller (i.e. subatomic) particles and a lot of empty space. At the centre of an atom is a 'core', the nucleus, made up of particles called **protons** which have a positive electric charge and **neutrons** which are uncharged. Around this central cluster of particles, in successive layers, are negatively charged particles, called **electrons**. The number of electrons is the same as the number of protons, so the atom does not have an electric charge overall because the positive and negative charges cancel each other. The electrons in an atom circle so fast and unpredictably that they form a cloud of negative charge around the central nucleus; these clouds (called electron shells) are usually represented in diagrams as circles (Figure 3.2).

A brief mention of **charge** – a concept that will crop up frequently throughout the course – is necessary here. We are all familiar with two types of electrical energy in everyday life: electric current or 'moving' electricity and static (non-moving) electricity. Electric current is the result of a stream of moving electrons. The effects of static electricity are seen when objects, such as hair or clothing, acquire a charge. The objects that have this electric charge do not look any different from before, but they do react with other objects differently. For instance, hair may stand on end as the similarly charged hairs repel each other. The only way you can tell that something is charged is to see how it reacts with other charged objects. An object can have a positive or a negative charge, or can be uncharged.

Objects with the same charge, i.e. both positive or both negative, repel each other. Those with opposite charges, one positive and one negative, attract each other. The positively charged protons in the nucleus of the atom attract and hold the negatively charged electrons. As you will see shortly, the principle of like charges repelling and unlike charges attracting is an important one in biological systems.

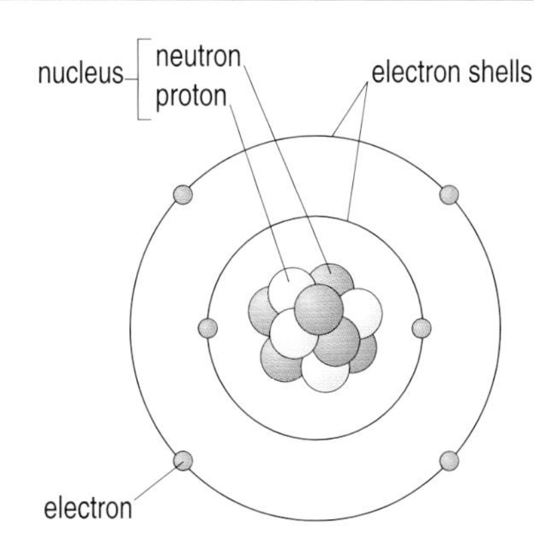

Figure 3.2 Simplified structure of an atom of carbon. The nucleus is made up of six neutrons and six protons (although only four are visible in this diagram).

Figure 3.2 shows the structure of just one particular type of atom – a carbon atom. All carbon atoms have six protons and six neutrons together in the nucleus and the positive charge of the protons is balanced by six electrons in two concentric shells outside the nucleus. The structure of, say, an oxygen atom (Figure 3.3) is different.

❑ Describe the atomic structure of oxygen.

■ An oxygen atom has eight protons and eight neutrons in its nucleus and eight electrons outside it – two in the inner shell (like carbon), and six in the outer one.

Carbon and oxygen are both examples of **elements**. Elements are defined as substances that contain only one type of atom; put another way, atoms of different elements differ from one another in their atomic structure, i.e. in the number of neutrons and protons in the nucleus and the number of counterbalancing electrons. Actually, for many elements the number of neutrons (but not the number of protons) may vary slightly; for example, most carbon atoms have six neutrons, whereas a small proportion have eight neutrons and an even smaller fraction have seven. These three atomic variants are called **isotopes**.

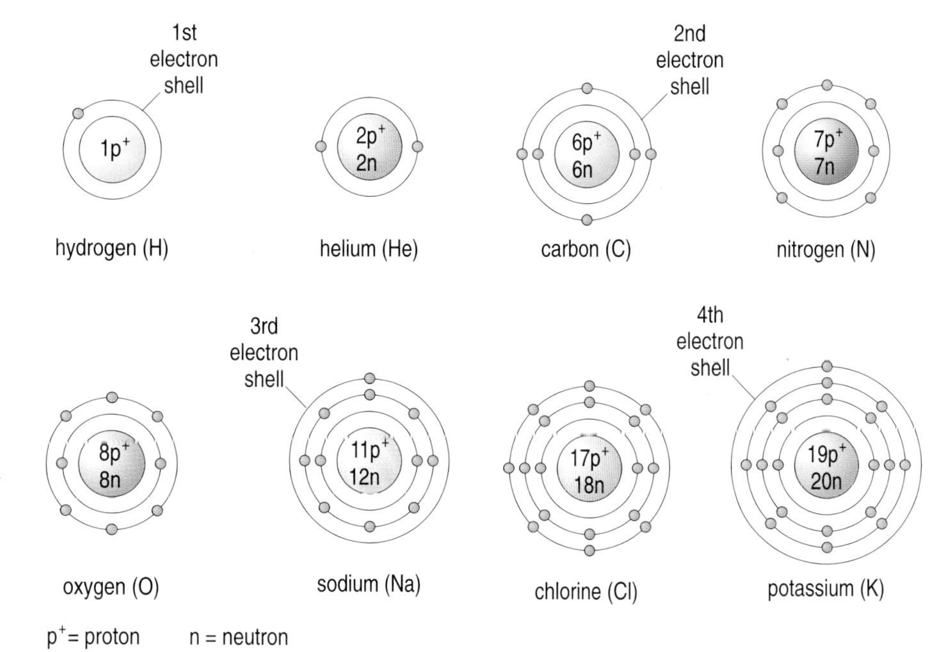

Figure 3.3 Atomic structures of various elements that are abundant in living matter. Notice that hydrogen, the lightest element, has no neutrons – its nucleus is a single proton. (The structure of the helium (He) atom is also shown.)

There are 92 naturally occurring elements, but living matter uses only a small proportion of these. In fact, organisms are made largely out of just four elements: hydrogen, carbon, nitrogen and oxygen. Hydrogen is the lightest element (i.e. made up of the smallest atoms: Figure 3.3) and exists as a flammable gas which is found in the Sun as well as on the Earth; in former times it was used to fill air-ships, with sometimes catastrophic consequences. The element carbon exists in nature as graphite (used for pencil leads), coal and diamonds, and occasionally as more complex structures. Nitrogen is the most abundant gas on Earth (making up about 80% of the air) and, unlike hydrogen, is very inert. Oxygen is the gas that makes up about 20% of the air; it is the element involved in burning – all methods used to extinguish fires involve excluding oxygen. As you will see later, oxygen fulfils a similar and vital role in living organisms in the 'burning' of food to provide energy.

❑ From your general knowledge, can you think of elements other than the four just mentioned that are found in living matter like ourselves?

■ Calcium is an element needed for strong bones and teeth; iron is needed for the proper functioning of the blood.

In addition, sodium and chlorine, the two elements of which table salt (sodium chloride) is composed, are needed for proper nerve and muscle action. Other elements such as the metals zinc, selenium, chromium and copper are required by the body in very small amounts and are therefore called **trace elements**. Many of these trace elements are poisonous if consumed in quantities significantly larger than those occurring naturally in the diet.

Each element is represented by a one- or two-letter symbol (Table 3.1). There is no need to learn all the element symbols but it will be useful to recognize those for the main ones found in living matter.

3.2.2 Chemical compounds

Elements are not commonly found in their pure form but in combination with each other to form **compounds**. (In fact, many of the elements were discovered by breaking down appropriate compounds.) Compounds are substances composed of two or more elements, i.e. types of atom. The most abundant compound in living matter (and indeed on Earth) is water, accounting for an average of 60% of total human body weight. Water is a compound of the gaseous elements hydrogen and oxygen. Table salt (sodium chloride) is another example; it is a compound of sodium (a metal) and chlorine (a pungent greenish gas). It is clear then that the properties of compounds can often be very different from those of their constituent elements. Just as the constituent units of elements are atoms, so many compounds are made up of **molecules** (but see ionic compounds below). For example, a molecule of water is made up of atoms of hydrogen and oxygen.

Table 3.1 The major elements found in living organisms and the symbols used to represent them. (Much of the information under 'Comments' will become clearer to you after you have read the chapter, and the remainder, when you have studied Books 2 and 3 of the course.)

Element and its symbol	Percentage of total body mass	Comments
oxygen (O)	65.0	constituent of water and organic molecules; needed for cell respiration, which produces adenosine triphosphate (ATP), an energy-rich chemical in cells
carbon (C)	18.5	found in every organic molecule
hydrogen (H)	9.5	constituent of water, all foods, and most organic molecules; contributes to acidity when it is positively charged (H^+)
nitrogen (N)	3.2	component of proteins and nucleic acids (DNA and RNA)
calcium (Ca)	1.5	contributes to hardness of bone and teeth; needed for many body processes, e.g. release of neurotransmitters, muscle contraction
phosphorus (P)	1.0	component of many proteins, nucleic acids, ATP; required for normal bone and tooth structure
potassium (K)	0.4	most abundant positively charged ion inside cells; important in conduction of nerve impulses and in muscle contraction
sulphur (S)	0.3	component of many proteins, especially the contractile proteins of muscle
sodium (Na)	0.2	most abundant positively charged ion outside cells; essential in blood and intercellular fluid to maintain water balance and important in conduction of nerve impulses
chlorine (Cl)	0.2	most abundant negatively charged ion outside cells
magnesium (Mg)	0.1	needed for many enzymes to function properly
iodine (I)	0.1	vital to production of hormones by the thyroid gland
iron (Fe)	0.1	essential component of haemoglobin (oxygen-carrying protein in blood) and of mitochondrial proteins involved in ATP production
aluminium (Al), boron (B), chromium (Cr), cobalt (Co), copper (Cu), fluorine (F), manganese (Mn), molybdenum (Mo), selenium (Se), silicon (Si), tin (Sn), vanadium (V), zinc (Zn)		these elements are called trace elements because they are present in minute concentrations

The processes in which different combinations of elements are made or destroyed are called **chemical reactions**. The outcome of any chemical reaction is determined by the number of electrons in the outermost electron shells of the participating atoms. This fact may seem a bit mysterious, but the following discussion should clarify things.

- ❑ Looking at Figure 3.3, what seems to be the maximum number of electrons that each shell can hold?

- ■ The shell nearest the nucleus holds only two electrons, the next holds a maximum of eight, as does the next.

Atoms form stable interactions with other atoms, i.e. they form compounds, by either filling their outermost shell with electrons to its limit or emptying it, either way ensuring that the outermost shell is complete. An example of this is seen in the formation of sodium chloride from sodium and chlorine (Figure 3.4).

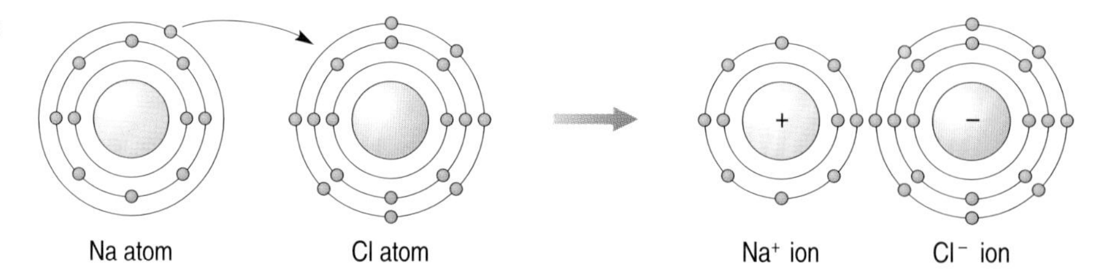

Figure 3.4 Formation of sodium chloride. An atom of sodium donates the single electron from its outermost shell, becoming a positively charged *sodium* (Na^+) *ion*, thereby filling up the outermost shell of the chlorine atom, which becomes a negatively charged *chloride* (Cl^-) *ion*.

- ❑ Why does the sodium atom now have an overall positive charge and the chlorine a negative charge? (You may need to refer back to Figure 3.3.)

- ■ Sodium has 11 protons in its nucleus and has lost a counterbalancing negatively charged electron by donating it to chlorine. Chlorine has 17 protons and in its interaction with the sodium now has 18 negatively charged electrons.

The 'charged atoms' formed in the above reaction are called **ions**. The sodium ion is written as Na^+ and the chlor*ide* (not chlor*ine*, now that it's in a compound) as Cl^-.

- ❑ The Na^+ ion has a positive charge and the Cl^- a negative charge. How then do they interact with each other?

- ■ Na^+ and Cl^- are held together by their opposite electrical charges.

This is an example of **ionic bonding**, also known as electrostatic interaction. As we shall see later, such interactions are very important in living systems.

Ionic bonds are very strong; in ionic solids, such as sodium chloride (NaCl), the oppositely charged ions are held together in a crystal lattice (Figure 3.5a). Although very high temperatures are required to melt such compounds (Figure 3.5b), they readily dissolve in water at normal temperatures. So where does the energy come from to prise the ions apart? As illustrated in Figure 3.5c, in aqueous solution each of the ions is surrounded by a shell of water molecules – the ions are said to be *solvated*. The energy made available by solvation is more than enough to separate the ions (otherwise they would not dissolve). Note that ionic compounds are not made up of discrete molecules – even in the rigid lattice shown in Figure 3.5a, we can't actually identify particular Na^+Cl^- ion pairs; all that we can say is that the number of Na^+ ions is balanced by an equal number of Cl^- ions.

Figure 3.5 (a) Lattice structure of a crystal of sodium chloride (NaCl) – the ions are fixed in space. (b) Liquid (molten) NaCl – the ions are mobile. (c) Solvated Na^+ and Cl^- ions in aqueous solution.

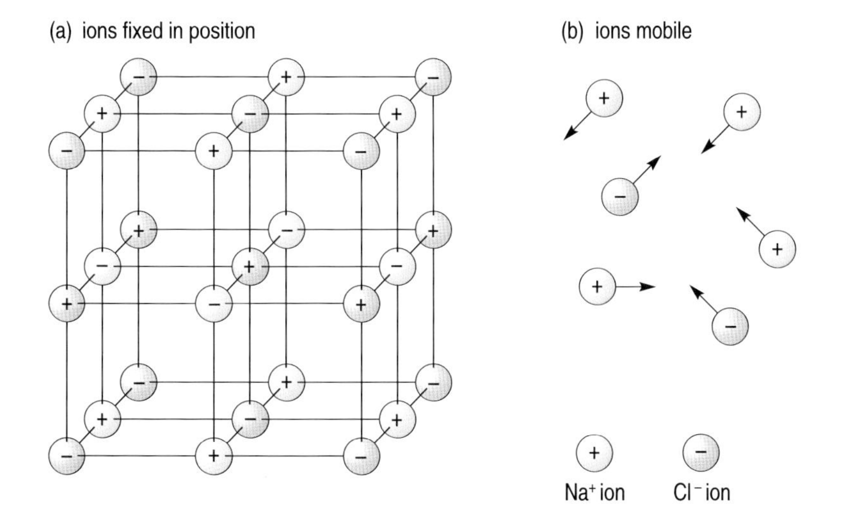

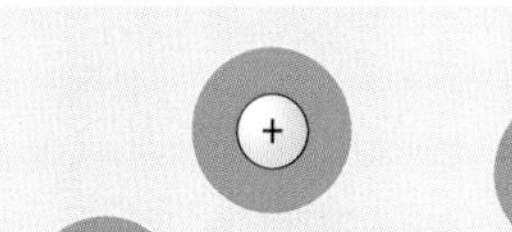

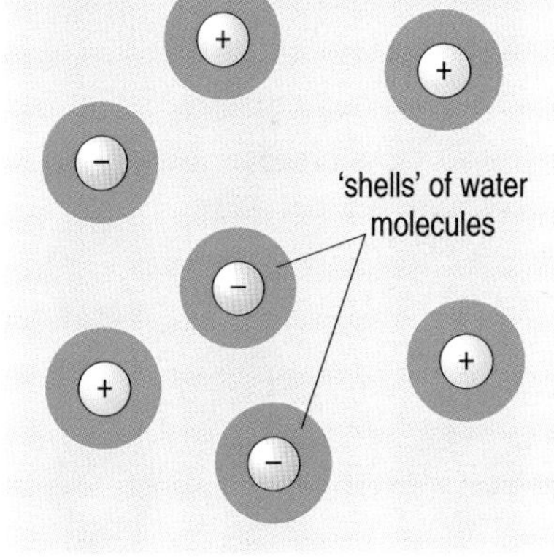

Many substances dissolve in water due to solvation. In fact, water is essential to life *because* it dissolves and holds many different substances in the same liquid. This allows chemical reactions to occur which are the basis for all living processes.

By no means all chemical reactions involve the donation and receipt of electrons to form ions. The other way in which the requirement for filled outer electronic shells can be met is by the *sharing* of electrons between the combining atoms. This is what happens in the formation of water from oxygen and hydrogen (Figure 3.6).

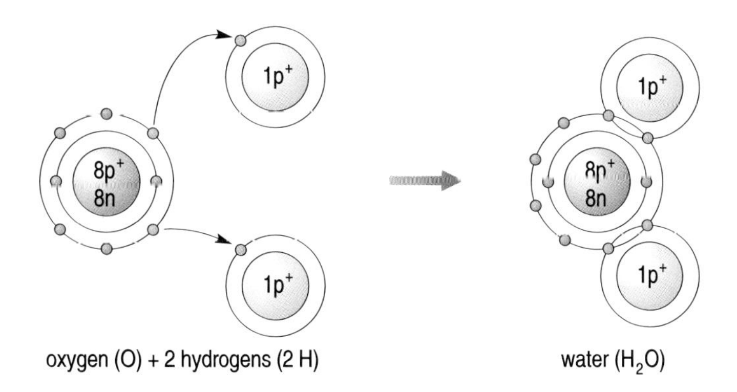

Figure 3.6 The formation of a molecule of water. The outer shell of the oxygen atom has six electrons and therefore has the capacity to hold two more, which are supplied by sharing the electrons from two hydrogen atoms. All three constituent atoms now have outer shells which are filled to capacity and therefore make up a stable molecule.

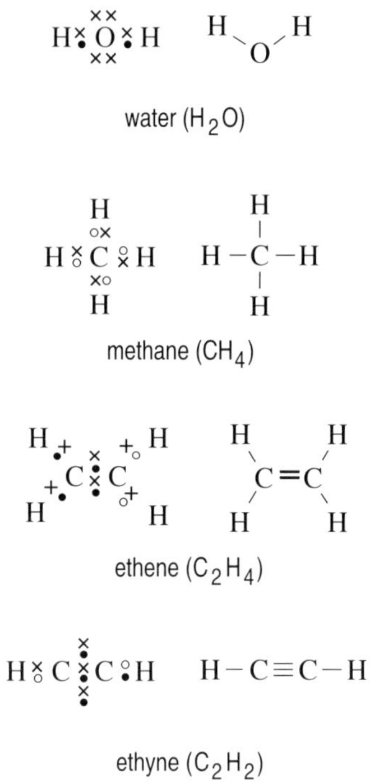

H×O×H H–O–H

water (H_2O)

methane (CH_4)

ethene (C_2H_4)

H–C≡C–H

ethyne (C_2H_2)

Figure 3.7 Examples of single, double and triple bonds in some simple hydrocarbons (compounds of carbon and hydrogen). The left-hand diagram of each pair shows the corresponding abbreviated electronic structure; here, the symbols used to denote the electrons are of no significance – they are merely used to show the origin of the electrons, i.e. which atoms in the molecule contribute the electrons to each bond. The right-hand diagrams are the corresponding structural formulae. The structure of water is also repeated here, using the same conventions.

This sharing of electrons forms a **covalent bond**. Atoms can share one pair of electrons between them to give a single bond, two pairs shared gives a double bond and occasionally three pairs are shared, giving a triple bond. You may wonder what it is about the constituent elements in a compound that determines whether the bonding is covalent or ionic. Put simply, it is the difference in the electron-holding power, or electronegativity, between the combining elements. In fact the electrons involved in covalent bonding are often not equally shared. Some atoms in combination do share electrons equally but others tend to pull electrons strongly towards their nuclei. In the case of the water molecule, the oxygen atom, being strongly electronegative, pulls more of the electrons towards it, giving it a partial negative charge whilst the hydrogen atoms, being much less electronegative than the oxygen, now have a partial positive charge. Because such covalent bonds have a polarity, i.e. a positive end and a negative end, they are called **polar covalent bonds**. As we shall see, this unequal distribution of charge within molecules has important consequences for the way in which they behave in biological systems.

Now we have an explanation for the process of ion solvation in aqueous solution which was mentioned above – the partial negative charge on the O atoms in water molecules is attracted to the positively charged ions (Na^+ in our example) and the partial positive charge on the H atoms is attracted to the negatively charged ions (here Cl^-).

Before we go on to consider the main types of molecules in the human body, we need to look at how molecular structures are represented. It would take too long to draw out the structures of all molecules (and would be impossible for very large molecules) so a kind of shorthand is used, called the **molecular formula**. For example, the molecular formula of water is written as H_2O, showing there are two atoms of hydrogen (H) and an atom of oxygen (O); likewise, a molecule of methane, CH_4, has four atoms of hydrogen and one of carbon (C) (Figure 3.7). A **structural formula**, which shows how the atoms are joined together, gives more information. Thus the structural formula of water shows that there are two single covalent bonds linking the constituent atoms. Double covalent bonds are represented by two parallel lines and triple bonds by three (Figure 3.7).

❑ Describe the types of bonds in the molecule (a hydrocarbon) that has the structural formula shown at the bottom left of the page.

■ It represents a molecule containing three carbon atoms and six hydrogen atoms. There are seven single covalent bonds altogether and one double bond.

You may have also noted that the carbon atoms form no more than four bonds and the hydrogen atoms only one.

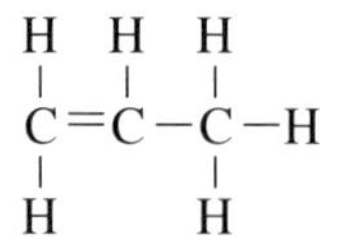

❑ Why do you think this is?

■ Carbon atoms have four electrons in their outer shell. So to complete that shell they must share *four* electrons from other atoms. That means each carbon atom can make four single covalent bonds, or two single bonds and a double bond or two double bonds, or a triple and a single bond. Hydrogen's outer shell has one electron and so needs to share *one* electron with another atom to complete that shell.

The number of bonds that an atom can make is called its **valency**.

❑ What is the valency of oxygen? (Look back at the molecular structure of water.)

■ Oxygen has six electrons in its outer shell and so needs only two more to complete its shell. Therefore it has a valency of two, hence the formula H_2O.

Obviously the structural formula given on the previous page is a flat representation of a three-dimensional molecule (called propylene, a substance used to make the plastic polypropylene). In biological systems the three-dimensional shape of molecules is very important for their functions, as we shall see later in this chapter. In Section 3.3 we take a closer look at some of the important molecules in living systems.

Summary of Section 3.2

1 Matter is made up of particles called atoms.

2 An atom has a central nucleus comprising neutrons and positively charged protons. Negatively charged electrons surround the nucleus.

3 An element contains only one type of atom.

4 Living things are mostly made up of the elements carbon, hydrogen, nitrogen and oxygen.

5 The trace elements are those required by the body in very small amounts.

6 Elements combine with each other to form compounds.

7 Covalent compounds are made up of discrete molecules; ionic compounds form a crystal lattice in which there are no discrete ion pairs.

8 In chemical reactions the outermost electronic shells of the participating atoms become filled by either the sharing or the transfer of electrons, resulting in covalent or ionic bonds respectively.

8 Polar covalent bonds are so called because the electrons are not shared equally, so the bond has a negative end and a positive end.

9 The molecular formula of a compound gives the number of each type of atom present.

10 The structural formula of a compound shows how the atoms are joined together.

11 The valency of an atom is the number of single bonds it can make with other atoms and can be predicted from its outer electronic structure.

3.3 Molecular components of living things

The water molecule discussed in the previous section is a very small molecule made of just three atoms and consisting of two elements. It is an example of an **inorganic** molecule. Most inorganic molecules do not contain any carbon atoms. *All* so-called **organic** molecules, however, contain carbon. The term 'organic' is a relic of the days when chemists thought that these compounds were derived only from living matter, or from matter that was once living. Although it was subsequently discovered that organic molecules could be made from non-living matter, the distinction was retained. Even today, chemistry is divided into organic and inorganic fields of study.

The molecules that make up living things range in size from the very small such as the water molecule to the extremely large molecules that are distinctive of living systems. These are called **macromolecules**. As we shall see in the following sections, biological macromolecules such as polysaccharides, proteins and nucleic acids are made from much smaller molecules (sugars, amino acids, nucleotides) joined together as very long chains (i.e. they are **polymers**). Table 3.2 lists some of the macromolecules that are present in living organisms, and also the small-molecule subunits of which they are composed.

All biological macromolecules and, of course, their small-molecule 'building blocks' contain carbon; life is carbon-based. Carbon atoms can form numerous, varied molecular structures, many more than can any other element. They can form long chains, join with many different elements and form ring structures. As we have seen by looking at the elements common to living matter, however, very great use is made of just a few elements, the bulk of the human body being composed of carbon, oxygen, hydrogen and nitrogen. Organic molecules built from these few starting elements provide the majority of the molecules necessary for life. The first type of biological molecules we shall look at are the carbohydrates.

3.3.1 Carbohydrates

You will probably already be aware that carbohydrates are an important component of the human diet, providing us with energy. The carbohydrates include small, soluble molecules called simple sugars or **monosaccharides** as well as large, insoluble molecules called polysaccharides (see Table 3.2). Monosaccharides are made up of carbon, hydrogen and oxygen atoms in the ratio of 1 : 2 : 1 and most of them have the molecular formula $C_6H_{12}O_6$.

❑ What information does this formula give us?

■ It shows that a monosaccharide molecule contains six carbon atoms, twelve hydrogen atoms and six oxygen atoms.

Table 3.2 Some of the macromolecular components of living things. (Information under 'Functions' and 'Examples' will be explained later in the chapter.)

Macromolecule	Small-molecule subunit(s)	Functions	Examples
Polysaccharides			
starch	glucose	energy store (plants)	potatoes, cereals
glycogen	glucose	energy store (animals)	present in liver and muscle
cellulose	glucose	structural: makes up cell walls in plants	used to make paper
*Lipids**			
fats and oils	glycerol and fatty acids	energy store	butter, lard, plant oils
phospholipids	glycerol, fatty acids and phosphate	make up cell membranes	all membranes
steroids	four fused rings of C atoms	component of cell membranes; act as chemical messengers	cholesterol, aldosterone, oestrogens, testosterone
Proteins			
globular	amino acids	help chemical reactions take place; help combat infection; transport small molecules	enzymes, antibodies, membrane proteins, haemoglobin
fibrous	amino acids	make up tissues that support body structures and provide movement	connective tissue proteins, muscle
Nucleic acids			
DNA	nucleotides	encodes hereditary information	chromosomes
RNA	nucleotides	helps decode hereditary information	messenger RNA

*As you will learn later, lipids are not strictly macromolecules but are large molecular aggregates.

Sugars containing six carbon atoms are called **hexoses** and include the monosaccharides glucose (well known as an energy source for the human body) and fructose (the sugar commonly found in fruit). Figure 3.8 shows the structural formulae of glucose and fructose.

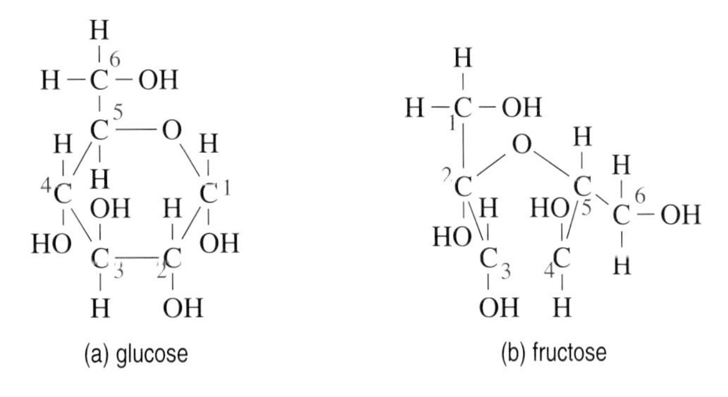

Figure 3.8 Structural formulae of glucose and fructose. (Numbering of the C atoms as done here is useful in describing the structures and chemical reactions of sugars.)

- ☐ What is the same about the structures of glucose and fructose and what is different?

- ■ Both are hexoses but the ring structures are different: glucose has six atoms joined in the ring and fructose has five.

In fact, although both hexoses have the same molecular formulae, i.e. have the same number and types of atoms, they are joined together slightly differently – hence the different ring formations. This is an example of **isomerism**; glucose and fructose are isomers (different forms) of each other. Isomerism is very common in biological molecules.

Single-ring sugars (i.e. monosaccharides) may be joined together into two-ring, soluble molecules called **disaccharides**. For example, sucrose (the sugar we all know as 'sugar') is made up of a molecule of glucose and one of fructose (Figure 3.9a). The disaccharide lactose is the sugar found in milk and is a combination of glucose with yet another hexose isomer, called galactose (Figure 3.9b). Both sucrose and lactose can be broken down (digested) in the body into their constituent monosaccharides, which can then be broken down further inside the cells to provide energy (the details of this process are dealt with in Book 3, Chapter 4).

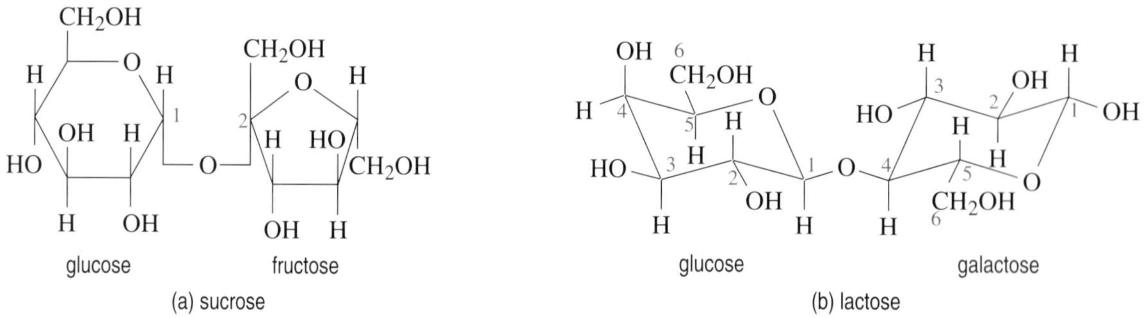

(a) sucrose

(b) lactose

Figure 3.9 Molecular structures of two common disaccharides, sucrose and lactose. The ring C atoms have been omitted, which is a commonly used convention. (Note that in (a) the bonds to the O atom linking the constituent hexoses are not really bent!) The representation used in (b) is more informative because it shows the actual shape of the molecule.

Monosaccharides can also be built up into long-chain, insoluble macromolecules called polysaccharides; polysaccharides can contain many thousands of monosaccharide units. The main polysaccharide in the human diet is a polymer of glucose, called starch, derived from plant material, e.g. potatoes, cereals, etc., in which it serves as a food (energy) store. Animals too have an energy storage polysaccharide, called **glycogen**, again a polymer of glucose units. Glycogen is stored in the liver and muscles and can be broken down into its constituent glucose subunits when the body requires glucose for energy. When there is a surplus of glucose in the bloodstream (as will be the case after a big meal) it can be stored as glycogen until it is required. (See Figure 3.10.) As you will learn later, a substance called insulin is a key player in the control of the balance between glycogen mobilization and glucose storage.

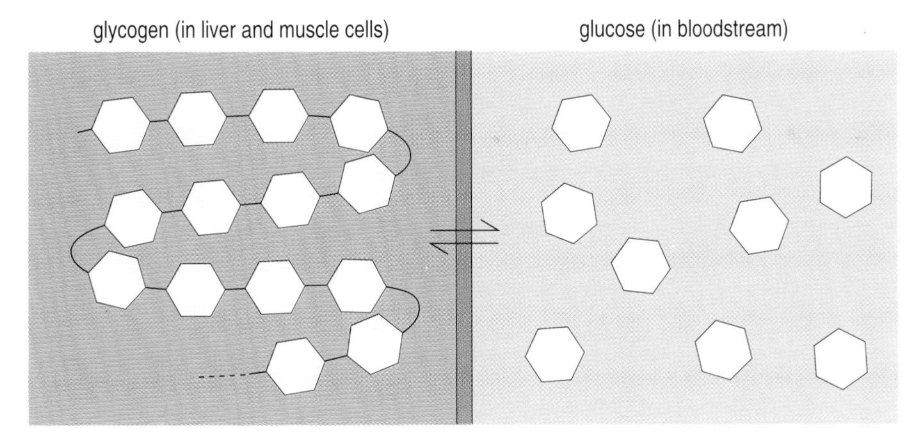

Figure 3.10 The interconversion of glucose and glycogen in the body.

3.3.2 Proteins

Another important group of biological macromolecules are the proteins. These are an essential component of the human diet and form 12–18% of the total body weight.

❑ What structures in the body are made up largely of protein?

■ Muscles – hence the perceived requirement of 'body-builders' for a high-protein diet. (In fact a typical Western diet contains more than ample quantities of protein to support significant rates of increase in muscle mass.)

As we shall see, proteins have many other functions apart from this fairly obvious structural one.

Chemically, proteins are made mainly from the elements carbon, hydrogen, oxygen and nitrogen. As with the polysaccharides, proteins are long-chain polymers, but this time the building blocks are not all the same but are a diverse group of molecules, called **amino acids** (so named because of the presence of the *amino*, $-NH_2$, and carboxylic *acid*, $-COOH$, groups of atoms in their structure; Figure 3.11a).

There are about 20 amino acids that commonly make up the proteins found in humans, eight of which must be included in the diet as they cannot be made by the body from other dietary components. All the other amino acids can be made by chemical reactions in the body. Note that the names of amino acids are usually written in a three-letter shorthand form (Table 3.3).

Table 3.3 The 20 common amino acids (three-letter abbreviations and full names). Nature of R groups: non-polar (yellow); polar but uncharged (mauve); positively charged (blue); negatively charged (red).

Gly	glycine
Ala	alanine
Val	valine
Leu	leucine
Ile	isoleucine
Met	methionine
Phe	phenylalanine
Ser	serine
Thr	threonine
Cys	cysteine
Asn	asparagine
Gln	glutamine
Tyr	tyrosine
Trp	tryptophan
Pro	proline
His	histidine
Lys	lysine
Arg	arginine
Asp	aspartic acid
Glu	glutamic acid

All the amino acids have a standard type of molecular structure: they vary only in the identity of the so-called **R group** (Figure 3.11a). The structure of the R group is crucial because it determines both the shape and chemical properties of an amino acid, which as we shall see, determines the structure of amino acid polymers, i.e. proteins.

How then is a protein formed from amino acids? Figure 3.11b shows the chemical reaction that can occur between two of the simplest amino acids, glycine (Gly) and alanine (Ala). After the Gly and Ala have joined together, via a **peptide bond**, to form the *dipeptide* glycylalanine (Gly-Ala), the groups at either end of the dipeptide (–NH_2 and –COOH) are free to undergo the same reaction with other amino acids. (In fact, in living systems the chain always grows by addition to the free –COOH group.) In this way a long chain of amino acids can be built up, i.e. synthesized, into what is called a **polypeptide** chain. The number of possible combinations of the 20 amino acids is enormous and later we shall see what it is that determines the actual *sequence* of amino acids in a polypeptide. Note that, although we have called an amino acid polymer a polypeptide here, the terms polypeptide and protein are often used synonymously. However, the distinction is useful in the case of those proteins that are made up of several polypeptide chains.

One familiar example of a protein is insulin which is made by a special group of cells inside an organ called the pancreas. As mentioned above, the function of insulin in the body is to regulate the level of sugar in the blood. (You will learn more about insulin and other regulatory substances in Book 2, Chapter 3.) Insulin is a relatively small molecule, made from only 51 amino acids, and was the first protein to have its sequence of amino acids determined – for which the head of the Cambridge research group reponsible, Frederick Sanger, obtained the Nobel Prize in 1958. This task took Sanger's group nearly six years to complete. However, the speed of sequencing amino acids in proteins has increased dramatically since then: it is now an automated process that can sequence several dozen amino acids overnight, using very small quantities of protein.

Figure 3.11 (a) The molecular structure of amino acids. (b) How two amino acids join together to form a dipeptide. Addition of another amino acid gives a tripeptide, and so on until the polypeptide chain is complete. (Notice that the formation of a peptide bond involves the elimination of a molecule of water.)

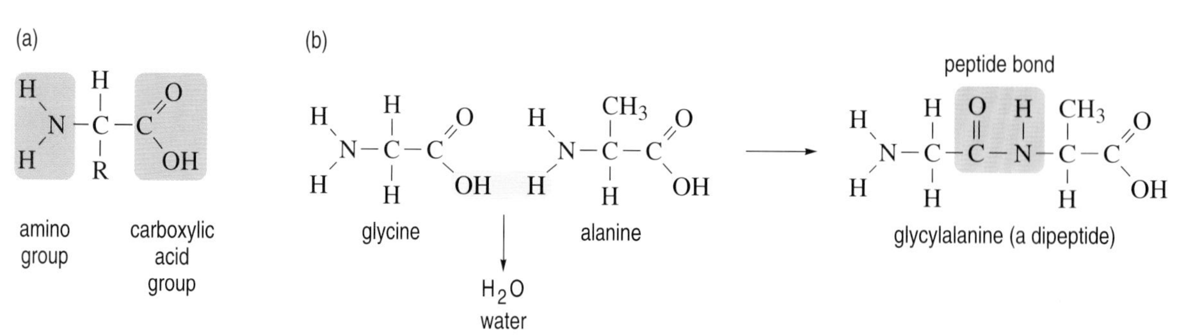

The enormous variety of amino acid sequences in proteins reflects their wide range of functions within the human body. Throughout this course you will encounter many different proteins performing a corresponding variety of functions. Some, like insulin, are *hormones*, i.e. chemical messengers secreted into the bloodstream. Others make up the overall *structure* of the human body, e.g. the muscle proteins actin and myosin, keratin found in skin and hair, and connective tissue proteins such as collagen which fill up the spaces between cells in the body (see Section 3.8.3). Other proteins form part of the immune system, *defending* the body from infection, e.g. the antibody proteins in the blood. Haemoglobin is a protein found in the blood that *transports* oxygen to the body cells. Another group of proteins *catalyse* (i.e. speed up) the many chemical reactions which take place in the body; members of this very important group are called **enzymes** (discussed in Section 3.4.1).

You are probably wondering how proteins can perform this huge variety of functions.

❑ What is different about each amino acid?

■ Amino acids differ from each other in the molecular structure, and therefore the properties, of their R groups (Figure 3.11a and Table 3.3).

Let's imagine a protein molecule surrounded by water molecules. The R groups that carry an electric charge, negative or positive (e.g. in Asp or Lys – Table 3.3) will tend to associate with water molecules easily – they are said to be **hydrophilic** ('water-loving'). Likewise the polar but uncharged R groups (e.g. Ser and Tyr) will also be hydrophilic.

❑ Why might this be so? (Think about the distribution of charge in the water molecule.)

■ You saw earlier that water molecules are polar, i.e. there are areas of negative and positive charge within the molecule. These areas will be attracted to the opposite electrical charges on the charged R groups and to the areas of opposite charge in the polar R groups.

The amino acids with non-polar R groups (e.g. Ala or Phe – Table 3.3) will tend not to associate with water molecules. In fact these so-called **hydrophobic** ('water-hating') groups will tend to cluster together in the interior of the molecule. The protein chain will thus fold up into a shape that maximizes the interaction of the hydrophilic R groups with the surrounding water molecules and at the same time maximizes the interaction of the hydrophobic R groups with each other inside the protein molecule, away from water (Figure 3.12). *The protein will therefore assume a shape determined by the sequence of its constituent amino acids.* Inevitably some charged R groups will find themselves forced to be in the interior of the molecule but pairs of these with opposite charges will attract each other and form ionic bonds within the protein chain which will further stabilize the structure.

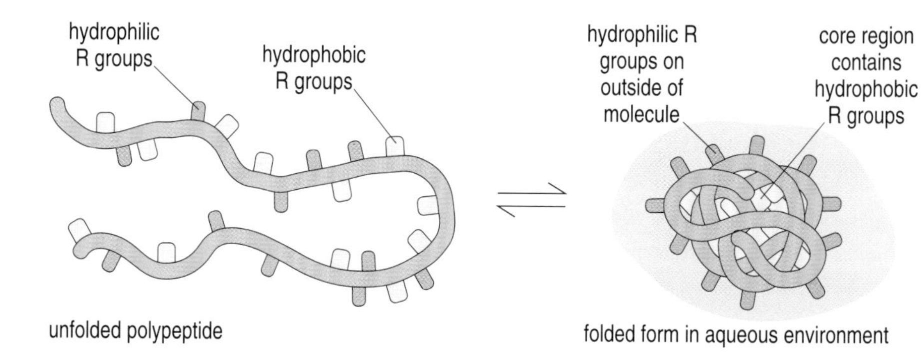

Figure 3.12 How a protein folds due to the interactions of its hydrophilic amino acid R groups with water and its hydrophobic amino acid R groups with each other.

There are two other types of interaction which are very important in the folding of protein chains. The first is called **hydrogen bonding**. Look back at the structure of the peptide bond in Figure 3.11b. You will see that it is made up of a C=O and an N—H group. Oxygen atoms, as we saw with the water molecule, are strongly electronegative – they tend to pull electrons towards themselves, resulting in a slight negative charge at the O, and a slight positive charge at the C. Nitrogen atoms are also electronegative, so there will be an excess of negative charge on the N and a small amount of positive charge on the H. This means that a C=O group of one peptide bond adjacent to an N—H group belonging to a peptide bond elsewhere in the chain will have a small but significant mutual attraction, referred to as a hydrogen bond (Figure 3.13a). Individually, hydrogen bonds are very much weaker than ionic or covalent bonds but in a protein molecule there will be a large number of these (particularly within sections of repeating structure: see below), resulting in a considerable stabilizing effect. Hydrogen bonds can also occur between polar R groups and the peptide bond groups (Figure 3.13b), between pairs of polar R groups and between a polar and a charged R group (Figure 3.13c).

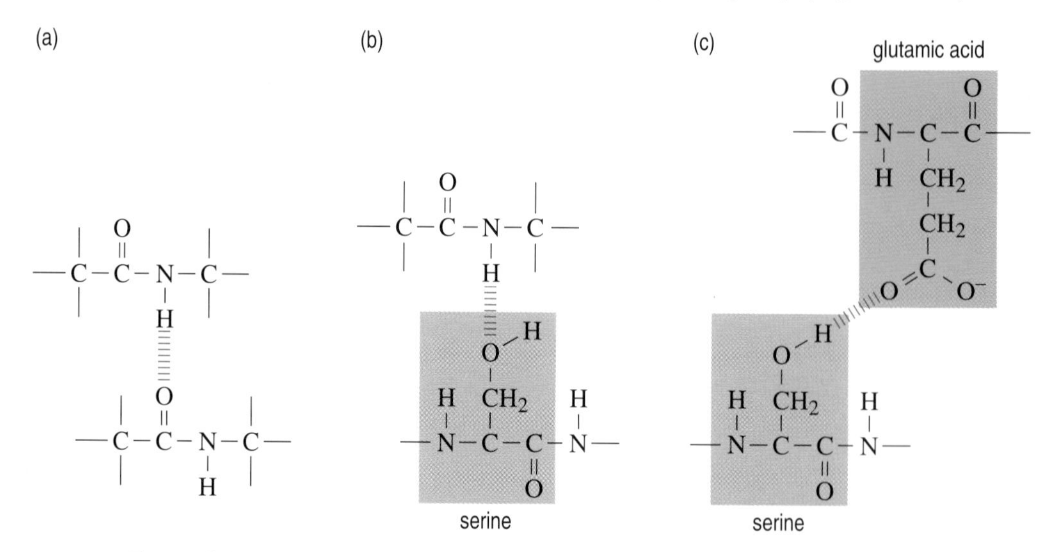

Figure 3.13 Hydrogen bonds (shown as hatched, red bars) can form between: (a) atoms in peptide bonds; (b) atoms in polar R groups and peptide bonds; (c) atoms in one polar and one charged R group. Note that because there is free rotation around single bonds, a representation of the peptide bond with the O and the H atoms on *opposite* sides of the C—N bond (as here) means the same as that with the O and H atoms on the *same* side of the C—N (as in Figure 3.11).

Hydrogen bonds are very numerous in biological systems.

❑ Why do you think this is so?

■ It is because in living systems there are many molecules that contain oxygen, nitrogen and hydrogen atoms and also because water, a combination of oxygen and hydrogen, is the liquid in which many of these organic molecules exist. Thus there are numerous opportunities for hydrogen bond formation within organic molecules, between organic molecules and water and between different organic molecules.

There is one last way in which the shape of a protein can be stabilized, and this is by **disulphide bridge** formation. As the name suggests, this involves covalent bond formation between two sulphur atoms of amino acids in different positions along the chain, giving an atomic 'bridge' across the molecule.

Methionine (Met) and cysteine (Cys) are the two common amino acids that contain sulphur atoms. Both can be found occasionally throughout most protein chains, but because of its particular chemistry, it is only the Cys which takes part in disulphide bridge formation. As the chain folds, two Cys amino acids may come into close contact, allowing the sulphur atoms in their R groups to react together to form a disulphide bond (bridge). Returning again to our insulin example, this little protein is made up of two polypeptide chains joined together by two (*inter*-chain) disulphide bridges; in addition, in one of the chains there is an *intra*-chain bridge too (Figure 3.14).

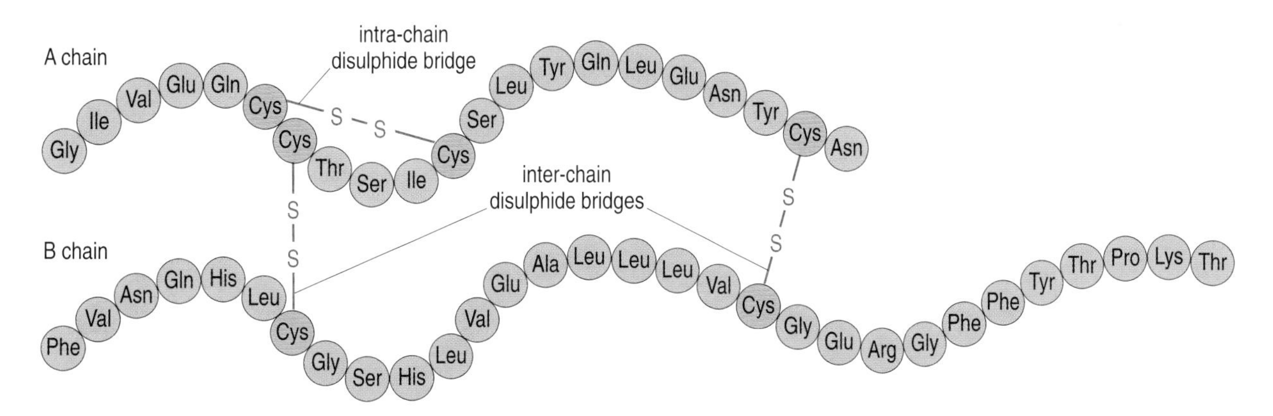

Figure 3.14 Simplified structure of insulin, a protein containing disulphide bridges.

Now that we have introduced the various forces that maintain the three-dimensional structure of protein molecules, we shall go on to consider the hierarchy of protein structure in a bit more detail.

The sequence of amino acids unique to a particular protein is called its **primary structure**. Any changes in the primary structure, even changing just one amino acid, can have great implications for the protein's three-dimensional structure and therefore its function. A well characterized example is provided by haemoglobin, the protein present in the red blood cells that is responsible for transporting oxygen from the lungs to the body cells. If, instead of glutamic acid (a negatively charged amino acid: see Figure 3.13c), there is a valine (a non-polar amino acid) at position 6 along the globin chain, the red blood cells will be sickle-shaped, instead of having the normal disc shape. This results in the characteristic sickle-cell disease syndrome shown in Figure 3.15.

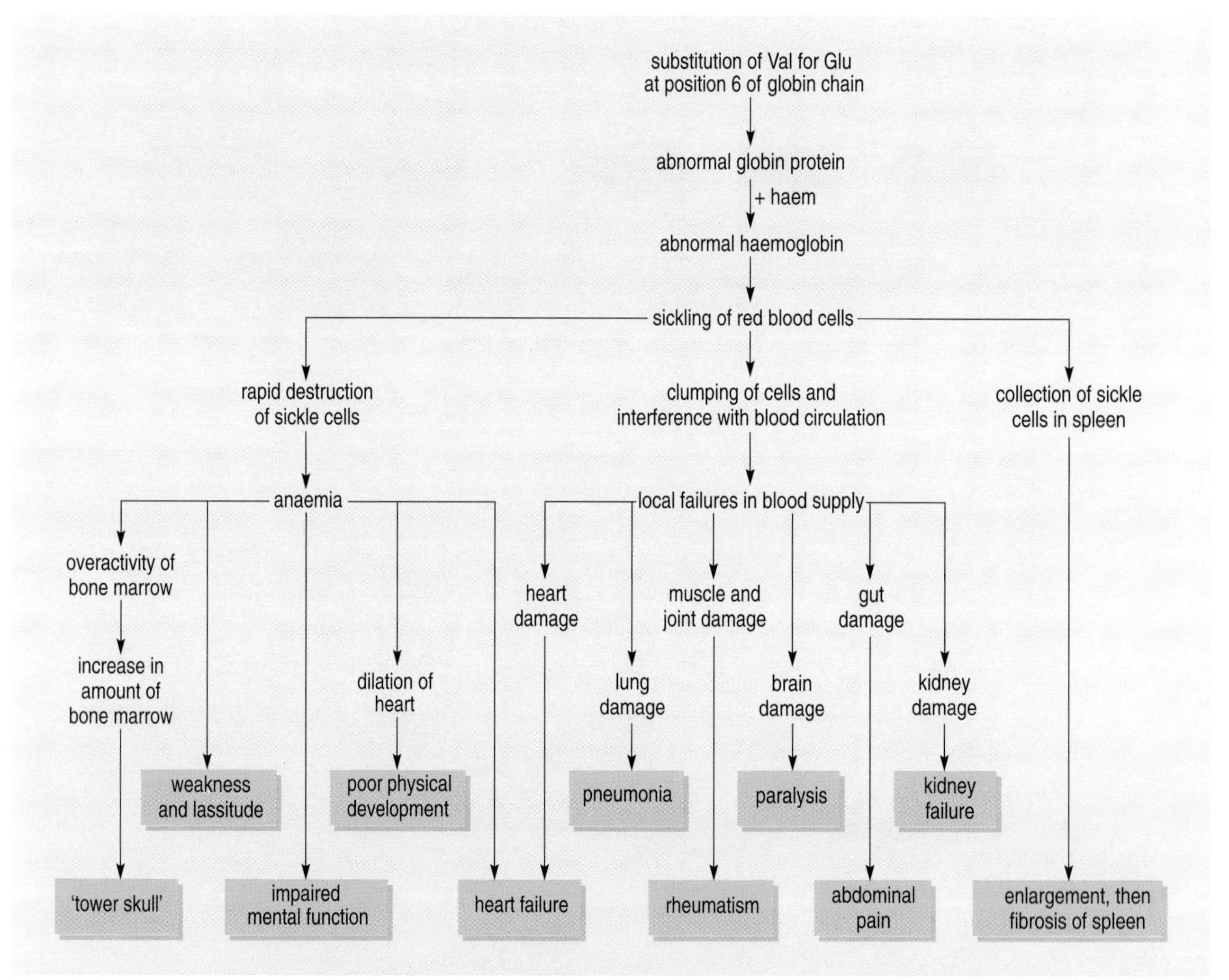

Figure 3.15 Origin and characteristics of sickle-cell disease.

The sickle-cell haemoglobin case is but one example of how the primary structure of proteins, and hence their three-dimensional structure, is finely tuned to their function.

Protein chains form two main types of structure: small and globular or long and fibrous (Table 3.2). With both types, there are two common chain folding patterns: the α (alpha)-helix (a spiral) and the β (beta)-sheet (see Figure 3.16). In the α-helix, the turns of the helix are held together by the peptide N—H to C=O hydrogen bonds described earlier. Similarly for the β-sheet, but here the bonds link adjacent separate chains, or parts of chains. Both these types of regular structure are referred to as the **secondary structure** of the protein.

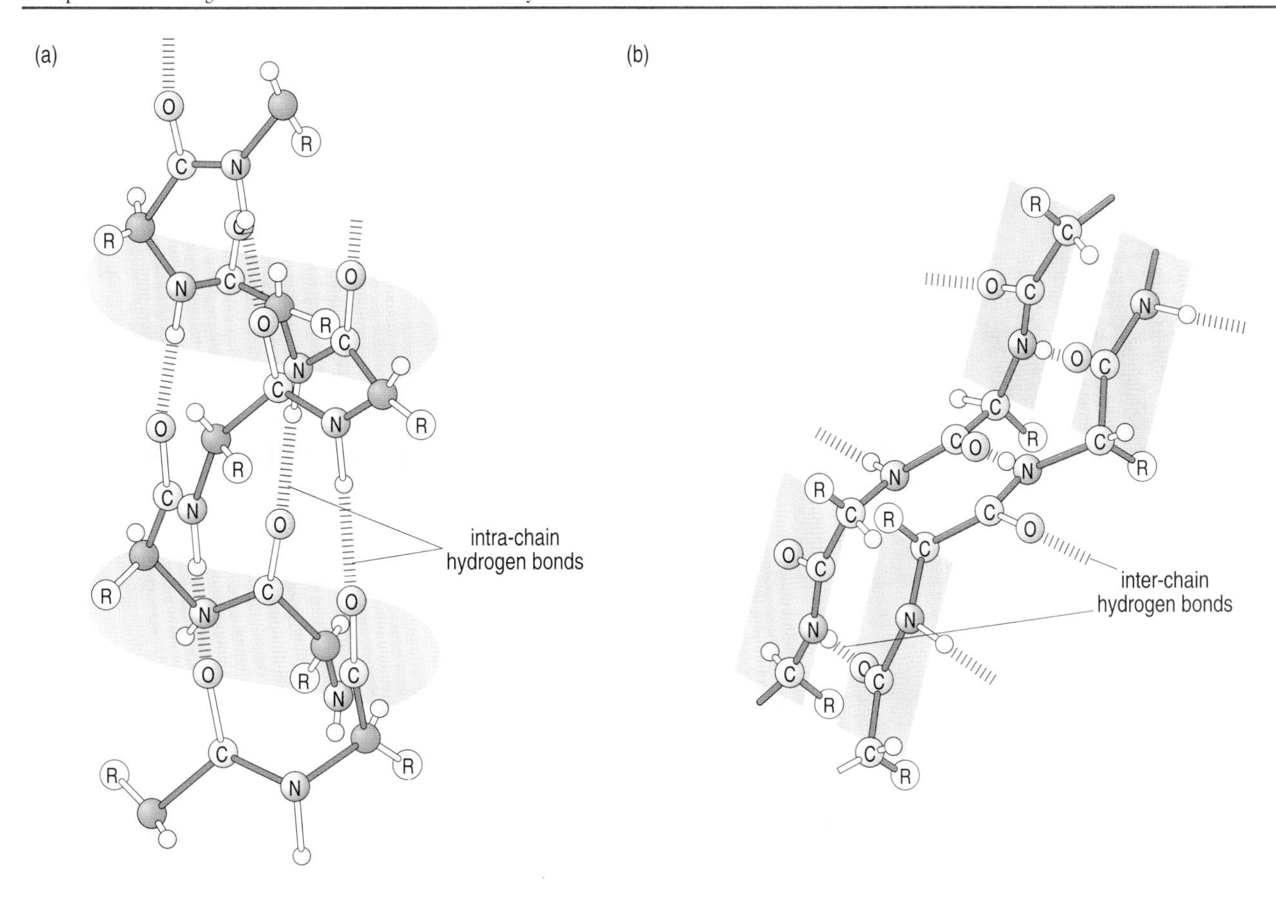

Figure 3.16 Protein secondary structures: (a) α-helix and (b) β-sheet. (The very small spheres are H atoms.)

The overall shape unique to a particular polypeptide chain is called its **tertiary structure**. We have already seen examples of how disulphide bridges can help to stabilize tertiary structure. As mentioned above, some proteins are made up of two or more separate polypeptide chains (also called subunits), e.g. haemoglobin which consists of four subunits. The shape assumed by a multisubunit protein is called its **quaternary structure**. Figure 3.17 shows the structure of an important two-subunit protein, triose phosphate isomerase – an enzyme that is present in all cells and which catalyses one step in the breakdown of glucose to provide energy. Notice that the enzyme is made up of two (identical) subunits and that in each there are regions with the regular secondary structures (α-helix and β-sheet) linked by sections of irregular chain.

❑ From the name of this enzyme, what type of reaction do you think is catalysed by triose phosphate isomerase? (*Hint*: think back to the discusion of sugar structures in Section 3.3.1.)

■ The interconversion of two isomers (these are both three-carbon sugar phosphate molecules).

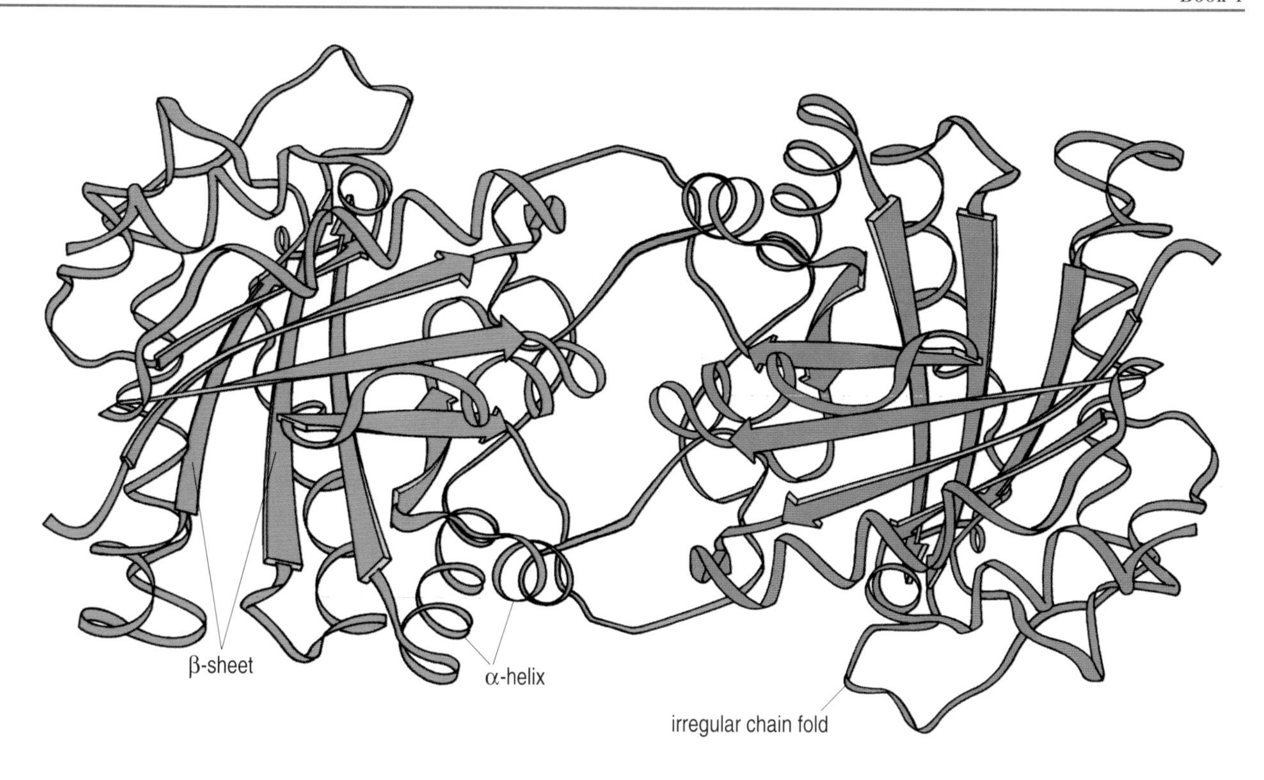

Figure 3.17 Structure of the enzyme triose phosphate isomerase. The two subunits are shown in different colours for clarity. Note that one is upside down with respect to the other. In this diagram, the common convention of representing lengths of β-sheet as broad arrows is used.

3.3.3 Lipids

We now go on to look at another important group of large organic molecules important to life, the **lipids**. In contrast to the polysaccharides and proteins discussed above, although lipids form large aggregates, they are not actually polymers. As with carbohydrates, lipid molecules contain the elements carbon, hydrogen and oxygen. However, lipids contain a smaller proportion of oxygen than do carbohydrates.

❑ What is the consequence of a lipid molecule having fewer oxygen atoms?

■ There will be fewer polar covalent bonds so less interaction with water molecules. Lipids are therefore hydrophobic.

The most well-known types of lipid are the fats and oils, collectively called **triacylglycerols** (also still known as triglycerides). Closely related chemically to fats and oils are the **phospholipids**, while there are also other lipids that are very different chemically, e.g. the **steroids** (such as cholesterol). We shall consider only the first two types of lipids, as steroids will be discussed later in the course.

Triacylglycerols are the most abundant lipids in the body; our capacity to store them is (unfortunately) almost unlimited. Body fat serves not just as an efficient energy source but also as insulation (the deposits under the skin) and as protection (around the major organs). Triacylglycerols also have a regulatory function in body metabolism.

Figure 3.18 shows the formation and structure of a triacylglycerol molecule. As you can see, it is made from three molecules of fatty acid (the 'triacyl') and one molecule of glycerol. The –COOH (carboxyl) group at the end of each of the fatty acid molecules reacts with a –OH (hydroxyl) group on the glycerol molecule.

- ❑ In the triacylglycerol molecule shown in Figure 3.18, why do you think one of the fatty acid molecules is drawn bent?

- ■ Because it has a double bond between two of the carbon atoms.

Unlike single bonds, which are very flexible, double bonds are rigid. Fatty acids without double bonds are referred to as saturated, those with one double bond are called *mono*-unsaturated, and those with many double bonds are called *poly*unsaturated. Because the chains of unsaturated fatty acids are bent, they take up more space. This looser arrangement means that unsaturated fats are less solid than are saturated ones, with their closely packed chains. Fats, which are fairly solid, contain a higher proportion of saturated fatty acids than the relatively unsaturated, and therefore more liquid, oils. You will have certainly come across these terms in food labelling and we return to their significance for health in Book 3, Chapter 7.

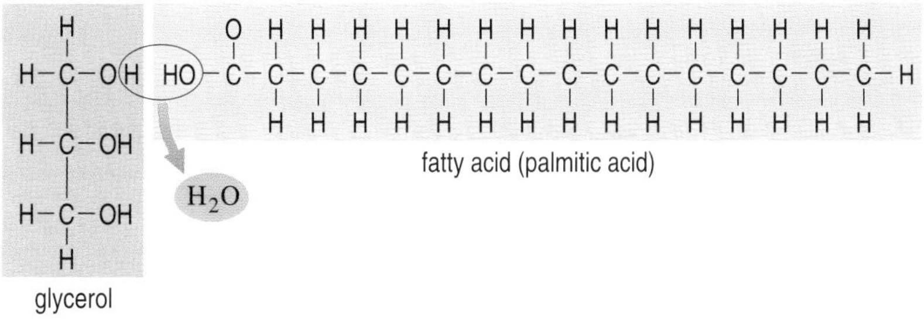

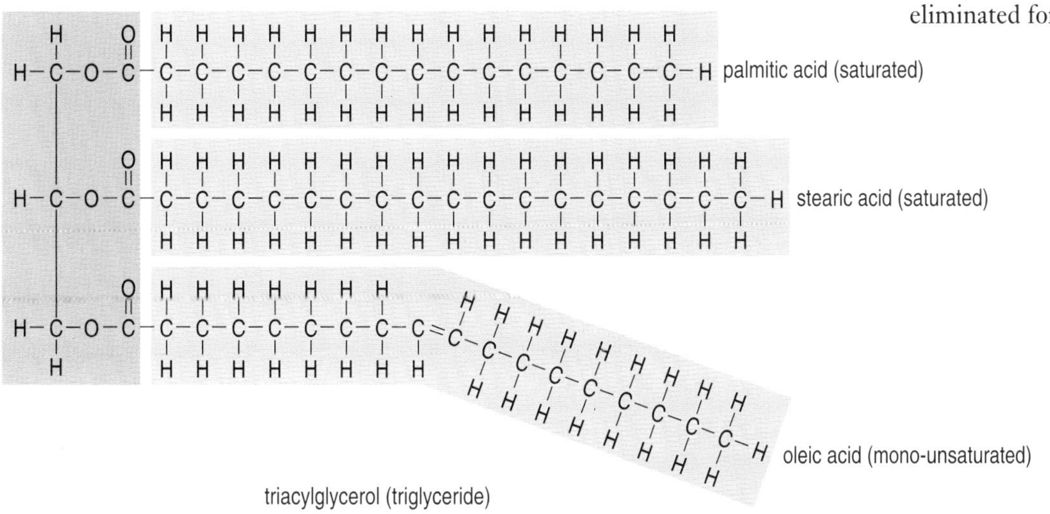

glycerol

fatty acid (palmitic acid)

H_2O

palmitic acid (saturated)

stearic acid (saturated)

oleic acid (mono-unsaturated)

triacylglycerol (triglyceride)

Figure 3.18 The formation of a triacylglycerol molecule by the successive addition of three molecules of fatty acid to a molecule of glycerol. (Notice that, in common with many types of biological synthetic reactions, e.g. peptide bond formation (Figure 3.11a), a molecule of water is eliminated for each bond formed.)

The phospholipids share structural similarities with the triacylglycerols in that they contain two fatty acids joined to a molecule of glycerol. However, instead of a third fatty acid, phospholipids have attached, as their name suggests, a group of atoms that includes a phosphorus (P) atom (Figure 3.19a). Phospholipids are the main structural component of biological membranes – the boundaries of cells and of many subcellular structures (see Section 3.5 later).

❑ In Figure 3.19a, there are electrical charges shown on part of the molecule whilst there are none on the fatty acid chains. How would you predict that each of these parts of the molecule will behave in water?

■ The charged part is hydrophilic and so will interact with the surrounding water molecules. The uncharged part is hydrophobic and, just as with protein folding, the hydrophobic groups will aggregate together, away from the water.

A phospholipid molecule is often represented as a hydrophilic 'head' joined to long hydrophobic 'tails' (Figure 3.19b). Using this simple convention, Figure 3.20 shows the structure of part of a phospholipid *bilayer*, a double layer of phospholipid molecules with the polar head groups on the outside and the non-polar tails on the inside. As you will discover soon, a biological membrane is rather more than just this passive phospholipid bilayer.

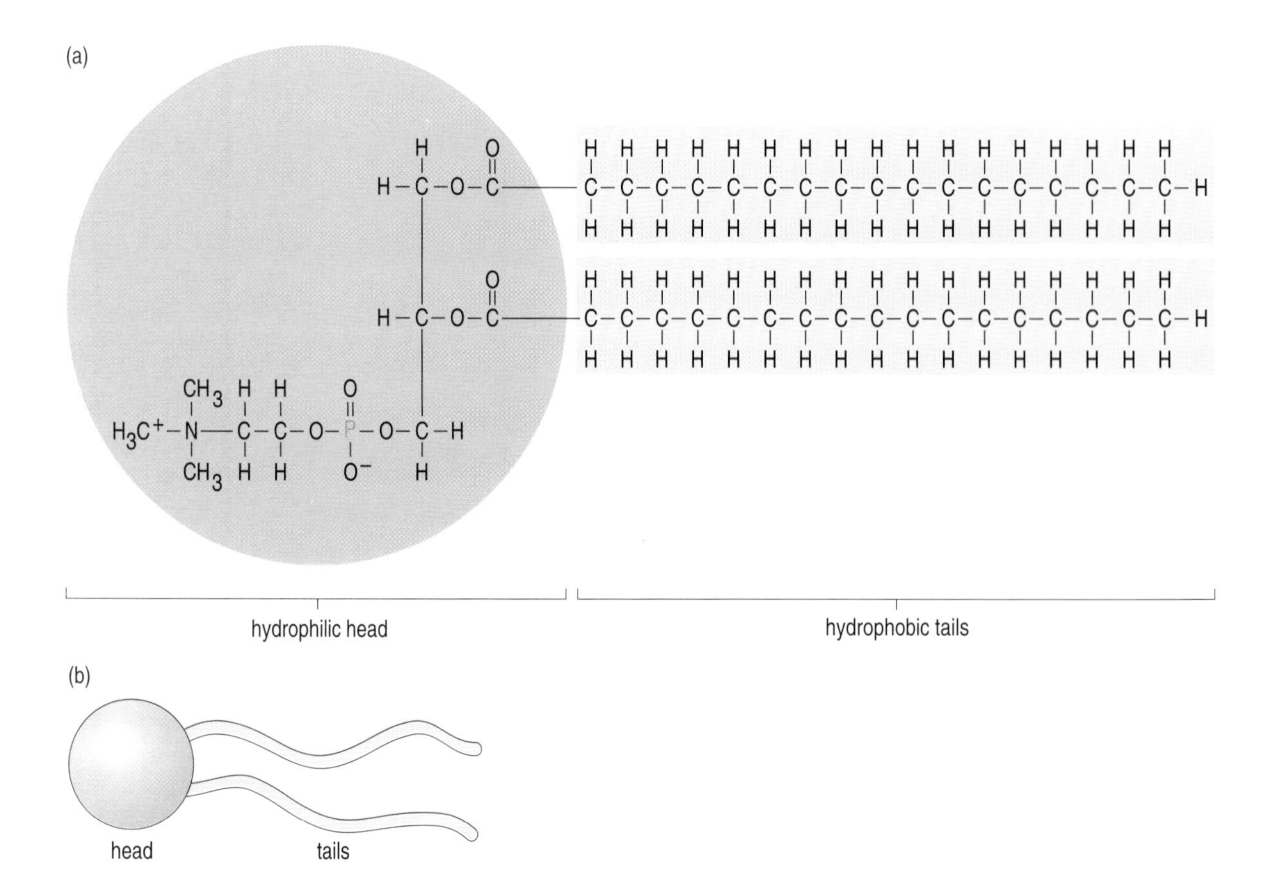

Figure 3.19 (a) Structural formula of a phospholipid molecule. (b) Diagrammatic representation of a phospholipid molecule.

There is one class of biological macromolecule that we have not yet considered, the nucleic acids. However, in order to appreciate fully the role of these important, informational macromolecules, it is necessary to become acquainted with both the chemistry (Section 3.4) and structure (Section 3.5) of living cells. We then turn our attention to the nucleic acids in Section 3.6.

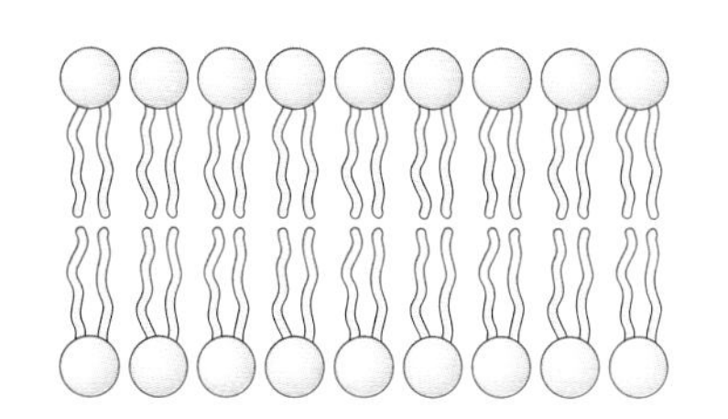

Figure 3.20 Structure of a phospholipid bilayer.

Summary of Section 3.3

1 Substances that do not contain carbon are referred to as inorganic; carbon-containing compounds are called organic.

2 Very large molecules are called macromolecules.

3 The carbohydrates are an important energy source, and include the soluble sugars (monosaccharides, e.g. glucose, fructose; disaccharides, e.g. sucrose, lactose), and insoluble polysaccharides (e.g. starch, glycogen).

4 The hexoses glucose and fructose are examples of isomers: they have the same molecular formula but the atoms are joined together differently.

5 Proteins have structural, hormonal, defence, transport and catalytic roles.

6 Proteins are macromolecules made up of amino acid building blocks.

7 The 20 amino acids found in biological proteins have different R groups.

8 Amino acids are joined together via peptide bonds.

9 Proteins fold so that the hydrophilic (water-loving) R groups are on the outside and the hydrophobic (water-hating) R groups are on the inside.

10 Hydrogen bonds help stabilize protein three-dimensional structure.

11 Cysteine-containing proteins may be further stabilized by disulphide bridge formation.

12 There is a hierarchy of protein structure: primary, secondary, tertiary and quaternary.

13 The primary structure of a protein determines its higher-order structure which in turn determines its function.

14 The α-helix and the β-sheet are the main secondary structural elements.

15 Lipids include: triacylglycerols (triglycerides), i.e. fats and oils; phospholipids; and steroids.

16 Lipids serve as an energy source, provide heat insulation, protect the major organs and some have a regulatory role.

17 Phospholipids make up the bilayer structure of cell membranes, with the polar phosphate 'head' groups on the outside and the non-polar lipid 'tails' on the inside.

3.4 Principles of cell metabolism

Having looked in some detail at protein *structure*, we shall soon be looking a bit more closely at the *function* of one particular category of proteins: the enzymes. This will serve to illustrate the importance of structure in determining function in living processes. But first we must set the scene by introducing some cell chemistry.

A huge number and variety of chemical reactions are going on in every cell of the body all the time; the sum total of all these reactions is called **metabolism**. The product of one reaction serves as the starting material (substrate) for one or more other reactions. Groups of sequential reactions are referred to as **metabolic pathways**; for example the first set of reactions involved in the breakdown of glucose to provide energy is called the glycolytic (from the Greek, meaning 'glucose-splitting') pathway. Reactions and pathways that result in breakdown of materials are referred to as **catabolic**. Those that result in the building up or synthesis of materials (e.g. the synthesis of muscle proteins in muscle cells, or of glycogen in liver and muscle cells) are called **anabolic**. As a general principle, catabolic pathways are energy-generating, while anabolic pathways are energy-consuming. The production and utilization of energy are linked by an important and universal molecule which serves as the 'energy currency' of the cell, **adenosine triphosphate** – usually known by its abbreviated name, **ATP**. Energy can be stored temporarily as molecules of ATP, via the addition of an inorganic (i) phosphate ion, P_i, to adenosine diphosphate, ADP, by coupling this reaction to an appropriate catabolic (energy-producing) reaction. The reverse reaction, in which ATP is split into ADP and P_i, thereby releasing the stored energy, is coupled to an anabolic (energy-consuming) reaction. Figure 3.21 shows the central role of ATP/ADP as the energy transfer system within cells.

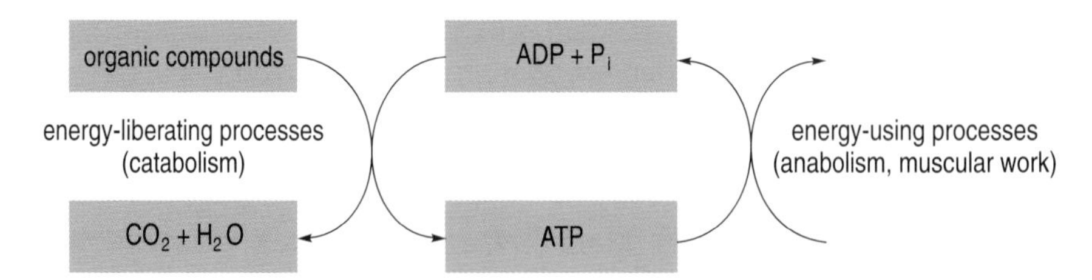

Figure 3.21 Energy economics of the cell: catabolic reactions produce ATP and anabolic reactions consume it.

Considering the catabolism of glucose ($C_6H_{12}O_6$) again, the end result is a substantial amount of energy (harnessed ultimately as ATP) and the simple waste products carbon dioxide (CO_2) and water (H_2O):

$$C_6H_{12}O_6 + 6O_2 \longrightarrow 6CO_2 + 6H_2O$$

If this reaction were to occur outside of the cell environment, e.g. if we oxidize (i.e. burn) a spoonful of sugar, a great deal of heat would be generated, and lost. However, in the cell the situation is very different – the

cell respiration reaction on the previous page merely summarizes a long sequence of reactions, each one catalysed by a specific enzyme, and each associated with a very small energy change, so that most of the energy is retained (as ATP), with a relatively small proportion being lost as heat. This important principle, which applies to all metabolic processes, is illustrated in Figure 3.22.

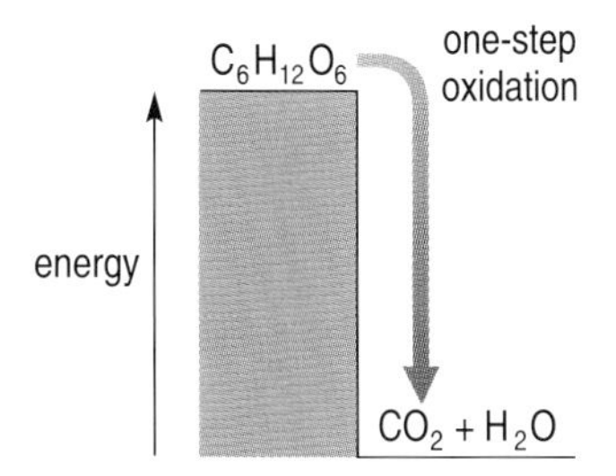

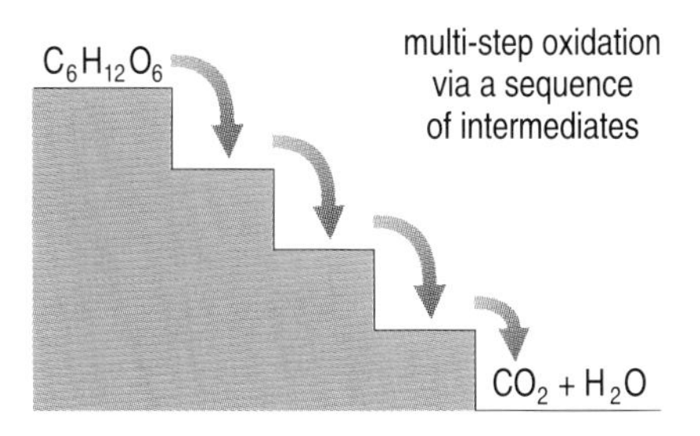

Figure 3.22 The catabolism of glucose in small steps leads to the production of the same amount of energy as in the one-step oxidation.

After that brief digression into the world of cell metabolism, you should now be able to appreciate that enzymes must be pretty remarkable proteins. In a 'typical' cell there are thousands of different enzyme-catalysed reactions all going on at the same time and all occurring at a rate that is appropriate for that particular cell at that particular time. There are 'core' sets of reactions that must occur in every cell, irrespective of its function, i.e. whether it is, say, a muscle cell, a nerve cell or a hormone-producing cell; these are often referred to as the 'house-keeping' reactions. Examples include the glycolytic pathway, the first stage in the breakdown of glucose, and the reactions by which the cell's structural macromolecules are synthesized. As well as those reactions common to all cells, cells of a particular type will have sets of enzymes and reactions *specific* to that type only. We return to the topic of cell specialization later.

You may be wondering how the *rates* of a cell's reactions are regulated to ensure that the right amounts of particular molecules are made at the right time and that the quantity of ATP produced (energy) is appropriate to the needs of the cell. As with the control of many *physiological* processes that you will meet later in the course, regulation of the cell's chemical reactions is achieved by **negative feedback** (also called end-product inhibition): the product of a metabolic pathway – be it a building block for synthesis or a molecule of ATP – when it exceeds a critical concentration, will 'switch off' an enzyme early in the pathway, thereby preventing further synthesis of the regulatory end-product (Figure 3.23). Consequently the concentration of the latter will decline as it is used up in other processes, thereby relieving the inhibition and so allowing the enzyme to start functioning again.

You will be able to understand how, in molecular terms, this inhibition occurs when we look in a bit more detail at the structure and properties of enzymes.

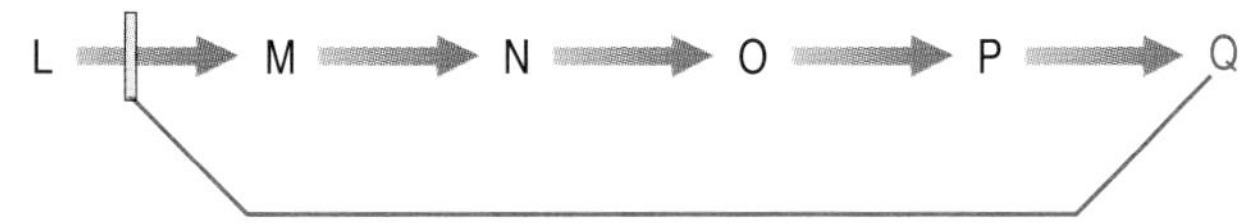

Figure 3.23 Negative feedback (end-product inhibition). The end-product of the pathway (Q) inhibits the enzyme that catalyses a reaction early on in the pathway – the conversion of L to M.

3.4.1 The role of enzymes

Enzymes are very specific **catalysts**: each one speeds up a particular chemical reaction without itself undergoing any net change in the process. The first thing to happen is that the enzyme (E) combines with a molecule of the substrate (S) to form an enzyme–substrate (ES) complex; then the transformation of substrate to product (P) occurs, P is released and the enzyme is free to combine with another molecule of substrate, and so on:

$$E + S \rightleftharpoons ES \longrightarrow E + P$$

(The two-way arrow denotes the reversibility of ES complex formation.)

You may be wondering how it is that an enzyme-catalysed reaction can be faster than an uncatalysed one using the same substrate and producing the same product(s). Put very simply, when the reaction proceeds via the enzyme-catalysed route (i.e. via an ES complex) less energy has to be put in to get the reactant moving along the route to product, so the reaction goes faster.

Each of the many thousands of enzymes present in a living organism has a unique three-dimensional structure which, as we have seen is the case for all proteins, is determined by the sequence of amino acids, i.e. the primary structure. At the surface of every enzyme is a relatively very small region, called the **active site** which is where the substrate binds and the catalysis occurs. The shape of the active site fits precisely the shape of the substrate molecule, rather like a lock matching a particular key (Figure 3.24). You should not be surprised to learn that the type of bonds holding the substrate

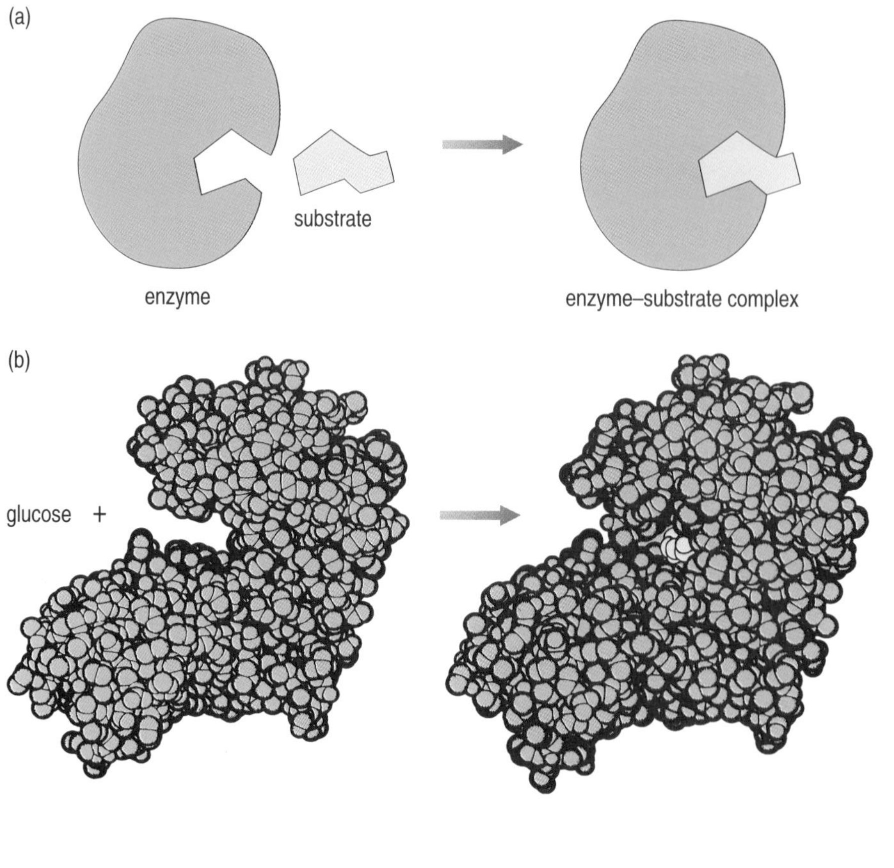

Figure 3.24 Binding of substrate at the active site to form the enzyme–substrate (ES) complex: (a) diagrammatic representation, with the active site shown as an indentation on the right side of the enzyme molecule, complementary in shape to the substrate molecule; (b) three-dimensional structure of the enzyme hexokinase binding a molecule of its glucose substrate. (This enzyme catalyses the first step of the glycolytic pathway.) Notice that the binding of glucose actually causes a small change in shape of the active site region, so the simple lock and key model of substrate binding is not wholly adequate!

to the enzyme molecule are those we have come across earlier: hydrogen bonds and electrostatic interactions with amino acid R groups within the active site. (Incidentally the same type of lock and key fit is responsible for the recognition of foreign material, i.e. antigens, by specific antibodies.)

Unlike simple chemical catalysts which are relatively robust, the proper functioning of enzymes requires precisely controlled conditions. Rates of enzyme-catalysed reactions increase with temperature up to a point, but then fall off dramatically. This is because, as temperature increases, the weak hydrogen bonds and hydrophobic interactions are broken, thereby allowing distortion and eventually destruction of the precise three-dimensional structure of the enzyme and with it the active site region.

The acidity of the solution also affects enzyme activity. High acidity will neutralize negative charges on the protein's surface and low acidity will neutralize positive charges (see Box 3.1). In both cases, the stabilizing interactions with water molecules will be reduced with a consequent distortion of the enzyme's three-dimensional structure, including that of the active site.

Finally, we are in a position to appreciate the molecular basis of the feedback control of enzymes. Just as the active site is specific for a particular substrate, so there is another site on regulatory enzyme molecules (called the **allosteric site**) in which the regulatory end-product fits precisely. Binding of end-product to the allosteric site causes a subtle change in the three-dimensional structure which is relayed to the active site (this may be many amino acids away in the protein chain), resulting in poorer binding of substrate (Figure 3.25). Thus the regulation of enzyme activity is a clear illustration of how flexible is the structure of an enzyme molecule and how crucial is the precise shape of the active site for proper functioning.

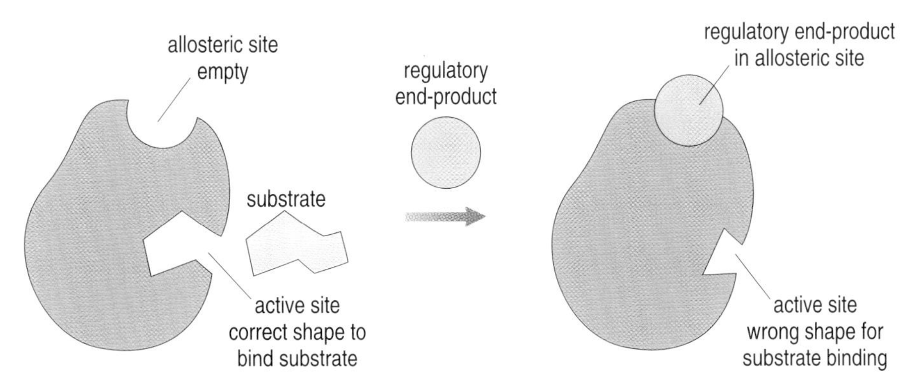

Figure 3.25 Allosteric regulation of enzyme activity.

Box 3.1 Acidity and alkalinity

An acid is defined as a substance that produces hydrogen ions (H^+) in solution. For example, hydrochloric acid (HCl) – the acid present in the stomach – dissociates thus:

$$HCl(aq) \rightleftharpoons H^+(aq) + Cl^-(aq)$$

where 'aq' means aqueous (in water) solution and the two-way arrow means the reaction is reversible. HCl is a very strong acid, i.e. it dissociates almost completely into its constituent ions. However, the acidic groups in most biological molecules are a lot weaker, dissociating to a much smaller extent. Liquid (l) water itself dissociates, but to a very small extent indeed (one molecule in 10 million) into H^+ and OH^- (hydroxide) ions:

$$H_2O(l) \rightleftharpoons H^+(aq) + OH^-(aq)$$

Because of the huge range of hydrogen ion concentrations that are possible, dealing in actual concentrations is very cumbersome. Therefore acidity of solutions is more usually expressed on a so-called *pH scale*, where a change in pH of one unit represents a factor-of-ten change in H^+. A pH value of 7 denotes a neutral solution (or pure water) and values lower than this progressively more acid solutions; values above 7, up to a maximum of 14, are for *alkaline* solutions. (For those who are interested, the explanation of the relationship between hydrogen ion concentration and the pH scale can be found in standard chemistry texts.)

In very acid conditions (high H^+ concentration and so low pH) *negatively charged* groups will tend to become neutralized, as shown below, the large concentration of H^+ 'pushing' the reaction from left to right:

$$—R^-(aq) + H^+(aq) \rightleftharpoons —RH(aq)$$

where $—R^-$ is a negatively charged amino acid R group and —RH is the uncharged form.

(It is a general principle of chemistry that if the concentration of any reactant changes, e.g. H^+ in the above equation, the reaction moves in the direction that tends to minimize that change.)

Conversely, when there is a low H^+ concentration (high pH) *positively charged* groups will tend to dissociate and so become neutral:

$$—RH^+(aq) \rightleftharpoons —R + H^+(aq)$$

where $—RH^+$ is a positively charged amino acid R group and —R is the uncharged form.

Summary of Section 3.4

1 The chemical reactions of the body are collectively called metabolism; in all cells, there are catabolic (breakdown, energy-producing) and anabolic (synthesis, energy-using) pathways.

2 ATP is the cell's energy currency.

3 The complete catabolism of glucose (to carbon dioxide and water) is called cell respiration.

4 Metabolic pathways are controlled by negative feedback (end-product inhibition).

5 Enzymes are biological catalysts; catalysis occurs at the active site and extremes of temperature or acidity/alkalinity interfere with enzyme activity via their effects on active site structure.

6 Feedback control involves binding to the allosteric site on a regulatory enzyme, which causes a small reversible change in active site structure.

3.5 Cell structure

It is now well-known that all living things are made up of similar subunits called **cells**. It was not until the invention of microscopes in the 17th century that the cellular structure of living material was discovered. Robert Hooke (1635–1703) is credited with giving cells their name, for as he examined a section (i.e. a thin slice) of cork with a microscope, he saw structures in that material which reminded him of the small, rectangular rooms in which monks lived, their cells. Cells are not all the same shape, nor do they perform the same functions; and this is not surprising considering the complexity in form and behaviour of living things. However, cells that do perform similar functions do resemble each other in structure.

Most cells are too small to be seen with the naked eye; in fact 100 human skin cells would stretch across only 1 millimetre. In the macroscopic world, measurements are made in metres, centimetres and millimetres. But when we move into the microscopic world, we need another set of measures to reflect the tiny scale of the cellular level. A convenient measure used is the micrometre (written as μm; 'μ' is the Greek letter mu). A micrometre is one-thousandth of a millimetre (mm), or one-millionth of a metre.

❑ What is the average diameter of the 100 skin cells that stretch across a millimetre?

■ Each cell is, on average, 10 μm in diameter.

A micrometre can be further subdivided into thousandths and each of these thousandths is called a nanometre (nm).

Pictures of cells taken with a light microscope show very little detail (Figure 3.26).

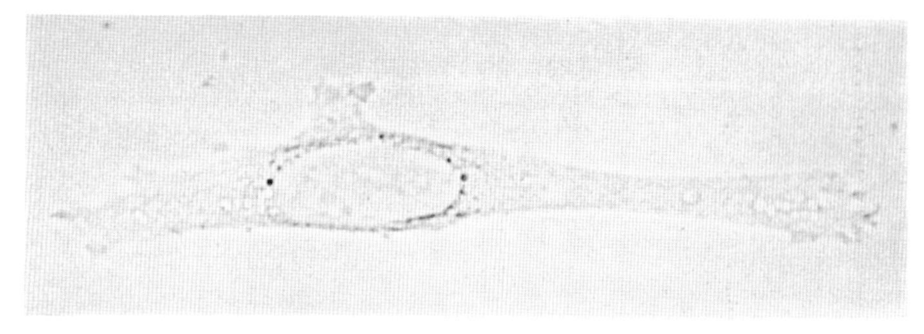

Figure 3.26 Human cell in culture seen under light microscopy (× 1 500). (Photographs of images from light microscopes are called photomicrographs.)

❑ What structures are visible in Figure 3.26?

■ Only the outer boundary of the cell, the cell membrane and a dark object in the centre of the cell, the nucleus, can be seen.

Any other structures inside the cell are hard to see, as the cells are nearly transparent; it's a bit like trying to see a contact lens in water. However, by using stains which colour particular parts of the cell (Figure 3.27a) or different types of lighting conditions (Figure 3.27b–d), light microscopy can reveal a little more internal structure.

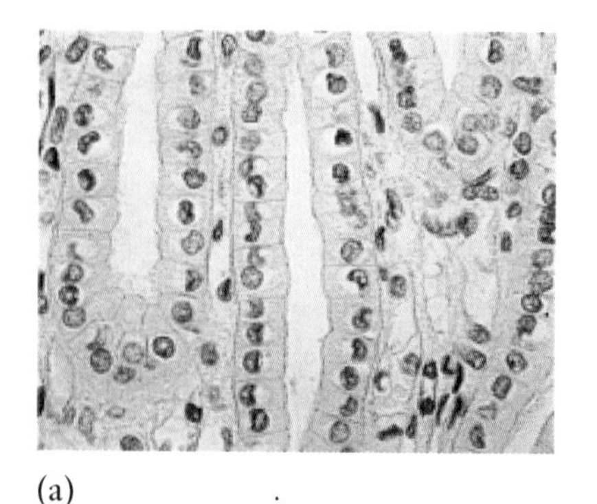

(a)

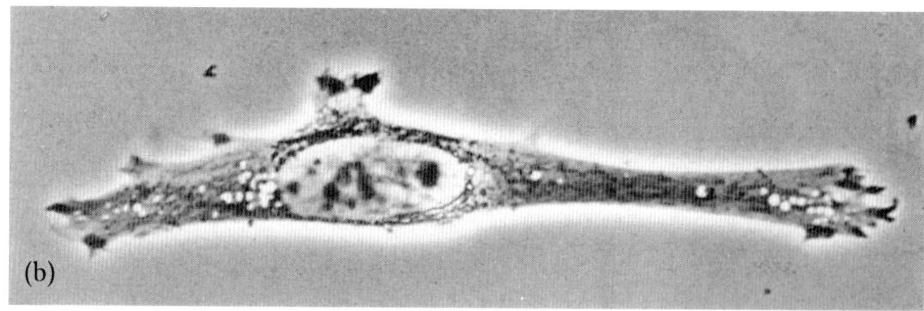

(b)

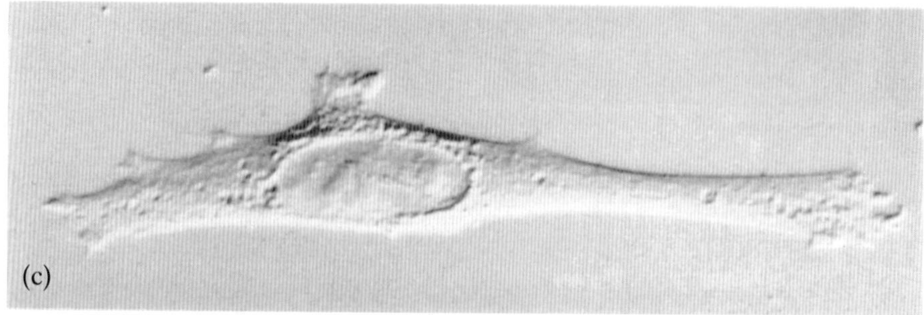

(c)

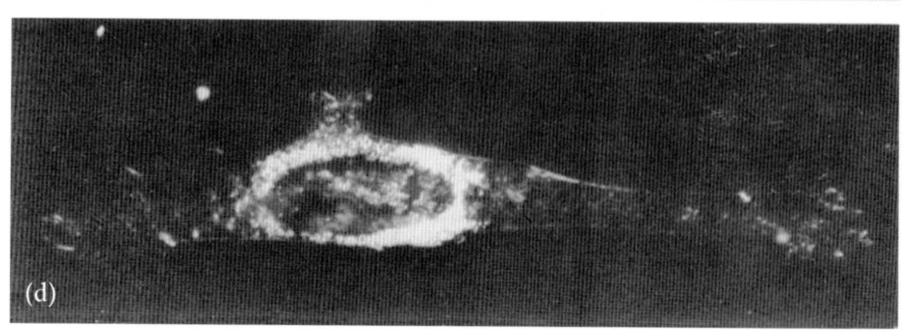

(d)

Figure 3.27 Photomicrographs of: (a) kidney cells in which the nuclei have stained blue (× 200); and of cultured human cells viewed under different lighting conditions (× 1 500) – (b) and (c) using 'interfering' beams of light; (d) using light directed at the cell from the side rather than from above or below.

It was not until the invention of electron microscopes that the full complexity of seemingly simple cell structures was realized (Figure 3.28). Electron microscopes were developed in the 1950s and use a beam of electrons with electromagnets, in place of a light beam with glass lenses. Specimens for inspection under electron microscopy have to be cut extremely thinly to allow the electron beam to pass through and the resulting image is focused onto a fluorescent screen. A vacuum has to be maintained inside an electron microscope as air disrupts the electron beam, so only dead material can be examined this way. Light and electron microscopy have extended the amount of detail that can be seen in biological material and Figure 3.29 shows the range of sizes of objects that can now be seen using these techniques.

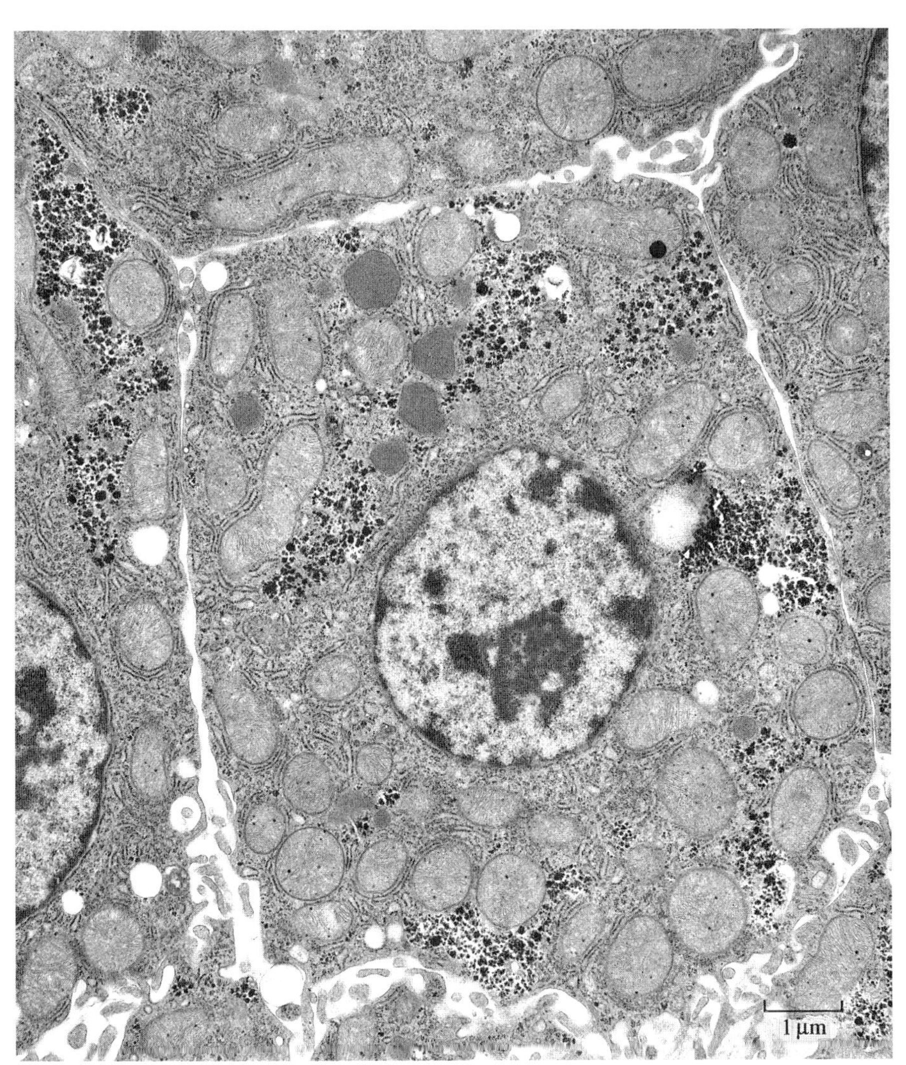

Figure 3.28 Cross-section of a liver cell seen using electron microscopy. (Images from electron microscopes are called electron micrographs.)

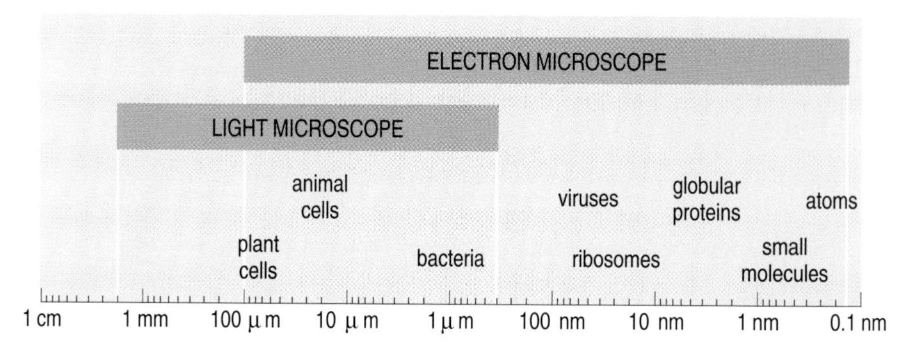

Figure 3.29 The range of biological material that can be seen using light and electron microscopes. (1 cm = 10 mm; 1 mm = 1 000 μm; 1 μm = 1 000 nm.)

The limitations of most types of microscopy are that they provide only a static view of cells in an artificial environment which is separated from the rest of the body. In addition, the preparation of biological material for microscopic examination can lead to the formation of **artefacts**; that is, structures which look like natural cell components, but are, in fact, the result of alterations in the cell's structure caused by the procedures used (chemical and physical) to prepare the specimen for microscopy. So caution must be exercised in the interpretation of micrographs of cells.

Let's now take a closer look at the cell and its components. The structure of a typical animal cell is shown in Figure 3.30.

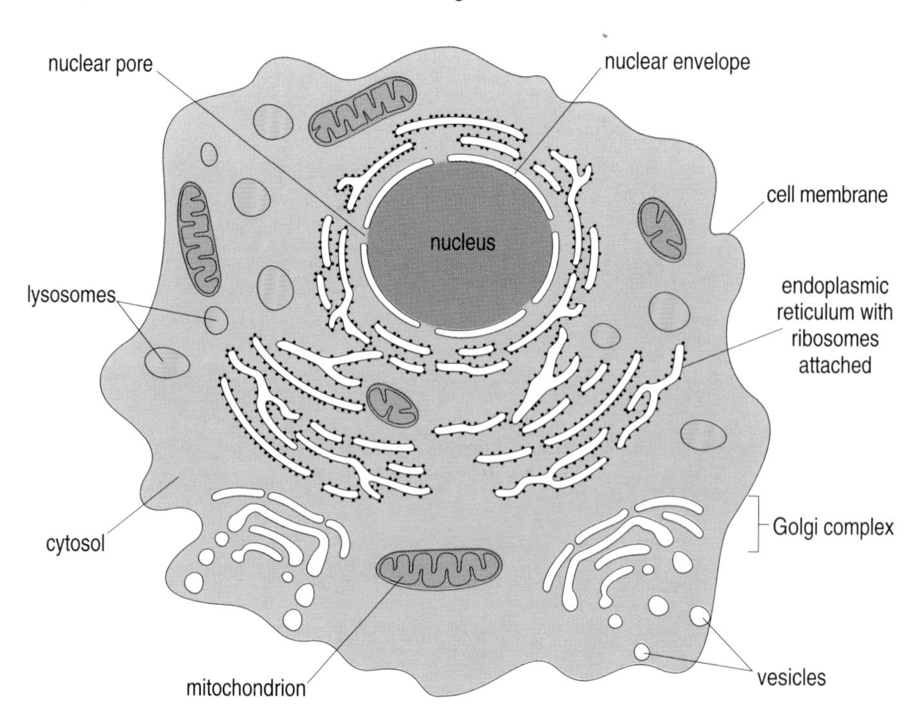

Figure 3.30 Cross-section diagram of a typical animal cell showing subcellular structures.

All cells are enclosed by a cell membrane (also sometimes called the plasma membrane) made up of a double layer of phospholipid molecules, together with other lipids and also proteins.

❑ How do phospholipid molecules interact with water molecules?

■ The hydrophilic head groups interact readily with the polar water molecules but the hydrophobic tails will minimize their interaction with water. The phospholipid molecules thus form a bilayer in which all the heads point out into the water and the tails point inwards together (Figure 3.31).

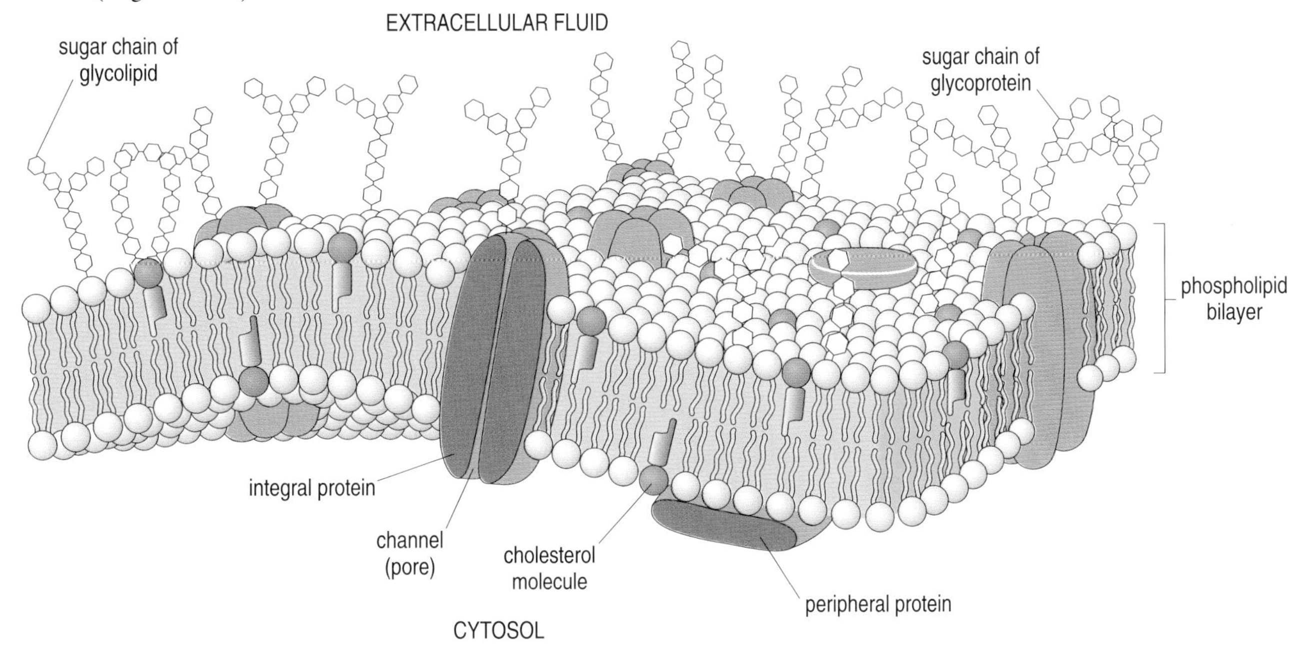

Figure 3.31 The structure of the cell or plasma membrane. Note that sugar molecules can attach to lipids (forming glycolipids) and to proteins (forming glycoproteins).

Far from being a rigid, fixed structure, the cell membrane behaves more like a fluid in which the phospholipid molecules move freely in the plane of the membrane, with their tails freely rotating and flexing. Cell membranes also contain molecules of the steroid cholesterol which, because of their flat, rigid structure, tend to decrease the fluidity of the membrane.

The cell membrane not only helps to maintain the cell's shape but also acts as a control on what substances pass into and out of the cell, thereby enabling the cell to maintain an internal environment that is different from its external environment. It is said to be *selectively permeable*; i.e. some substances can cross it freely whilst others are excluded from the cell or can only pass across the membrane with the expenditure of energy.

This selective permeability of the cell membrane is due to the presence of numerous membrane proteins which are associated with the membrane lipids. Some of these are just on the surface of the membrane, either on the inside or the outside; these are the *peripheral proteins*. Other membrane proteins penetrate the membrane completely (*integral proteins*). These integral proteins perform many functions vital to the cell. Some act as pores or channels allowing substances to pass through the cell membrane by **diffusion** (Box 3.2). Others transport substances across the membrane using energy (from ATP) in so doing; such energy-consuming transport processes are collectively called **active transport** (Box 3.3). Water moves into and out of cells by the passive (i.e. non-energy-requiring) process of **osmosis**, the amount and direction of net water movement being determined by the total concentration of dissolved substances (*solutes*) inside as compared to outside the cell (Box 3.4).

Box 3.2 Diffusion

Consider two solutions of different solute concentrations separated by a membrane which is freely permeable to the solute. In this situation there is said to be a *concentration gradient* across the membrane. After a while the solute concentrations either side of the membrane will have equalized. The process by which this uniformity of concentration is achieved is called diffusion – the solute moves from the concentrated to the dilute solution, i.e. down the concentration gradient. For charged solutes, another factor besides the concentration gradient is important in determining the direction of net diffusion. An unequal distribution of charges across a membrane (see Box 3.3) gives rise to an electrical potential gradient or *voltage* across the membrane (often referred to as the *membrane potential*). The concentration gradient and the electrical potential gradient can be combined and the net driving force or *electrochemical gradient* for a charged solute can be calculated. For cell membranes, the inside is usually negative with respect to the outside. (You will learn about the significance of this gradient for the conduction of nerve impulses in Book 2, Chapter 3.)

Diffusion across biological membranes is often mediated by specific channel proteins which allow the solute to move through tiny membrane pores, or by carrier proteins which bind to the solute on one side of the membrane and release it on the other side. These processes are collectively referred to as *facilitated diffusion*.

Box 3.3 Active transport

Substances required by a cell that are at a higher concentration inside than outside can only be taken in by active transport, which is membrane transport that requires energy. Active transport is always mediated by carrier proteins. The actively transported solute is 'pumped' up its electrochemical gradient using metabolic energy, either by the breakdown of ATP to ADP and P_i or by linking solute transport in one direction to ion transport in the opposite direction or a combination of both, e.g. the pump that maintains the concentration gradient of Na^+ (sodium) and K^+ (potassium) ions across cell membranes. The concentration of K^+ is 10–20 times higher inside cells than outside and the reverse is true for Na^+ ions. These concentration differences are maintained by the operation of an *Na^+-K^+ ATPase* – a special protein which is both a carrier for Na^+ and K^+ ions and an enzyme that breaks down ATP to ADP and P_i. This membrane protein pumps two K^+ ions into the cell and three Na^+ out for each molecule of ATP used.

Box 3.4 Osmosis

If the total solute concentration on one side of a membrane that is permeable to water but not to solutes is different from that on the other side, then *water* will move across the membrane from the dilute solution into the concentrated solution. This passive movement of water that is determined by a gradient in solute concentration is called osmosis. The type of membrane that permits osmosis is referred to as *semipermeable*. A familiar example of osmosis is the swelling of dried fruit placed in water, caused by water entering the cells of the fruit. Conversely, crisp, fresh lettuce leaves immersed in salt water wilt because water moves from the dilute solution inside the lettuce cells into the more concentrated salt solution. (You will meet examples of osmosis from human physiology in Book 3.)

Proteins that are partly exposed on the outside of the cell membrane can act as receptors, i.e. recognition sites for molecules outside the cell. Molecules that interact with specific receptors are called ligands. Receptor–ligand binding is similar to the binding of an enzyme to its specific substrate, discussed in Section 3.4.1, but here the result is not chemical transformation of the bound molecule but the triggering of a particular process inside the cell. Receptor–ligand binding plays a major part in many biological processes and will crop up throughout the course. Membrane proteins can also act as cell identity markers, enabling cells to recognize other cells as similar or different. Again, a 'lock and key' fit, this time between a protein receptor and a region of the marker protein, is involved. As shown in Figure 3.31, many membrane proteins and lipids have sugar molecules attached to them; these are called **glycoproteins** and **glycolipids** respectively, and the sugars are often responsible for the specificity of receptor–ligand binding.

The fluidity of the cell membrane is illustrated by the fact that a tear in the membrane will 'self-heal' because the phospholipid molecules flow back to fill in the discontinuity, just as water flows back over the hole made by a diver entering the water. The fluid membrane also allows the cell to ingest large solid particles or large volumes of liquid by forming extensions called pseudopods ('false feet') around the material being ingested. The pseudopods then rejoin, thus capturing the material inside the cell. This process is called **endocytosis**. If the material taken into the cell is solid it is called **phagocytosis** ('eating cell') (Figure 3.32). As you will learn in Section 3.9.2, specialized phagocytic cells (phagocytes) are responsible for cleaning up wounds. Phagocytes are also important in protecting the body against infection, by engulfing and digesting invading microbes. The ingestion of liquid by cells is called **pinocytosis** ('drinking cell'). The reverse process can happen, with large amounts of material being exported from the cell (e.g. the secretion of hormones into the blood or of digestive enzymes into the digestive tract); this activity is called **exocytosis**.

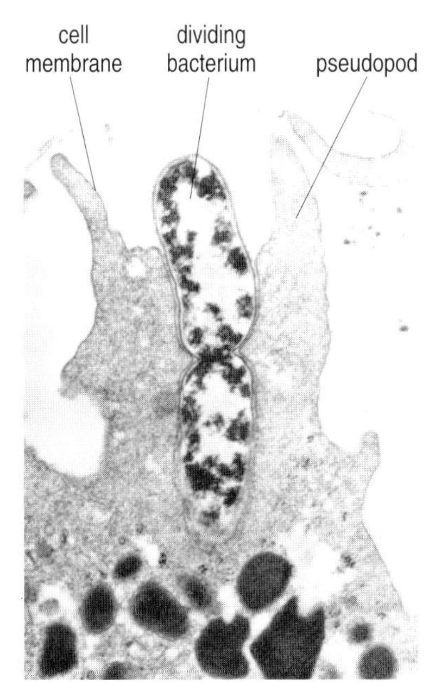

Figure 3.32 Cell performing phagocytosis. Here, the material being ingested is a bacterium in the process of dividing into two.

The main body of the cell is filled with a viscous, transparent gel (75–90% of which is water) called the **cytosol**. The gelatinous nature of the cytosol is due to its macromolecular constituents, mainly proteins and polysaccharides.

Various ions and small molecules, e.g. K^+, Na^+, Cl^-, sugars and amino acids, are also dissolved in the cytosol. It is here that many (but by no means all) of the chemical reactions of the cell occur. The term **cytoplasm** is frequently used to describe the material inside cells. It is *not* synonymous with cytosol but refers to all the cell contents outside the nucleus; in other words, cytosol + cytoplasmic structures = cytoplasm.

Running through the cytosol are threads that make up a protein scaffolding called the **cytoskeleton** (made mostly from the structural proteins actin and tubulin) which maintains the cell's shape and helps to move material around the cell, rather like a conveyor belt. This last activity is particularly important when the cell is reproducing and so needs to coordinate the movement of vast amounts of material; we return to the process of cell reproduction later.

The large dark object seen in light microscopy that is often towards the centre of the cell is the nucleus. The nucleus is bounded by a *double* layer of membranes, the **nuclear envelope**, which is perforated by **nuclear pores**. It is within the nucleus that the cell's 'instructions' are kept. How information is stored by the nucleus and distributed to where it is used in the cytoplasm is one of the most important issues of modern biology, which we shall explore in Section 3.6. Before then, let's look at what other structures can be seen in the cell, outside the nucleus.

Within the cytosol are subcellular membrane-bound structures, collectively called **organelles** ('little organs'). As with the organs in the body, each of the organelles has a specific job, all of them working together in a properly functioning cell.

The first type of organelle we shall consider is the **mitochondrion** (plural, mitochondria). Mitochondria are sausage-shaped structures bounded by two layers of membrane: the outer membrane is smooth while the inner one is folded into leaf-like extensions called **cristae** (Figure 3.33). Mitochondria are often referred to as the power-houses of the cell; for these are the organelles that carry out most of the catabolic reactions in the breakdown of glucose and other fuels, i.e. cell respiration, producing large quantities of ATP in the process. The catabolic enzymes are dissolved in the fluid inside the mitochondrion while the synthesis of ATP is carried out by precisely ordered arrays of special enzymes and other proteins embedded in the adjacent cristal membrane.

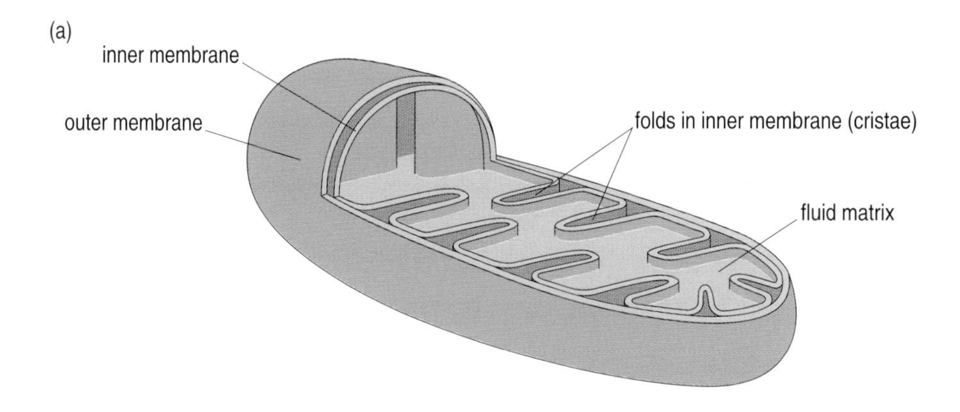

Figure 3.33a Cut-away diagram of a mitochondrion.

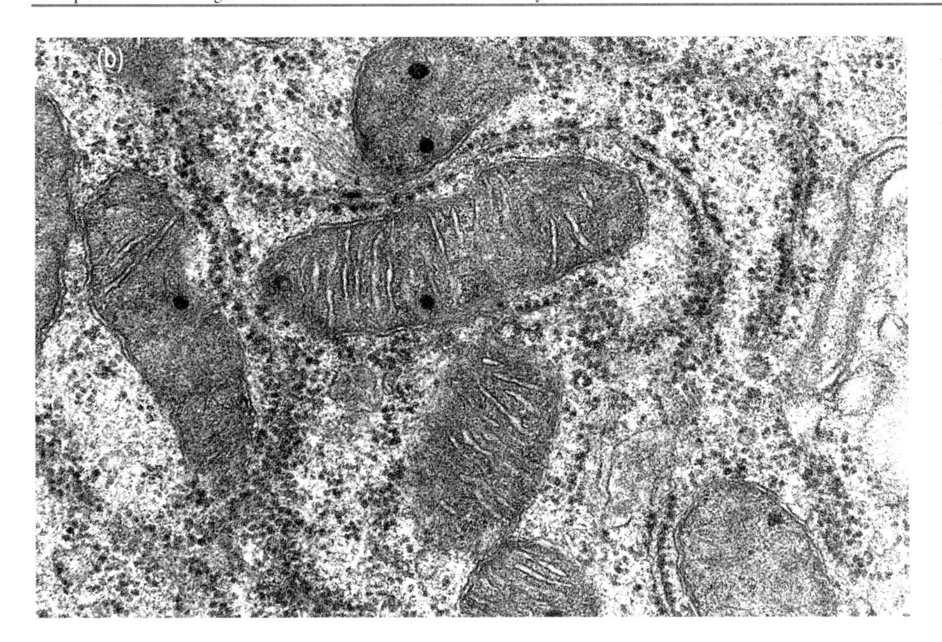

Figure 3.33b Electron micrograph of mitochondria in a liver cell (× 45 000).

❑ What could you conclude about the activities of a cell that contained numerous mitochondria?

■ That the cell requires a lot of energy to perform its functions. Examples include muscle cells, nerve cells and sperm.

In fact, increased levels of physical fitness are associated with increased numbers (and also size) of muscle mitochondria.

Another organelle in the cell, which is structurally – and functionally – much simpler than the mitochondrion, is the **lysosome**. Lysosomes are membrane-bound 'bags' of digestive enzymes.

❑ Under what circumstances might cells need to break down material rapidly using lysosomes?

■ If there is cell debris which needs clearing up as part of the process of wound healing or if there is a microbial infection.

Cells engulf the material to be digested (cell debris or microbe) via the process of phagocytosis (Figure 3.32). The ingested material, now enclosed by a small portion of the cell membrane, is transported to and fuses with the lysosome where it is digested into small units which are then released into the cytosol to be used in the cell's metabolism.

❑ What types of large molecules will be the substrates of the lysosomal digestive enzymes and what will be the small-molecule products?

■ Polysaccharides will be digested to their constituent simple sugars (monosaccharides), proteins to amino acids and fats to fatty acids and glycerol. (In addition, nucleic acids will be broken down to nucleotides – see later.)

Lysosomes also help recycle old and used organelles, as the cell – like the rest of the body – renews and repairs itself. Human liver cells, which are very active metabolically, renew half their contents every week in this way.

The cytosol is permeated by a system of membranous sacs called the **endoplasmic reticulum** (usually abbreviated to ER). Attached to the surface of much of the ER are numerous small structures, the **ribosomes** which give the ER a rough appearance in electron micrographs. Some of the ribosomes occur free in the cytosol but the majority are membrane-bound. (Notice the numerous ribosomes in Figure 3.33b.) As we shall discuss soon, ribosomes use the information supplied by the nucleus to make proteins. As you would expect, cells that synthesize and export large quantities of protein have lots of rough ER.

❑ What type of cells have you already come across that make and export protein?

■ The insulin-producing cells of the pancreas, which secrete the protein hormone insulin into the bloodstream.

The last intracellular structure to be mentioned is the **Golgi complex**. The Golgi was named after its discoverer, the Italian Camillo Golgi, who in 1898 (long before electron microscopes) used the very skilful process of staining the cell with a silver compound to reveal structures in the cytosol. The Golgi complex is a stack of flattened membranous sacs in which proteins manufactured by the ribosomes are processed and packaged for export out of the cell in secretory vesicles – these vesicles pinch off from the Golgi sacs, move away from the Golgi and then fuse with the cell membrane, thereby secreting the protein from the cell (Figure 3.34).

❑ What is the name of the process whereby material is released from the cell in this way?

■ Exocytosis.

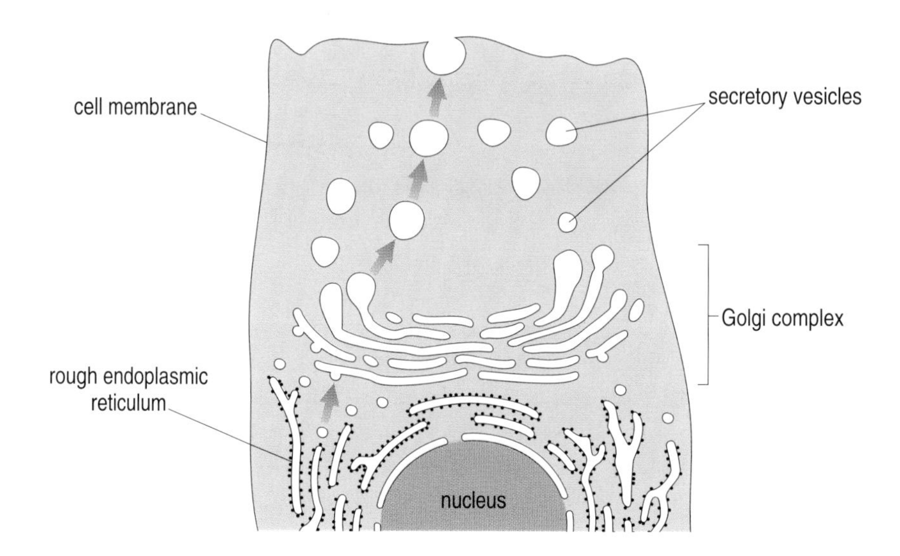

Figure 3.34 The transport of proteins from the endoplasmic reticulum through the Golgi, then into secretory vesicles and out of the cell.

The processing of newly synthesized proteins, which occurs during their passage across the successive layers of Golgi sacs, usually involves the attachment of short chains of sugars to the R groups of particular surface amino acids. This is just one of the functions of protein processing in the Golgi; addition of sugar chains to proteins in the Golgi is also important as a way of 'tagging' proteins to determine their destination *within* the cell.

Summary of Section 3.5

1 All living things are made up of units called cells.

2 Even with developments in staining and lighting conditions, light microscopy can show only a limited amount of internal cell structure.

3 Electron microscopy shows the fine structure of cells much more clearly.

4 Because of artefacts, caution has to be exercised in the interpretation of micrographs.

5 The cell membrane is a selectively permeable barrier, controlling the cell's internal environment. It is composed mainly of phospholipids, with cholesterol and various proteins; many membrane lipids and proteins have sugar chains attached.

6 The roles of membrane proteins include solute transport, ligand binding and cell–cell recognition.

7 The fluid structure of the cell membrane allows it to engulf and ingest material (via endocytosis); either solids (phagocytosis) or liquids (pinocytosis).

8 The cytoplasm is the fluid contents of the cell (the cytosol) plus all the structures (organelles, membranes, ribosomes, cytoskeleton) that are outside the nucleus.

9 The nucleus is surrounded by a double membrane, the nuclear envelope, which is studded with pores.

10 The cell organelles include the mitochondrion, which carries out cell respiration, and the lysosome, which is involved in intracellular digestion and recycling of materials.

11 The endoplasmic reticulum (ER) is a membrane system within the cell; many of the ribosomes (structures involved in protein synthesis) are attached to the ER.

12 The Golgi complex is another intracellular membrane system – it is involved in protein processing and export.

3.6 Information transfer in the cell

3.6.1 Genes and chromosomes

We now return to the special role of the nucleus in the cell. If a cell that is in the process of dividing to become two cells is viewed under the light microscope after appropriate staining, threadlike structures called **chromosomes** can be seen inside the nucleus (Figure 3.35). (These structures were first observed by Walther Flemming in 1882, in the nuclei of dividing salamander cells.) However, when the cell is just going about its day-to-day metabolic business, the nucleus appears to have little or no internal structure. Chromosomes are made up of a material called **chromatin**, which is an association of a macromolecule called deoxyribonucleic acid (**DNA** for short) with a special group of proteins, the *histones*. In a non-dividing cell, the chromatin is unravelled, existing as long strands of DNA wound at intervals around molecules of histone (Figure 3.36a). However, to become chromosomes, the chromatin undergoes several levels of condensation, as shown in Figure 3.36a–e. If all the DNA in just one human cell were to be unravelled it would measure about 1.7 metres. So the compacting is pretty impressive!

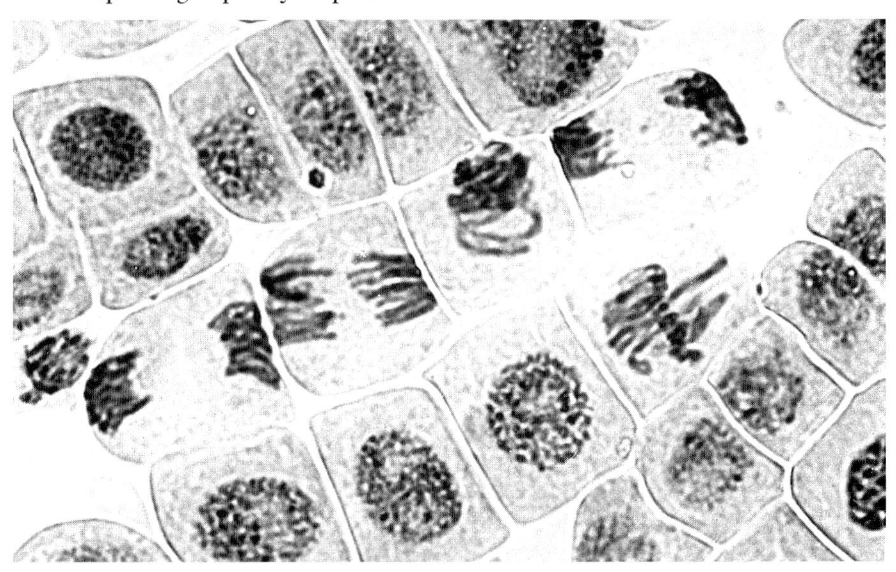

Figure 3.35 Chromosomes seen under a light microscope in dividing root-tip cells of broad bean. (The cells shown have been 'frozen' at different stages of the division process, which is described later.)

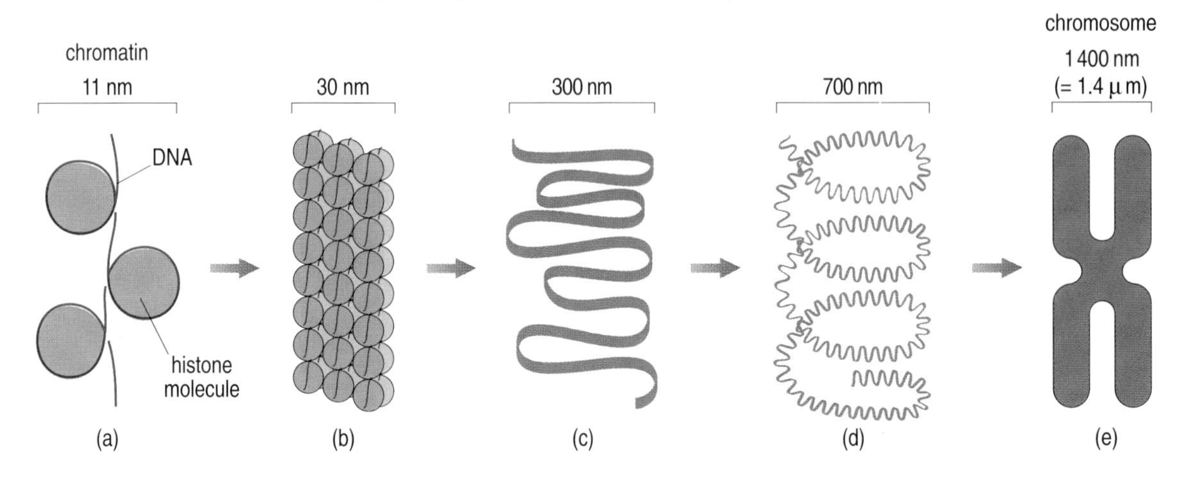

Figure 3.36 (a) Unravelled chromatin (DNA–histone complex); (b–e) Successive stages in the condensation of chromatin into the compact structures called chromosomes.

Later (in Section 3.8.2) we shall have more to say about the important process of cell division, but it is now time to look at the function of DNA in more detail. The DNA in the nucleus holds all the information required to direct the activities of the cell, including making accurate copies of itself.

So, if DNA holds this information, the cell's instructions, what form do the instructions take and how are they executed? This section sets out to answer both of the questions.

Back in the 1860s, Gregor Mendel, an Austrian monk, performed some interesting breeding experiments with peas. What Mendel concluded from his experiments was that the 'instruction' to produce peas of, say, a particular colour is transmitted from generation to generation in the sex cells, one instruction in each of the egg and the pollen (in animals, egg and sperm). However, in Mendel's time, little was known about cell structure, so the physical nature of what he referred to as 'particles of inheritance' remained a mystery for many years.

However, with improvements in the techniques used for the study of cell structure (cytology), the details of the cell division that gives rise to the sex cells (a process called *meiosis* – see Chapter 4) were eventually worked out. A key observation was that the behaviour of chromosomes during meiosis matched that of Mendel's proposed 'particles of inheritance', i.e. their number is halved in the production of sex cells, and restored at fertilization. The obvious conclusion therefore was that the chromosomes are the hereditary material.

Mendel's 'particles' are now called **genes**. From breeding experiments using laboratory organisms, we know that genes are arranged linearly along chromosomes, i.e. a gene is a section of DNA. The question now is: what is it about a particular length of DNA that determines its information content?

Samples of DNA had been isolated from salmon sperm by Johann Miescher as far back as 1889, but the chemical techniques of the time did not allow the examination of the structures of such large biological macromolecules. By the mid-1950s, the chemical constituents of DNA had been established, but it appeared a very unlikely candidate for the hereditary molecule as its composition showed very little variation whether it was extracted from bacteria, plants, humans or other animals. However, an important observation was made in 1944 by Oswald Avery and colleagues who showed that DNA extracted from disease-causing (virulent) bacteria caused harmless ones to become virulent, and to remain so through successive generations. This confirmed that DNA is an informational molecule and is inherited.

Chemically, DNA is made up of repeating molecular units called **nucleotides** (Figure 3.37a). A DNA nucleotide itself has three components: a **base**, a sugar called deoxyribose (Figure 3.37b) and a phosphate group. The bases are ring-shaped molecules (made up of C, H, N and O) with N atoms as well as C atoms forming the ring(s). There are four different bases in DNA, two with a double-ring structure called *adenine* and *guanine* and two with a single-ring structure, *cytosine* and *thymine*. Figure 3.37c shows the skeleton structures of the two types of base but not any attached atoms and groups. Bases are usually referred to by just their initials, A, G, C and T. Figure 3.37a shows a small section of a single DNA strand: a chain of alternating sugar and phosphate groups (the sugar-phosphate backbone), with a base attached to each sugar (deoxyribose).

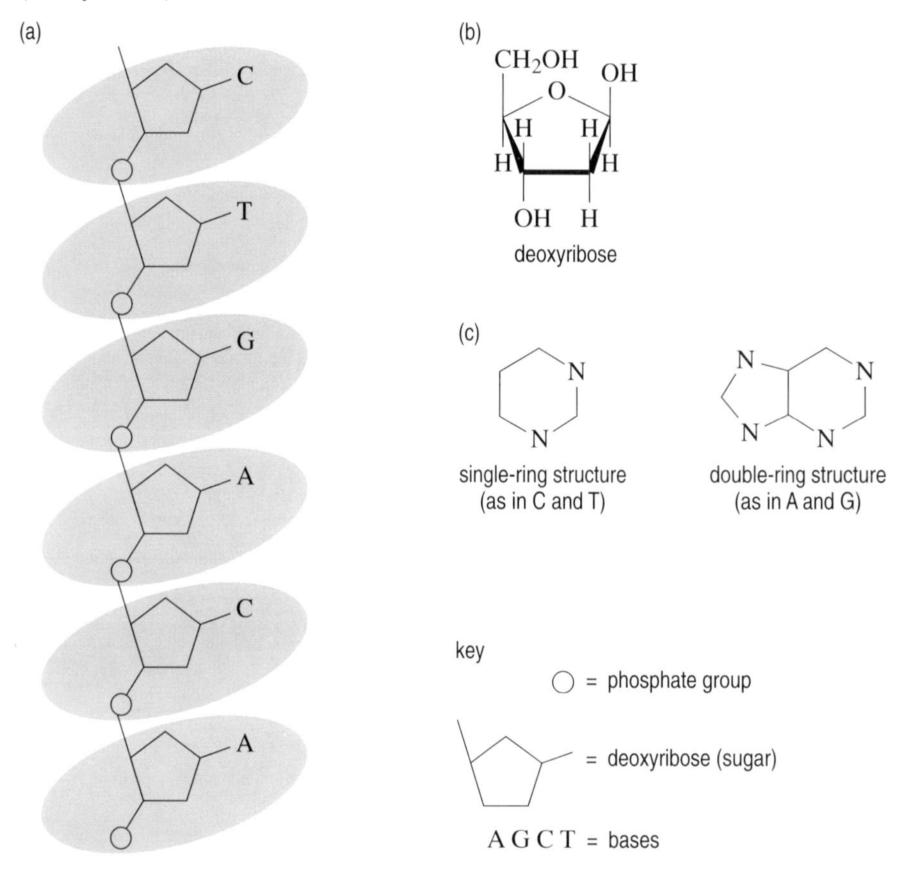

Figure 3.37 Simplified structure of DNA: (a) section of a DNA strand made up of six repeating units, called nucleotides (shaded ovals); (b) molecular structure of deoxyribose; (c) skeletons of the four bases (the ring N atoms but not the C atoms, are shown).

Whatever the source of DNA, chemical analysis showed that there was always as much guanine as cytosine and always as much adenine as thymine present. How could a molecule with so little variation carry the varied information an organism needs? Further evidence about DNA structure came from studies on crystals of pure DNA. X-rays passed through these crystals are bent (diffracted) and produce a pattern on special film, from which the molecular structure of DNA was deduced. This work (done by Rosalind Franklin and Maurice Wilkins at King's College, London in the early 1950s) showed that DNA exists as *two* strands wound together in a spiral or helical shape. Using this X-ray evidence and the chemical evidence already available and building actual models of proposed DNA structures, James Watson and Francis Crick found that if they put the sugar and phosphate groups together in two long strands like the sides of a ladder, then the bases would fit across the ladder like the rungs. The neatest arrangement was to have adenine always opposite thymine and guanine opposite cytosine (Figure 3.38a, b). As this meant that a two-ring base was always opposite a one-ring base, the sides of the ladder were kept at an equal distance from each other. By twisting the ladder into the helix suggested by the X-ray work, they found a structure that not only explained the finding that there was always as much guanine as cytosine and thymine as adenine, but also immediately suggested a way for DNA to carry and duplicate information in its structure. Recently, the double-helical structure of DNA has actually been seen at the atomic level, a final proof that DNA really has this shape (Figure 3.38c, d).

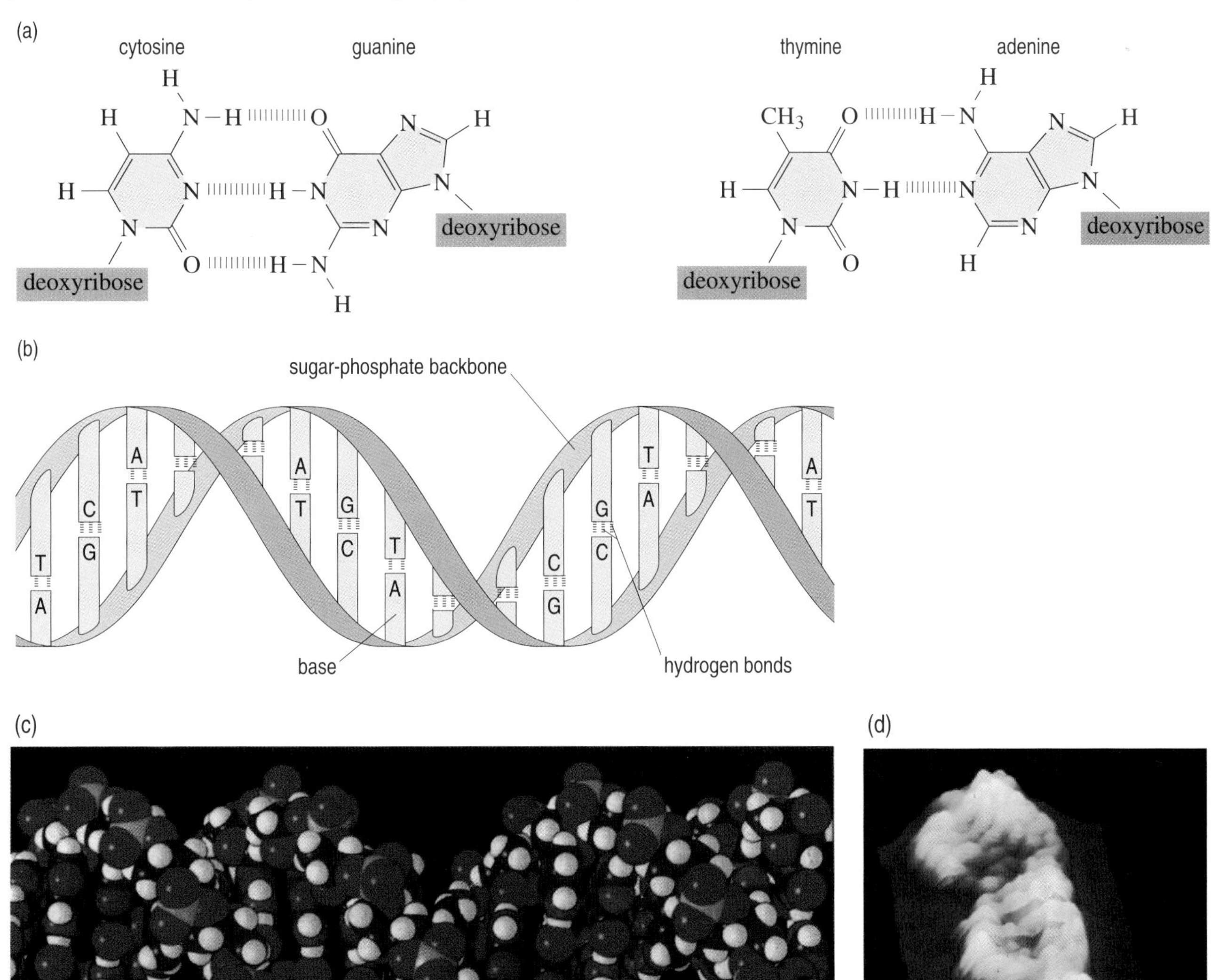

Figure 3.38 Structure of the DNA double helix: (a) hydrogen bonding between cytosine (C) and guanine (G) and between thymine (T) and adenine (A) base pairs; (b) diagrammatic representation of the double-helical structure (each red bar represents a hydrogen bond); (c) space-filling model of the DNA double helix, which indicates the relative sizes and positions of the constituent atoms; (d) electron micrograph of DNA. (One turn of the double helix = 3.4 nm.)

How then does the structure of DNA enable it to carry information? The only thing that changes along the DNA double helix is the sequence of bases. Also, by looking at the sequence along one of the strands, you could deduce the sequence of the other, *complementary* strand.

❑ Given that the sequence of bases along one strand of a DNA double helix is AACTGGAT, write out the base sequence of the complementary strand.

■ The sequence is TTGACCTA.

The pairs of complementary bases are held together by hydrogen bonds (Figure 3.38a).

❑ How many hydrogen bonds will there be in the section of DNA double helix above?

■ From Figure 3.38a, each A–T pair is held together by two hydrogen bonds and each G–C pair by three bonds, so there will be $(5 \times 2) + (3 \times 3) = 19$ hydrogen bonds altogether.

Thus the macromolecular DNA double helix is held together by a very large number of hydrogen bonds. In addition, there are hydrophobic interactions between the stacked base rings which make a significant contribution to the stability of this structure.

The base-pairing hydrogen bonds and base-stacking forces can be fairly easily overcome. So when another copy of the DNA is required by the cell, in preparation for cell division, the DNA double helix begins to unwind and split down the middle and each strand serves as a *template* for the synthesis of another, complementary strand. (Because of access problems – for the required enzymes and raw materials – the process of DNA synthesis can only occur after the DNA double helix has unwound.) Thus two new DNA double helices are formed, each having one strand from the 'old' original DNA double helix and one newly synthesized strand; in other words, DNA synthesis is *semi-conservative*. The process of making new DNA molecules is called **replication**. Note that DNA replication takes place in the nucleus, using nucleotides dissolved in the cytosol.

❑ How do these DNA building blocks get from the cytosol to where they are needed in the nucleus?

■ By passing through the pores in the nuclear envelope.

The process of DNA replication involves a complex and well coordinated set of enzyme-catalysed reactions. DNA synthesis begins when a cell receives a signal to reproduce, resulting eventually in the formation of two new cells from one. (The process of cell reproduction, called *mitosis*, is discussed later in this chapter.) The important outcome of replication is that the sequence of bases in DNA is faithfully copied from one molecule to the next. This sequence of bases is the vital information that DNA carries from one generation of cells to the next. The DNA base sequence is like a code – it must then be deciphered into a form that can be interpreted via the cell's metabolism.

❑ What other type of macromolecule have we met which has a very precise sequence of repeating units?

■ Proteins.

As we shall see, it is the sequence of bases in a stretch of DNA, a gene, that determines, i.e. is the code for, the sequence of amino acids in the corresponding protein. In other words, the information carried by DNA is *expressed* as the primary structure, hence as the three-dimensional structure and thus via the *function* of proteins.

❑ Does this give you a clue as to how DNA information is linked to cell metabolism?

■ Metabolic reactions are catalysed by the class of proteins called enzymes – thus DNA information is interpreted by directing the synthesis of cellular enzymes (and, of course, all the other proteins of the cell).

The above picture is much simplified. By no means all the cell's DNA has a coding role. Also, there must be some control on which stretch of DNA, which gene, is expressed and when, and for how long. As you can appreciate, the copying of bases in DNA has to be accurate because this base sequence determines the sequence of amino acids in cellular proteins. As we have seen in the sickle-cell haemoglobin example, even a slight change in this amino acid sequence can cause changes in the protein structure that can make that protein not function correctly (or even not function at all). A change in DNA base sequence is called a **mutation**. There are many inherited disorders which are now known to be the result of an alteration in DNA base sequence, and in some cases (e.g. sickle-cell disease, cystic fibrosis) the corresponding protein has actually been identified.

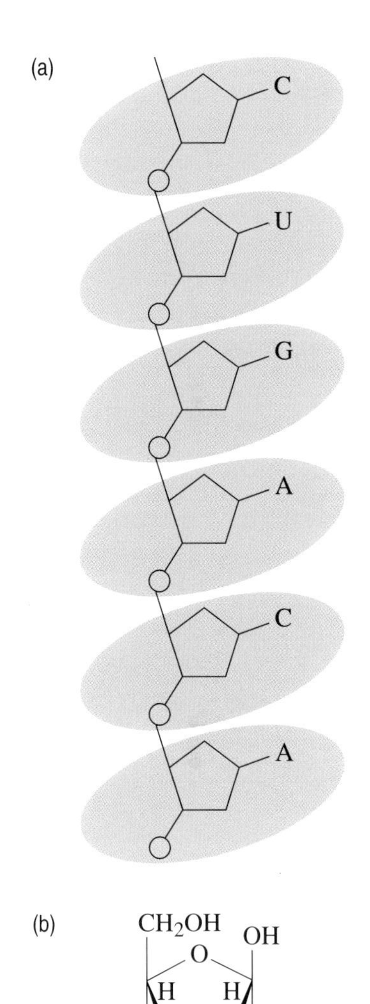

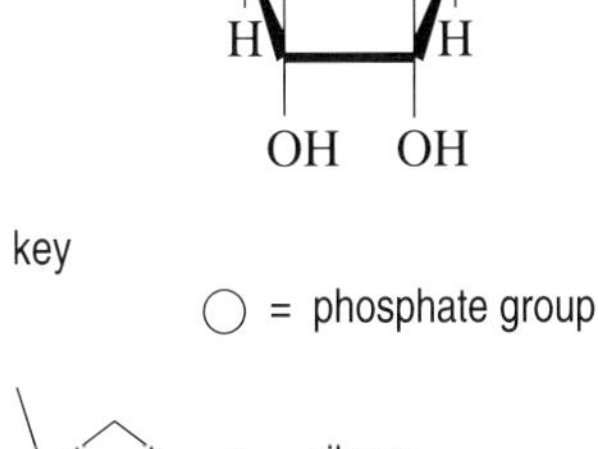

Figure 3.39 Simplified structure of RNA: (a) section of an RNA strand made up of six nucleotides (shaded; ovals); (b) molecular structure of ribose.

3.6.2 From DNA to protein

We shall now go on to describe the actual sequence of events by which the DNA message is interpreted as a specific protein product.

Considering for a moment the layout of the cell: DNA is in the nucleus but it is at the ribosomes that proteins are built, so there has to be a way of getting the information encoded in the DNA from the nucleus to where it is used. You may not be surprised to learn that it is not DNA itself which leaves the nucleus, carrying the information, but another type of nucleic acid, called ribonucleic acid or RNA (Figure 3.39); in fact it is a particular class of RNA, **messenger RNA** (**mRNA** for short).

❑ Compare Figure 3.39 with Figure 3.37. What differences can you see between an RNA and a DNA molecule?

■ In RNA, the sugar is not deoxyribose but a slightly different one called ribose (hence the difference in names). Also, in RNA the base *uracil* (U) is present instead of thymine.

Don't be too concerned with these rather subtle chemical differences between DNA and RNA. Another difference – which is significant to the function of these two types of nucleic acids – is that DNA is double-stranded, whereas RNA is mostly single-stranded.

Unlike DNA, mRNA molecules are not around in the nucleus all the time but are made specifically to order, using a stretch of DNA, a gene, as a template. Just as when DNA is replicated, the process of making a complementary mRNA strand starts with a part of the DNA double helix unwinding its two strands. Just as in the making of a new DNA molecule, the right bases are built into the growing mRNA chain because only G pairs with C and A with U. Thus a sequence of mRNA bases complementary to the DNA bases on one of the strands is built up; this process of synthesizing mRNA on a DNA template is called **transcription** (Figure 3.40). Note that only *one* of the DNA strands acts as a template; this one is called the *coding* strand. The other, which has the same sequence as the mRNA product (except, of course, that in the mRNA the base T becomes U) is called the *non-coding* strand.*

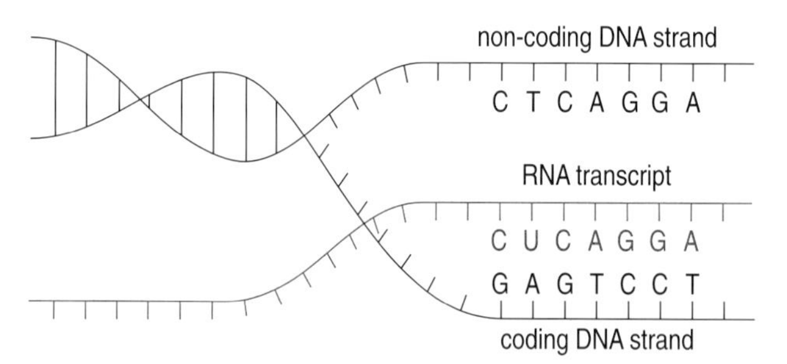

Figure 3.40 The transcription of a DNA base sequence into an mRNA base sequence.

When transcription is complete, the mRNA molecule detaches from the DNA coding strand, allowing the DNA to rewind itself. The number of copies of mRNA made from a specific gene depends on the activity of gene regulatory proteins that either enhance mRNA transcription or inhibit it, depending on the need of the cell for that specific mRNA.

The process of mRNA synthesis in organisms other than simple ones, such as some bacteria, is not quite this straightforward. In simple bacteria, the transcription product is a functional mRNA molecule. However, in other organisms, humans included, although the whole gene is transcribed,

* In some texts you may find that the terms coding and non-coding are, rather confusingly, used the other way round, i.e. template strand = non-coding strand, and strand with same sequence as corresponding mRNA = coding strand.

sections of the primary transcription product (RNA) have to be cut out and the ends rejoined (spliced) to make the functional mRNA molecule (Figure 3.41). The parts of the mRNA molecule that are removed have been copied from sections of the gene called **introns** and those that are retained in the mRNA come from stretches of the DNA called **exons**.

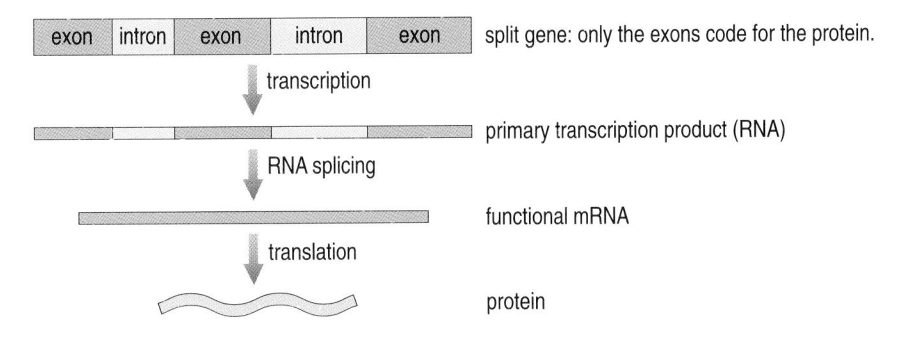

Figure 3.41 Transcription in most organisms involves the removal of parts of the newly transcribed mRNA molecule, a process called RNA splicing.

Why is there this extra step? It is known that most genes contain many introns and exons and that the same gene transcribed in different tissues may be processed in different ways to produce different mRNA molecules and hence different proteins. An example is shown in Figure 3.42.

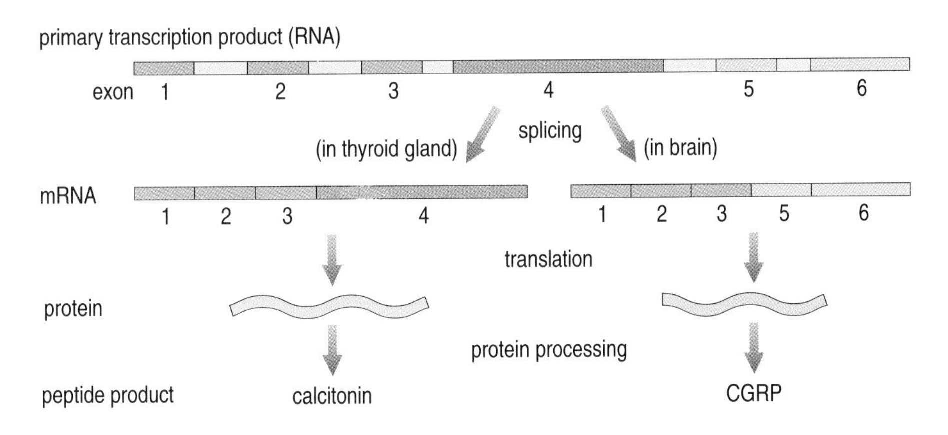

Figure 3.42 Alternative processing of the calcitonin/CGRP gene. In the thyroid gland, exons 1, 2, 3 and 4 are spliced together, giving the mRNA coding for the hormone calcitonin. In the brain, exons 1, 2, 3, 5 and 6 are used, giving the mRNA that codes for CGRP, a regulatory peptide in the brain.

After processing, newly synthesized mRNA molecules leave the nucleus, through the pores in the nuclear envelope and out to the ribosomes. Ribosomes are structures composed of RNA (another class of RNA, ribosomal RNA or rRNA) and protein, and are made up of a small subunit and a larger subunit. Ribosomes become attached to the mRNA as it emerges from the nucleus and can be seen, strung along mRNA molecules, in electron micrographs (Figure 3.43). Now that the ribosomes are in place, **translation** of the coded sequence of mRNA bases into a protein amino acid sequence can begin.

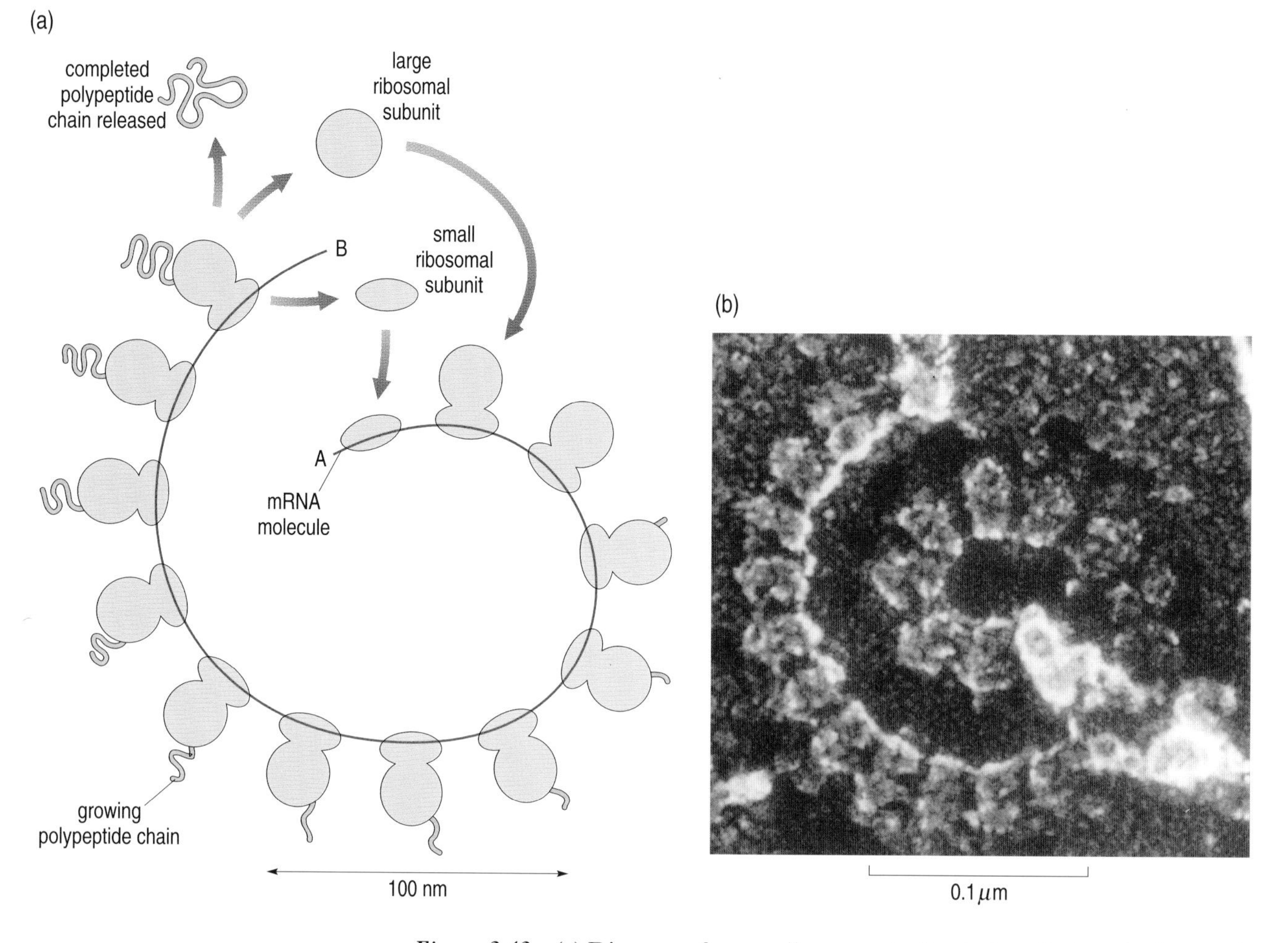

Figure 3.43 (a) Diagram of many ribosomes (collectively called a polyribosome) attached to one mRNA molecule. The ribosomes are all moving along the mRNA from end A towards end B. (b) Electron micrograph of the structure shown in (a).

The mRNA message is 'read' three bases at a time; each set of three bases is called a **codon** and a given codon corresponds to a particular amino acid.

❑ Why do you think it takes a three-base sequence to specify one amino acid? (Compare the number of bases with the number of amino acids which have to be coded for.)

■ There have to be enough codons to specify all 20 of the amino acids. A two-base code would only provide $4 \times 4 = 16$ permutations; a three-base genetic code provides $4 \times 4 \times 4 = 64$.

So there are many more codons available than there are amino acids – consequently most amino acids are coded for by several different codons: the code is said to be *degenerate*. The full genetic code, which was worked out by experiments using synthetic mRNAs back in the 1960s, is shown in Table 3.4 (which you need not memorize!).

Table 3.4 The genetic code: mRNA codons and the corresponding amino acids.

Amino acid	mRNA codons					
alanine (Ala)	GCA	GCC	GCG	GCU		
cysteine (Cys)	UGC	UGU				
aspartic acid (Asp)	GAC	GAU				
glutamic acid (Glu)	GAA	GAG				
phenylalanine (Phe)	UUC	UUU				
glycine (Gly)	GGA	GGC	GGG	GGU		
histidine (His)	CAC	CAU				
isoleucine (Ile)	AUA	AUC	AUU			
lysine (Lys)	AAA	AAG				
leucine (Leu)	UUA	UUG	CUA	CUC	CUG	CUU
methionine (Met)	AUG					
asparagine (Asn)	AAC	AAU				
proline (Pro)	CCA	CCC	CCG	CCU		
glutamine (Gln)	CAA	CAG				
arginine (Arg)	AGA	AGG	CGA	CGC	CGG	CGU
serine (Ser)	AGC	AGU	UCA	UCC	UCG	UCU
threonine (Thr)	ACA	ACC	ACG	ACU		
valine (Val)	GUA	GUC	GUG	GUU		
tryptophan (Trp)	UGG					
tyrosine (Tyr)	UAC	UAU				

So how then is a codon actually translated into an amino acid? To answer this question we need to introduce one more participant in the translation process. This is yet another, relatively small type of RNA, called **transfer RNA** or **tRNA** (Figure 3.44). (Both rRNA and tRNA are produced in the same way as mRNA, i.e. by the transcription of stretches of DNA, genes, in the nucleus, but unlike mRNA they are not translated.) As the name implies, the transfer RNAs carry amino acids to the ribosome to be incorporated into the protein chain. There are many different tRNA molecules in the cell, each one specific for the particular amino acid carried at one end and each having a specific sequence of three bases, the **anticodon**, which is complementary to a particular mRNA codon. Having introduced all the players, we can now look at the mechanism of translation.

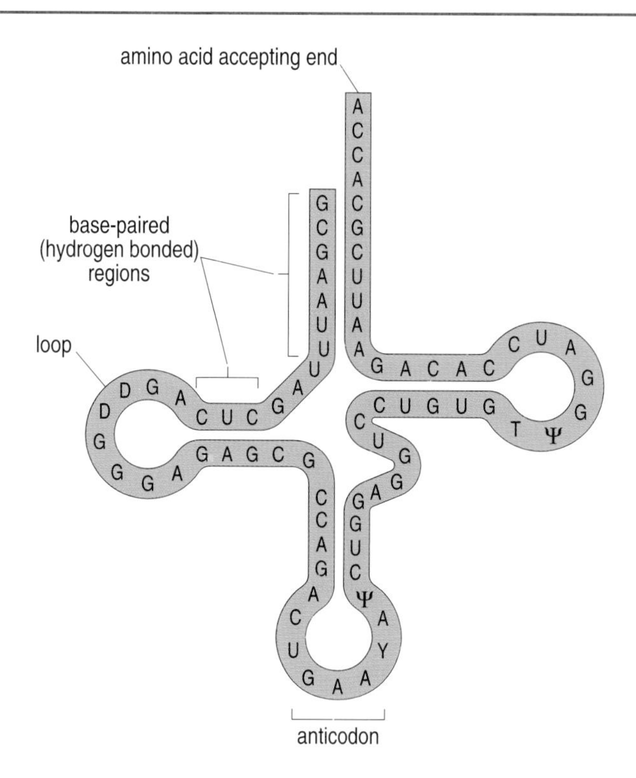

Figure 3.44 The molecular structure of a transfer RNA molecule. Notice that the anticodon has the sequence AAG, which is complementary to the mRNA codon UUC, so this particular tRNA carries the amino acid phenylalanine – Table 3.4. (D = dihydrouracil and Ψ = pseudouracil, bases closely related to uracil.)

In Figure 3.45 (stage 1) you can see that the ribosome becomes attached to the mRNA molecule at a specific codon, AUG. This is known as the start codon because it is where translation begins.

❑ What will be the first amino acid in the protein chain?

■ From Table 3.4, the mRNA codon AUG corresponds to the amino acid methionine (Met), so all newly synthesized protein chains start with Met.

Thus AUG allows the binding of a Met-specific tRNA molecule carrying its methionine 'cargo' (Met-tRNA^{Met}, for short), which has the complementary anticodon (Figure 3.45, stage 2).

❑ What is the sequence of the tRNA^{Met} anticodon?

■ UAC.

A ribosome can hold two tRNA molecules at once. Suppose, as in Figure 3.45, that the next mRNA codon is GGA. According to Table 3.4, this codon will direct the binding of Gly-tRNA^{Gly}. The next step is peptide bond formation, i.e. the protein chain starts to grow: the carboxyl (–COOH) group of the Met reacts with the amino (–NH_2) group of the Gly, giving the dipeptide Met-Gly (stage 3, and see also Figure 3.11b earlier), still attached to tRNA^{Gly}, and the free tRNA^{Met} is then released to renew its cargo of Met. As illustrated in stages 4–5, the ribosome moves on, making room for a third tRNA loaded with its amino acid, and a second peptide bond is formed as before (stage 6). The process continues with the chain of amino acids growing in the amino (N) to the carboxyl (C) direction until the codon signalling 'stop translating' (UGA or UAA – not listed in Table 3.4) is reached, and the completed protein chain, which has already begun folding into its precise three-dimensional shape, can then be released from the ribosome (Figure 3.43a).

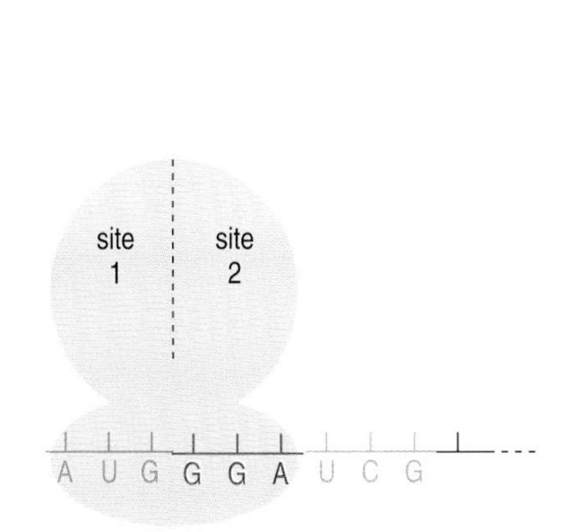

1 A ribosome attaches to the left-hand end of an mRNA molecule.

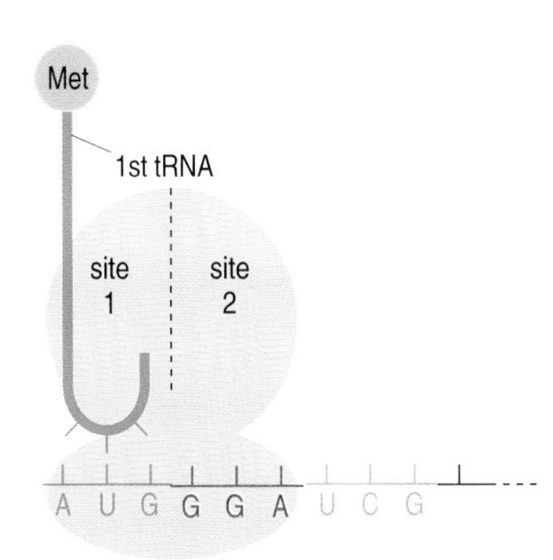

2 The N-terminal (first) amino acid, methionine, attached to its tRNA as Met-tRNA^{Met}, binds to site 1 of the ribosome and the tRNA^{Met} anticodon then pairs with AUG (mRNA codon for methionine).

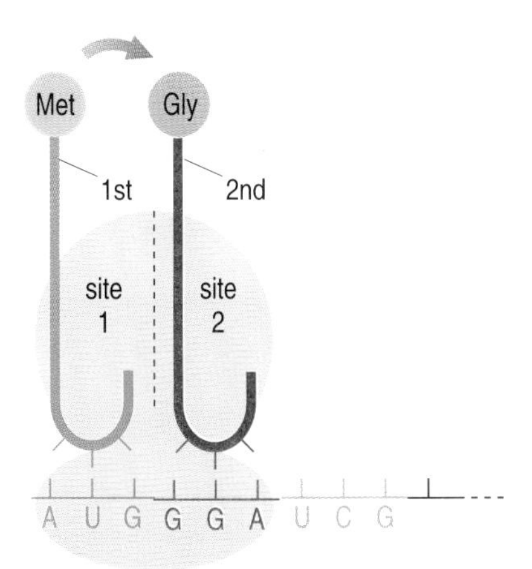

3 Gly-tRNA^{Gly} binds to site 2. Then the N-terminal Met migrates to form the dipeptide Met-Gly.

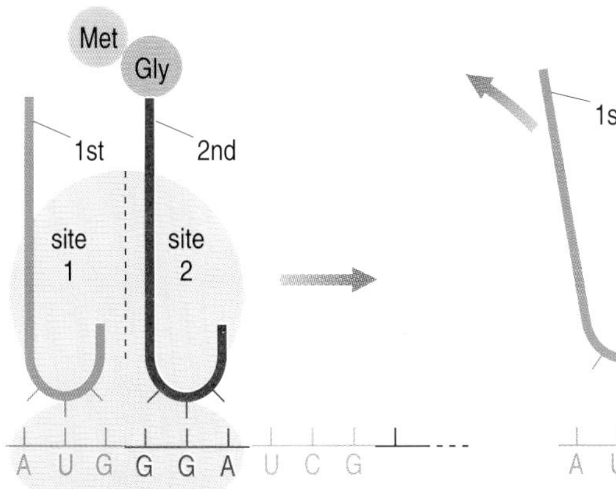

4 The ribosome moves along the mRNA by one codon.

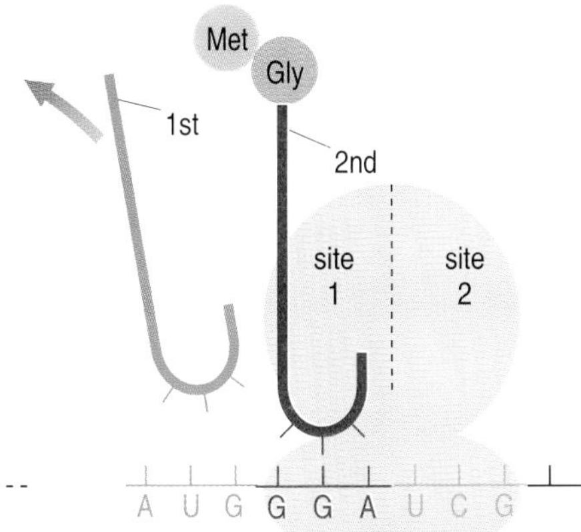

5 The unloaded tRNA^{Met} leaves and the now empty site 2 is ready to receive the next amino acid-tRNA – in this case Ser tRNA^{Ser}.

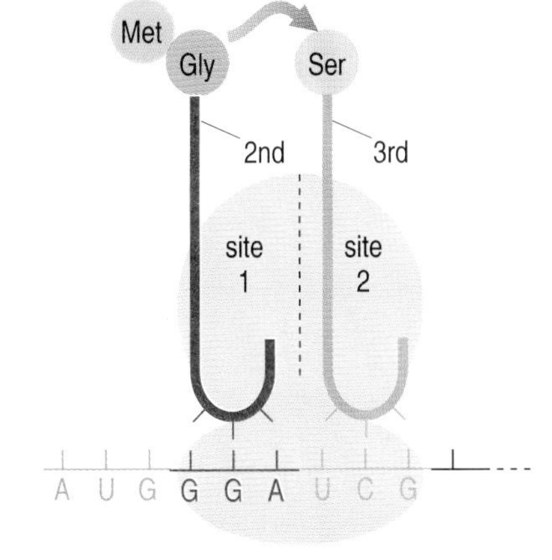

6 The Met-Gly dipeptide migrates to form the tripeptide Met-Gly-Ser.

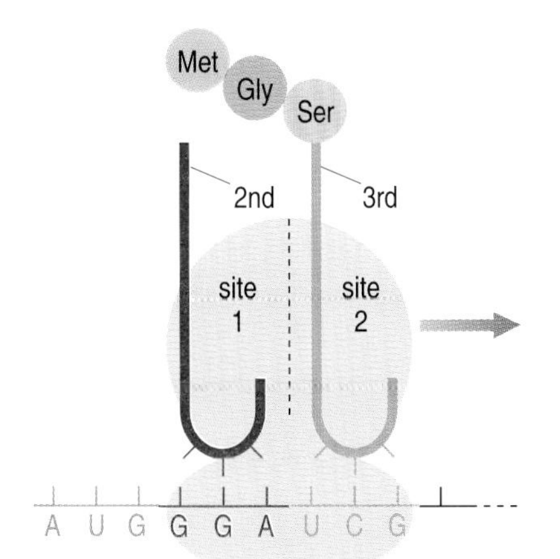

7 The ribosome moves along the mRNA by one codon.

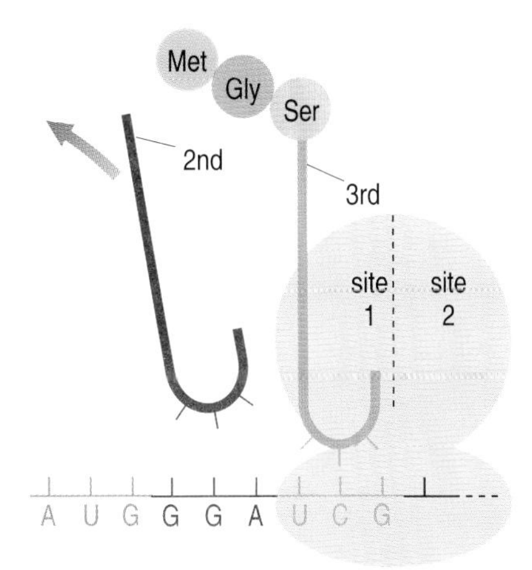

8 Site 2 is once again empty; the unloaded tRNA leaves (This is a repeat of stage 5.)

Figure 3.45 The stages of translation.

❑ Suppose that a mutation has occurred in a gene which changes the last base of the UGA codon in the corresponding mRNA from A to G. What do you think will be the effect of this?

■ The stop codon (UGA) will be changed to UGG, which codes for the amino acid tryptophan (Trp). It is likely therefore that the chain won't be released from the ribosome, i.e. a functional protein will not be produced.

The fate of the now free mRNA molecule will depend on the needs of the cell. If enough of the protein it encodes has been produced, it will be broken down into its small-molecule constituents; alternatively, it may undergo further rounds of translation before finally being degraded.

Thus we come to the end of our description of the complex sequence of events, which began with a DNA molecule in the nucleus and ended with a newly formed protein in the cytosol.

Summary of Section 3.6

1. Chromosomes are made of chromatin (a DNA–histone complex); they condense and so become visible prior to cell division – but in cells that are not dividing they are unravelled and therefore invisible.
2. Genes, the particles of inheritance, are arranged linearly along the chromosomes; a gene is thus a section of DNA.
3. DNA is a sequence of nucleotides; a nucleotide is base-sugar-phosphate unit. There are four different bases in DNA and it is the base sequence that determines the information content of the DNA.
4. DNA exists as a double helix of two complementary strands; X-ray crystallography confirmed this structure.
5. DNA replication is semi-conservative, each strand serving as a template for the synthesis of the complementary strand.
6. Genes are expressed in the synthesis of proteins; a mutation, i.e. an alteration in base sequence, can result in an altered protein product.
7. The first stage of expression is the transcription of the gene to the corresponding messenger RNA (mRNA); only one strand of the DNA (the coding strand) is used as template.
8. mRNA is chemically slightly different from DNA – in particular, the base thymine (T) is replaced by uracil (U).
9. In most organisms, the primary transcription product is 'edited' before translation – sections (introns) are removed and the remainder (the exons) are then spliced together.
10. The second stage in expression is translation of the mRNA into the corresponding protein; this occurs at the ribosomes in the cytosol.

11 Each mRNA codon corresponds to a particular amino acid; transfer RNA (tRNA) molecules translate each codon in sequence via a complementary anticodon at one end and the specific amino acid carried at the other. The result is a protein chain with an amino acid sequence dictated by the corresponding mRNA base sequence, and so ultimately by the base sequence of the gene from which the mRNA was transcribed.

3.7 Cells into tissues

Our discussion so far has focused on the internal workings of a 'generalized', isolated cell. We now move on to look at cells that have become specialized and work together to perform a particular function. All cells have certain characteristics in common and these have been the subject for much of this chapter. They all have a nucleus (though some – red blood cells – lose it as they mature), a cell membrane and cell organelles. The different groupings of cells can be classified according to structure and function (Table 3.5). A group of cells performing the same function is called a tissue. Figure 3.46 shows a sample of different human tissues as seen under light microscopy.

Table 3.5 Cells classified by function. (Note that cells can have more than one function and therefore be classified into more than one group, e.g. some cells of the immune system are also blood cells, some support cells are also contractile and hormone-producing cells can be classified as epithelial.)

Cell group	Examples	Functions	Special features
epithelial cells	lining of digestive tract and blood vessels, outer layer of skin	as barrier, absorption, secretion	tightly bound together by cell junctions
support cells	fibrous support tissue, cartilage, bone	organize and maintain body structure	produce and interact with extracellular material
cells of adipose (fat) tissue	under skin, around organs	heat insulation, physical protection	contain very little cytoplasm – most of cell volume is fat
contractile (muscle) cells	skeletal muscle, heart muscle, smooth muscle	movement	contain protein filaments which mediate contraction
nerve cells	brain and spinal cord	direct cell communication	communicate with each other and with other cell types by releasing chemical messengers (fast-acting communication)
germ (sex) cells	eggs, sperm	reproduction	have half the normal chromosome number
blood cells	circulating red and white cells	oxygen transport, defence	haemoglobin in red cells binds oxygen, white cells destroy bacteria
cells of immune system	lymphoid tissues (lymph nodes and spleen)	defence	recognize and destroy foreign material
hormone-producing cells	thyroid and adrenal glands, pancreas	indirect cell communication (via bloodstream)	secrete chemical messengers into blood (slow, long-lasting communication)

Cells of specific types are said to be **differentiated.** Every cell of the body has the same set of genes on the chromosomes in the nucleus; some of those genes – those responsible for the cell's 'house-keeping' functions – are active, i.e. **expressed**, in all the cells, while the expression of others is tissue-specific – genes that are switched on in one tissue are suppressed in others. This **differential gene expression** gives rise to the wide range of different cell types, some of which are shown in Figure 3.46. Such differential expression of particular genes may also be time-dependent, with different genes expressed at different stages of the cell's and the individual's life cycle.

All the cells in a particular tissue function together to perform a specialized function, e.g. **connective tissue** which, as the name suggests, is the packaging tissue filling up the space around the body's internal structures and **epithelial** (covering) **tissue (epithelium)** which lines cavities (e.g. Figure 3.46a) and covers various structures – including, of course, the whole body, as the outermost layer of the skin.

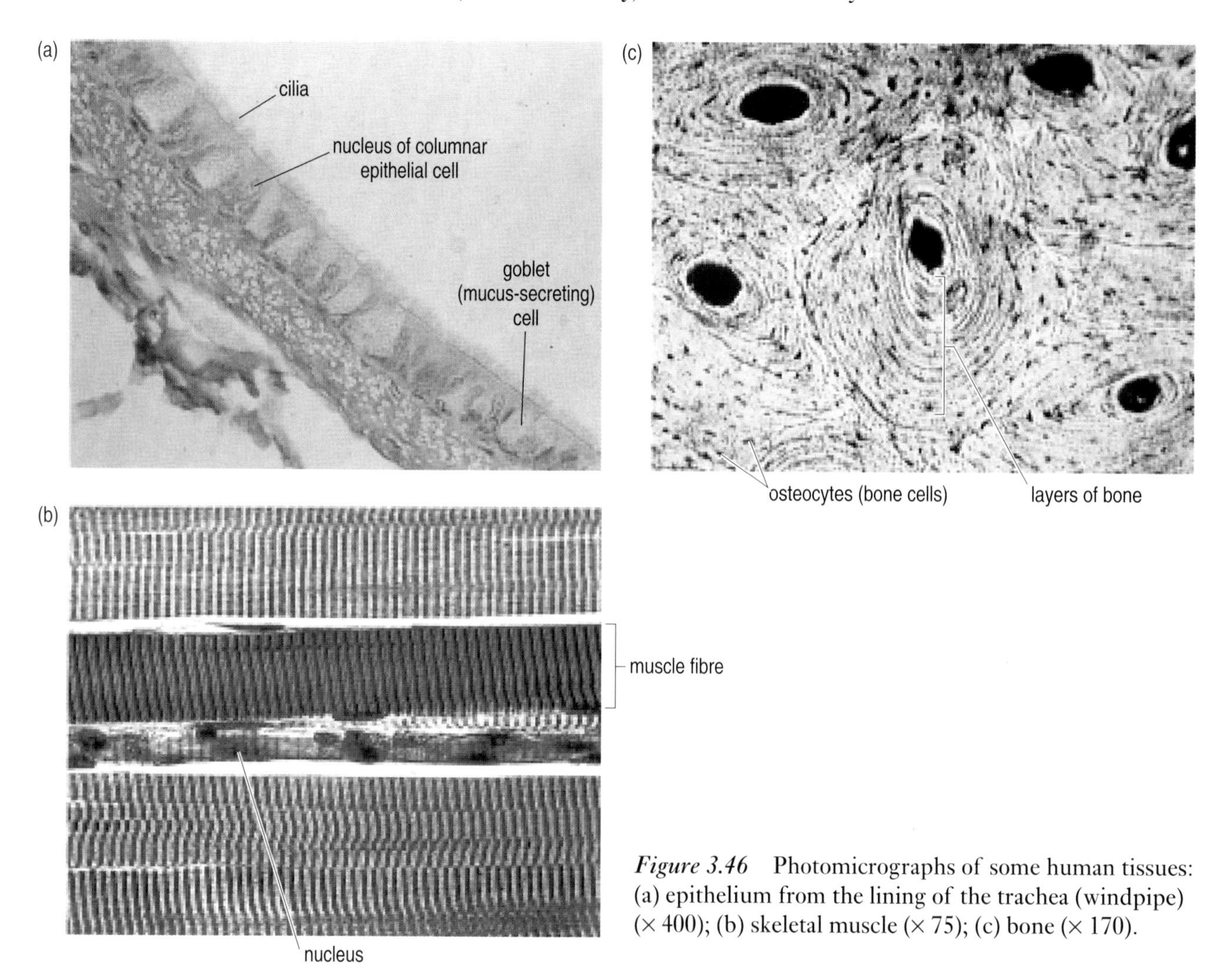

Figure 3.46 Photomicrographs of some human tissues: (a) epithelium from the lining of the trachea (windpipe) (× 400); (b) skeletal muscle (× 75); (c) bone (× 170).

Cells of many tissues are held together by a variety of **cell junctions** which also allow cells to communicate with each other. *Tight junctions* connect neighbouring cells together tightly enough to form a fluid-tight seal. Epithelial cells are often joined together by tight junctions and so can act as a barrier to fluid loss. Cells in tissues that are subject to friction and stretching such as the outer layers of the skin, muscle tissue in the heart, the neck of the womb and the epithelial lining of the digestive tract, are commonly joined by different types of *anchoring junctions*. Anchoring junctions use the cell's cytoskeleton to join cells to one another or to extracellular material (material outside of the cell). A third type of junction is a *gap junction* where adjacent cell membranes come extremely close to each other but do not actually touch. Proteins called *connexins* span the gap, forming fluid-filled channels through which small molecules can pass directly from the cytosol of one cell into the cytosol of a neighbouring cell. Cells that are cancerous do not have gap junctions and so do not communicate well with the cells around them, growing and reproducing in an uncontrolled and uncoordinated way.

Figure 3.47 shows how collections of cells form tissues that are then formed into **organs**. An organ is a collection of distinct groups of tissues which performs a specific function. Examples of the major organs are: the liver, which controls the blood glucose level (Section 3.3.1) and also destroys or detoxifies harmful chemicals (e.g. alcohol) entering the bloodstream; the kidneys, which remove waste products from the blood, producing urine; and the heart which pumps blood to the lungs and around the body. Organs can be grouped into **organ systems** (or body systems) in which all the component organs work together. For example, the kidneys, the bladder and tubes called ureters which connect these organs make up the urinary system. The heart and the body-wide system of blood vessels through which the blood pumped by the heart is circulated, is the cardiovascular system. Ultimately in this hierarchy of structure and function we reach the whole human body, in which all the systems work together to produce a functioning person. Throughout the rest of this course, the structure and function of the human body will be studied at all these different levels, with emphasis on their interdependence and interaction. Figure 3.47 summarizes the complete hierarchy of biological organization and concludes the first part of this chapter.

In Part II we focus our attention on a particular and very familiar collection of tissues, the skin, and its response to injury, i.e. the healing process. As you will see, this necessarily requires the production of new cells, an activity that is central to the net growth and repair of all tissues. We shall therefore look in some detail at the process of cell division, or mitosis, which we introduced in Section 3.5.1.

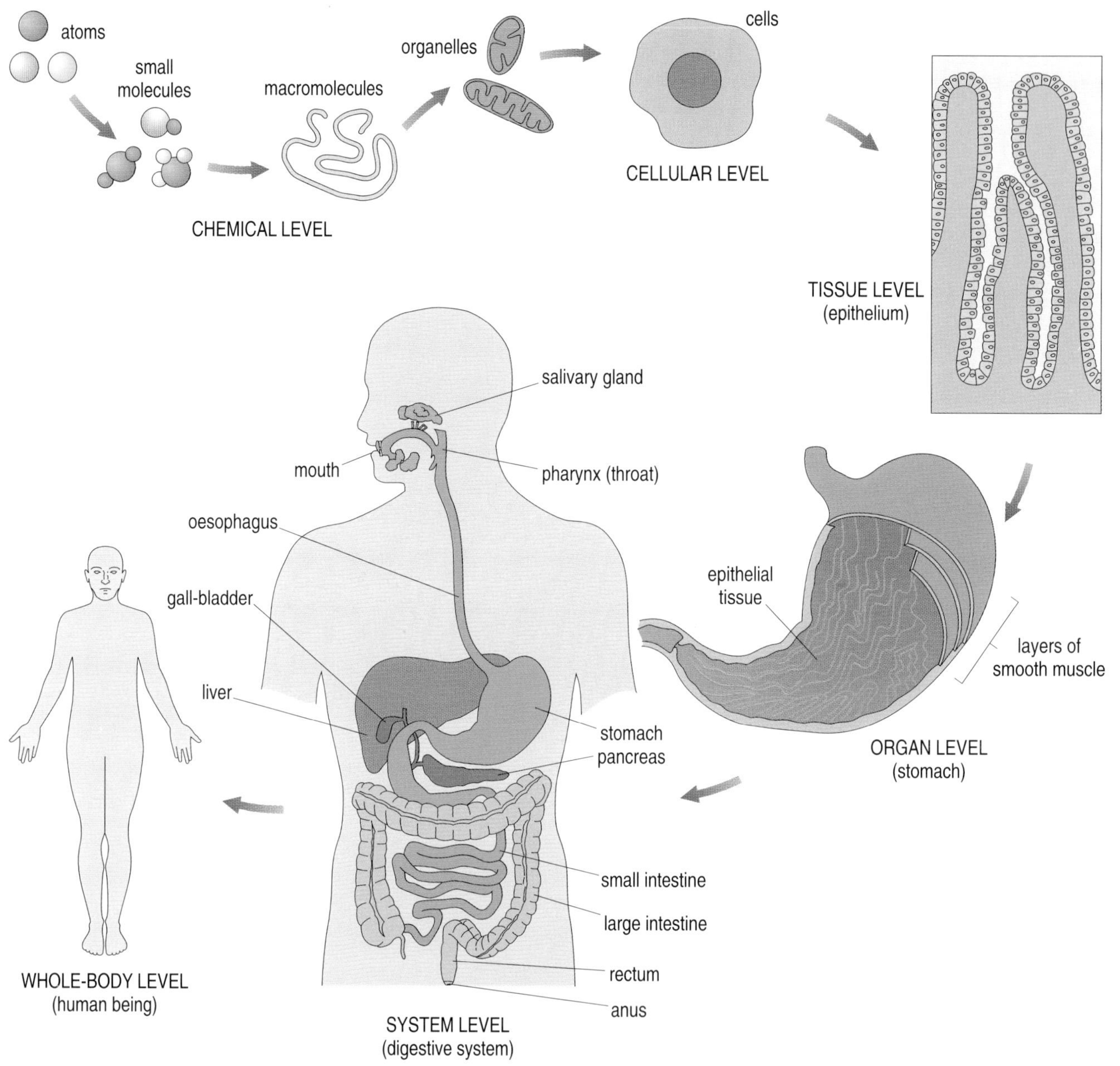

Figure 3.47 Levels of biological organization: atoms, molecules, organelles, cells, tissues, organs, systems, whole body.

Summary of Section 3.7

1 Tissues are groups of differentiated cells performing the same function (e.g. epithelial tissue and connective tissues).

2 The differences between tissues are due to differential gene expression; this can be time-dependent as well as tissue-dependent.

3 Cells in tissues are held together by various types of cell junctions.

4 Organs are collections of tissues performing a particular function (e.g. liver, kidney).

5 In organ systems the activities of the component organs are integrated (e.g. the urinary and cardiovascular systems).

PART II TISSUES IN ACTION

3.8 The healing process

Healing is by no means limited to the special emergency measures which the body calls upon to restore order to tissues after damage from, say, a skin wound, a sprained ankle, or a broken bone. Healing means making whole, a process which goes on in organisms all the time. Organisms maintain their structure and form rather as a fountain does: there is a continuous flow of material through the form so that its stability to disturbances comes from a highly dynamic nature, not from static forces such as those that maintain the stable form of, say, a chair or a car. Even our bones are dynamic; if they were not, fractures wouldn't heal. What we see happening in tissues after wounding is just an amplified version of the tissue recycling or turnover which is taking place all the time in the body. (As you know, there is constant turnover at the subcellular level also, with organelles and other cell components being broken down and resynthesized all the time.) In this part of the chapter the objective is to take a look at various aspects of the healing process by examining in detail the structure of the skin and how this structure is restored after wounding. This will take us through the different levels of organization of the body – organs, tissues, cells, and molecules – leading to an understanding of how each of these levels contributes to the integrity of the whole. An essential insight that will emerge is that it is not just what cells and tissues are made of, but more importantly, it is the dynamic *relationships* between the various components that hold the key to living order.

3.8.1 Structure of the skin

The outermost layer of the skin is epithelium. As mentioned in Section 3.7, epithelium forms the surface layer of nearly all the organs of the body, both inside and out. The skin has some important characteristics with which we are familiar. It has a tough surface, is elastic and pliable, hurts when pinched or cut, and has remarkable healing properties, as required of a barrier that separates the body from its physical, chemical and biological environment. When a section of skin is examined under a microscope that magnifies about 100 times, it has the structure shown in Figure 3.48. Starting at the surface (top) and working down, you see several layers and types of cell that are divided into **epidermis** (the epithelial tissue) and **dermis**. These two layers are separated by an undulating single layer of cells shown as little units with dots in the middle, which are the cell nuclei. The highly corrugated structure of the junction between the epidermis and the dermis results in a very strong union that resists shearing forces that might otherwise separate the layers as the skin rubs against objects.

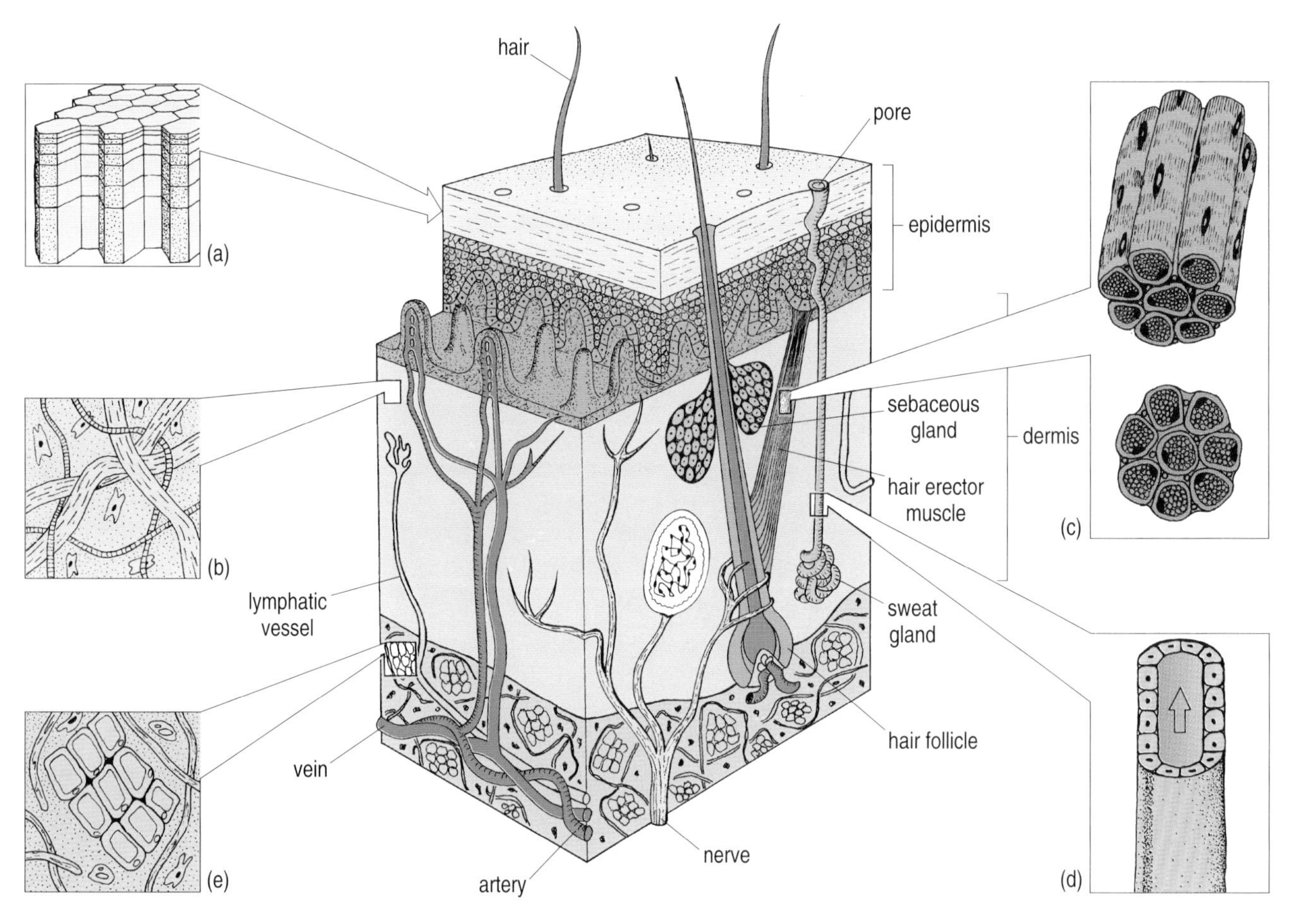

Figure 3.48 Structure of the skin, showing the various types of cells and tissues: (a) epithelial cells in epidermis; (b) connective tissue; (c) muscle; (d) epithelial cells lining the duct (tube) which carries sweat to the surface of the skin. Also shown are (e) the fat cells (adipose tissue) just beneath the dermis; this tissue provides insulation and also acts as an energy reservoir.

Look at the distribution of the different components in Figure 3.48 – hairs, blood vessels, nerves, muscle – and then try the following questions.

❑ If a scratch removes only the upper layers of the epidermis, will it bleed?

■ No. The blood vessels are confined to the dermis, so a cut needs to go through the epidermis before bleeding occurs.

❑ Since the nerves do not extend to the epidermis, why do we feel the light touch of a feather on our skin?

■ We feel it through the hairs, which when moved can activate the associated nerves.

❑ Do you have any areas of skin that are insensitive to touch? Why?

■ If you have had a fairly deep wound that removed both epidermis and dermis, including the hairs, then the patch of healed skin can be without hairs or nerve fibres because these fail to regenerate, so the patch will be insensitive. Also, heavily calloused regions of the skin become insensitive because the epidermis becomes so thick that a light touch fails to activate any nerves in the dermis.

The epidermis has some distinctive features that illustrate the principle that the body is continuously engaged in the healing process to maintain its structure. Inset (a) of Figure 3.48 is a diagrammatic view of the way epidermal cells are organized into hexagonal columns known as **epidermal proliferative units** (EPUs) – but note that these are not actually visible as discrete units. The structure of a single EPU is shown in Figure 3.49. There are several layers, described as basal, spinous, granular and cornified. Cells in the basal layer are attached to a non-cellular, polysaccharide structure, the *basement membrane* (which they produce and maintain), lying just above the dermis. These cells grow and divide, as shown by the cells on the right where the two sets of chromosomes are separating to the two newly formed cells, a stage in the process of mitotic cell division (see Section 3.8.2 below). Cells leave the basement membrane to become part of the spinous layer, where they flatten and spread, the nuclei also changing shape. At the same time, these cells begin to produce the protein keratin, which gives the cell mechanical resistance and impermeability. The cells are all attached together by tight junctions (see Section 3.7). Although these junctions are broken down and reformed as the cells move relative to one another, they are very important in maintaining the structural integrity of the epidermis.

Cells in the different layers are distinguished from one another mainly by their content of keratin, which increases towards the surface, as well as by their shape. At the top, in the cornified layer, the cells have become thin desiccated plates filled with keratin and joined together at their edges by very strong tight junctions. They have lost their nuclei and so die, getting sloughed off the surface of the skin as the body goes about its business of interacting with the world. Much of the dust in your house comes from these scales of keratinized skin; most of the rest of it is hair, which is also made of keratin.

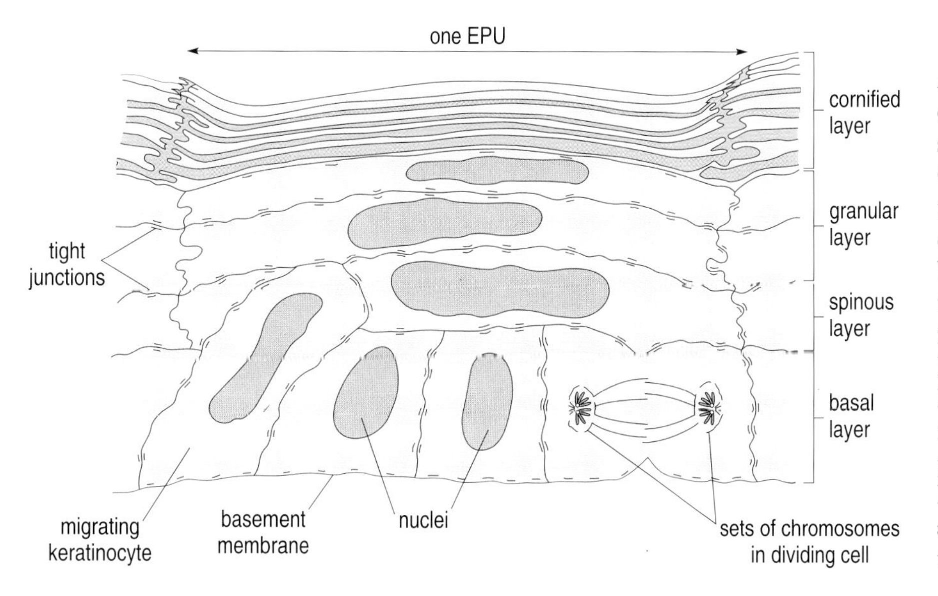

Figure 3.49 Section through an epidermal proliferative unit (EPU), showing the layered structure. Cells in the basal layer, attached to the basement membrane, grow and divide. Cells are displaced from the basement membrane and move into the spinous layer, where they flatten and differentiate, producing keratin. The cells (called keratinocytes) are pushed upwards from layer to layer, flattening progressively and producing more keratin until they lose their nuclei and become the hardened scales of the cornified layer.

A surface view of the epidermis (five EPUs) is shown in the top part of Figure 3.50. The bottom part of the figure is a cross-section of three EPUs, with a basal cell dividing in the one on the right. The hexagonal shape of the EPUs is a result of mechanics and geometry: the most stable arrangement of units on a surface, given that the units reduce the extent of their boundaries to a minimum (a strategy that minimizes use of materials and energy), is a hexagonal array.

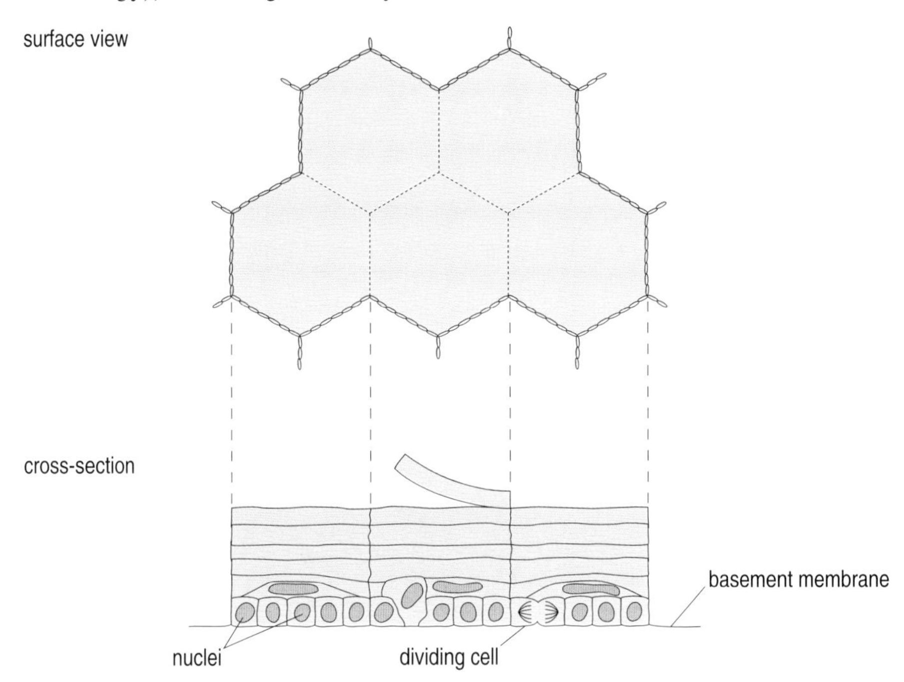

Figure 3.50 (Top) A surface view of several EPUs showing their characteristic hexagonal shape; (bottom) three EPUs in cross-section. (Only the nuclei of the basal and spinous cells are shown.)

3.8.2 Cell division

The process at the foundation of most tissue maintenance is cell growth and division, whereby one cell gives rise to two. This type of cell division is called **mitosis**. Cells lost through wear and tear are continuously replaced, normally at a rate that just balances loss. Otherwise, the result would be excessive cell production, leading to a tumour; or deficiency, resulting in tissue wasting. There are of course tissues of the body in which, in adult life, no cell division occurs – most notably the brain and the rest of the nervous system. Nerve cells that wear out and die are not replaced, so there is a continuous decrease in their number after a certain point in the life of the individual (when division of nerve progenitor cells stops). But most of the tissues of the body are renewed continuously by their dividing cell populations. In the epidermis, this actively dividing population is the basal layer. The following account describes cell division with particular reference to the epidermis, but bear in mind that the same sequence of events occurs in *all* the dividing cells of the body.

In most of the basal cells, no chromosomes are visible in the nucleus: all that can be seen with a light microscope, looking at cells that have been stained, is that the nucleus appears much denser than the surrounding cytoplasm (as in Figure 3.27a). Within the nucleus are the chromosomes, 46 of them (23 pairs) in a human cell. The cells of the basal layer grow, and

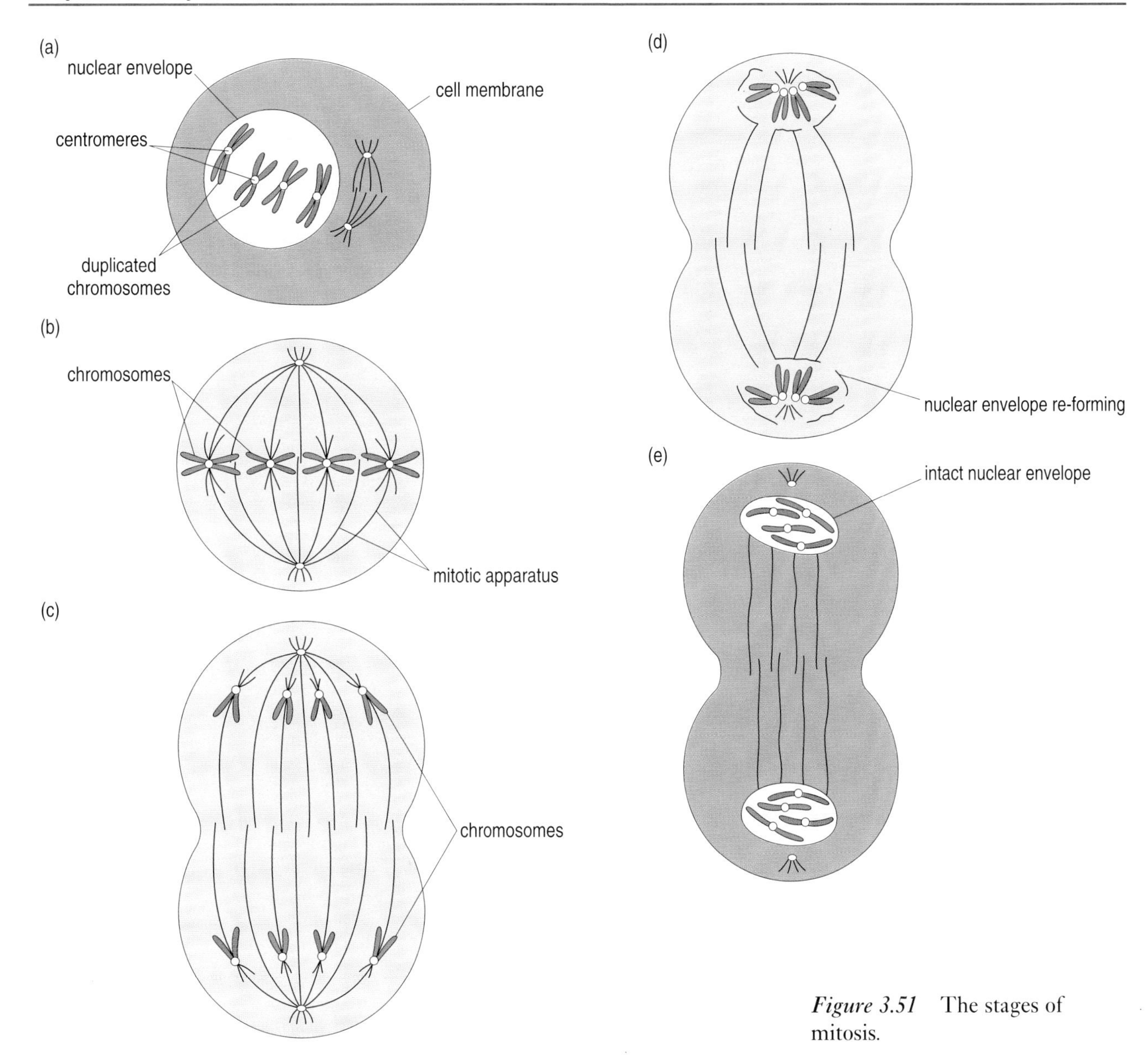

Figure 3.51 The stages of mitosis.

at a certain stage of growth the DNA of the chromosomes starts to replicate, as described in Section 3.6. This DNA replication process takes several hours.

Once a basal cell has duplicated its chromosomes, it is well on its way to making two cells, each of which will get a complete set of chromosomes. The sequence of events resulting in the partitioning of the duplicated chromosomes into two nuclei is shown schematically Figure 3.51 (for simplicity, only four of the 46 chromosomes are shown here). The duplicated chromosomes, at this stage still joined together at their centromeres, first condense into compact structures within the nucleus (Figure 3.51a). These are made visible under light microscopy by staining, as described above. The nuclear envelope then disintegrates and the chromosomes gather at the centre of the cell (Figure 3.51b) called the equator, by analogy with the Earth. A special mitotic apparatus (also called a mitotic **spindle**), made of protein, is involved in separating one copy of each chromosome from its duplicate and moving them to opposite poles (ends) of the cell (Figure 3.51c–d). A new nuclear envelope is formed around each of the sets of chromosomes and the cell then divides into two (Figure 3.51e), a process called **cytokinesis**. Thus mitosis is now complete – one cell has become two, each with a chromosome content identical to that of the original cell.

The whole process, from the growth of a newly formed cell through DNA replication and chromosome duplication to the end of mitosis, is called the **cell cycle**. Of course, all the other constituents of the cell described in Section 3.3 also increase in quantity during the growth phase, so that both new cells have a complete allocation of the materials and structures that make up a living cell. Attention is focused on the chromosomes and the genetic material in this process because of the importance of an equal partitioning of genes to both progeny cells in order that they retain the capacity to grow and to differentiate normally. However, it is important to realize that the total organization of the cell is required for this to occur. What reproduces is the whole cell and this has to be understood in terms of the dynamic interactions of *all* its constituents. This integrated system is more than the sum of its constituent parts, none of which would be able to perform as they do without the presence of the other constituents and their organized interactions. Also, to understand why a cell in a tissue such as the epidermis carries on growing and dividing rather than differentiating into a specialized cell such as an epidermal keratinocyte, we need to take account of its environment, e.g. the location of the basal cell layer in the tissue and the properties of the basement membrane, its position relative to the nutrients that come to it from the blood vessels in the underlying dermis and in relation to differentiating cells above it. These provide conditions and signals that influence cell behaviour in the basal layer, determining how rapidly cells grow and divide, conditions that can be disrupted.

3.8.3 The dermis

Now look back at the structure of the dermis in Figure 3.48. We have already noted the presence of nerves and blood vessels in this tissue and their absence in the epidermis.

❑ What strikes you as a sharp contrast between the cellular organization of the dermis and that of the epidermis?

■ There are fewer cells and much larger amounts of connective tissue in the dermis than in the epidermis.

Inset (b) of Figure 3.48 gives a schematic view of the structure of the loose connective tissue of the dermis, which has the appearance of a disorganized tangle of fibres with cells interspersed among them. The thicker threads are fibres of the structural protein collagen which have great mechanical strength and elasticity. These fibres form bundles that run in characteristic directions in different parts of the body. You can determine their direction by pinching your skin together and finding out which way it folds most easily into fine wrinkles. After a certain age, no pinching is necessary; the wrinkles are just there. This is because the elasticity of collagen and other fibres gradually decreases with age.

The natural folds of the skin are known as cleavage lines, and they are important to a surgeon: an incision that separates parallel bundles of collagen fibres without rupturing them heals with a fine line, whereas an incision severing and disrupting them produces a broad scar.

❑ Which way do the collagen fibres run on your torso?

■ Around it, i.e. transversely.

❑ So which is the best direction for an incision to remove an appendix?

■ A transverse incision.

Within the connective tissue of the dermis are located cells of different type and the various structures such as sweat glands, hairs with associated muscles and sebaceous (oil-producing) glands, vascular tissue (fluid-conducting vessels – arteries, veins, lymphatics) and nerves. The sweat glands secrete sweat onto the surface of the skin, getting rid of some of the body's wastes and at the same time keeping the skin cool. The greasy substance secreted by the sebaceous glands lubricates the skin, keeping it supple and preventing it from drying out. As in the epidermis, the structures of the dermis are dynamically maintained: hairs grow, muscle cells get damaged and are replaced, vascular tissues rupture and are repaired, and collagen gets remodelled in response to the stresses and strains on the skin.

As you can see in Figure 3.48, inset (b), there is more to connective tissue than just collagen bundles. Thinner fibres are entangled with the collagen. These are elastic fibres, largely made up of the protein elastin, which increase the strength of connective tissue. Also present are many types of non-fibrous protein that make up the material between cells and fibres – known as the **extracellular matrix** (ECM). Cells called **fibroblasts** ('fibre cells') move about in the dermal matrix, producing collagen and other constituents of the ECM, which they are continuously remodelling, thereby maintaining a distinctive chemical and structural environment which itself acts on the cells, keeping them in a particular state. When this is disturbed, as we consider next in relation to wounding, the changed environment of the cells stimulates them to act in ways that naturally restores the previous state of dynamic order, which is described as a *steady state* (a balanced condition maintained by continuous restoration of components which are always wearing out and being replaced). This process of maintaining a constant condition by a dynamic balance between opposing forces is an example of **homeostasis**, a basic principle of biological order which you will encounter in different contexts throughout this course.

Summary of Section 3.8

1 Healing is an expression of the normal maintenance activities of the body, based on a balance between production and destruction of components.

2 The epidermis is maintained by a dynamic balance involving cell division, cell movement and cell differentiation, the production of extracellular materials and cell death. These processes are spatially organized in epidermal proliferative units, with dividing cells in the basal (bottom) layer and progressively differentiated, keratinized cells in layers up to the surface, where dead cells are removed through wear and tear.

3 Cell division (mitosis) involves chromosome replication and their equal partitioning to the two new cells as well as duplication and partitioning of all other cell constituents, with the result that both progeny cells inherit the entire dynamic organization characteristic of the living state.

4 The dermis is loosely packed connective tissue containing collagen and elastin fibres within an extracellular matrix produced and maintained by fibroblasts. Within the dermis are sweat glands, hairs with associated muscle fibres and sebaceous glands, vascular tissue and nerves.

3.9 Wound healing

3.9.1 Tissue damage and pain

The first response to a skin wound is the rapid reflex action which gets the body away from the cause of the damage, and the experience that follows quickly is that of pain (there is much more on this in Book 2, Chapter 3). This signals physical disruption. In the skin, the nerves are responding to an altered pattern of relationships within the tissue and with the environment: high pressure (causing bruising), high temperature (causing burning) or cutting (causing bleeding), with exposure of normally covered nerve endings.

The next stage of the wound response is provisional (i.e. temporary) recovery of integrity by the processes of blood clotting to cover the exposed area. If the wound involves considerable blood loss that is not quickly staunched, then mechanisms within the body that rapidly restore blood pressure and blood volume are brought into play, so that the heart continues to function normally. These wound response mechanisms depend upon regulatory processes operating over the body as a whole, not just local ones occurring in the skin. Depending on the severity of the wound, different levels of the body's regulatory activities are brought into operation. The large-scale or higher-level mechanisms of maintaining the body's normal state will be discussed in detail in Books 2 and 3. Here the focus is on the healing processes within the skin itself.

3.9.2 Provisional recovery: inflammation and clean-up

As soon as blood vessels are damaged and blood is released into the wound area, the blood-clotting mechanism is triggered. A protein in the blood plasma (the liquid part of the blood) changes from a soluble form known as fibrinogen to insoluble fibres of fibrin which become tangled together with blood cells and form a dense mat, a fibrin clot. This is a provisional extracellular matrix that is not unlike that of connective tissue with its tangled fibres, but it is made of different proteins and is much denser. As it dries out, this clot forms a scab. In the neighbourhood of the wound, blood vessels dilate in response to other substances released from disrupted cells, notably a compound called histamine. The dilation makes the vessels leaky so that more blood escapes into the wound region and contributes to clot formation. The dilated vessels can also carry an increased volume of blood to the wound region.

❑ What are the external signs of this increased blood flow to the wound area?

■ Redness and swelling (inflammation).

This is the typical inflammatory response: swelling, redness, and also tenderness. Blood cells get trapped within the fibrin network, particularly the so-called inflammatory cells (white blood cells and macrophages, a special kind of phagocytic cell). These cells play important roles in healing, by digesting dead cells and tissue remnants and by combatting bacterial infection. They do this by engulfing cell fragments and bacteria (Figure 3.32) and then digesting them.

❑ What is this engulfing process called and which particular organelle is responsible for intracellular digestion?

■ The material is taken in by phagocytosis and broken down in the cell's lysosomes.

White cells and macrophages also carry out this mopping-up operation within the blood itself, ingesting and digesting damaged and dying blood cells, so that they are continuously involved in this tidying-up process. (And of course, when they die their fragments are ingested by other white cells and macrophages in the blood.)

3.9.3 Regeneration and repair

As the fibrin clot forms and dries, producing the temporary protective cover of the scab in a few hours, other processes are initiated which bring about tissue regeneration. Blood cells of a particular type called platelets release a protein called platelet-derived growth factor (PDGF), which stimulates cell division in epidermal, dermal, and blood cells (the white cells and macrophages involved in clearing up cell debris and engulfing bacteria). In the dermis, the fibroblasts are also stimulated to divide and become more active. They migrate into the wound area, guided by the fibrin network, and differentiate into both collagen-synthesizing cells and cells containing filaments of actin, a protein which gives these cells contractile properties like those of muscle. These contractile cells form in a ring around the wound margin, connecting to one another and to the fibrous components of the extracellular matrix. Their active contraction draws the regenerating dermal tissue inwards over the wound, a process known as **wound contraction**. The fibroblasts produce new collagen and other materials that are laid down as the margin of the wound closes, producing a new extracellular matrix for the regenerating dermis.

In response to PDGF, epidermal cells also become more mobile, migrating over the surface of the closing dermal ring. They secrete materials that become the extracellular matrix of the epidermis, which itself facilitates and guides cell migration. A mere 48 hours after wounding, the surface of the wound may already be covered by a thin layer of epithelial cells. For the next week or so, this grows into the multilayered epithelium of normal skin.

The healing tissue is particularly vulnerable to damage during this phase. The regenerating epithelium is initially a thin layer unable to offer any protection against environmental forces. The same is true of regenerating connective tissue, which does not yet have sufficient strength or elasticity and also contains numerous small, thin-walled blood vessels. At this stage a wound may reopen or bleed as a result of traumas that would be insignificant in normal circumstances. Such damage may delay the healing process or even prevent full recovery if it recurs frequently.

Within the dermis, hair follicles, sebaceous glands, and sweat glands can also regenerate if the wound is not so deep that all remnants of these organs are removed. However, after severe wounding these may not regenerate so that the scar tissue lacks these normal skin components. The new connective tissue has a dense network of blood vessels. This is known as granulation tissue because to the naked eye it has the grainy appearance characteristic of a healing wound (due to the many tiny, newly-formed blood vessels within it). The wound is now closed, but the new tissue is still different in many ways from normal skin.

3.9.4 Restitution: remodelling and transformation

The last phase of wound healing involves a process of restitution – the reorganization and remodelling which produces a tissue that is as close as the regenerative process can get to the original structure and properties of the skin. Within the granulation tissue of a wound, the remodelling process begins with the disappearance of many of the (now superfluous) small blood vessels by contraction and closure. Some of the vessels remain to serve the new tissue, those that deliver blood developing a strong, muscular wall and becoming little arteries (arterioles) while the veins (which return blood to the heart) retain a thinner, softer wall.

❑ What do you think the skin looks like at this stage of healing?

■ The closing of blood vessels is visible to the eye as a gradual fading of the initially dark purple surface of the wound and the restoration of normal skin colour.

The extracellular matrix of the newly formed dermis has a rather higgledy-piggledy arrangement of collagen fibres, which were laid down quickly by the migrating fibroblasts as the wound closed. These are remodelled by the cells of the dermis so that the fibres become reoriented in the direction of the tension lines of the skin, reforming the cleavage lines described earlier. The elasticity of the tissue is restored by the close packing of the other constituents of the extracellular matrix and the reorganization of the fibres. The epidermis takes on the characteristics of the mature tissue with the reformation of the stacked layers of cells in the hexagonal epidermal proliferative units, though there are often disturbances to the regularity of this arrangement in scar tissue.

The overall result is the restoration of the tough, impermeable, elastic properties of the skin within a period of 2–4 weeks. As stressed earlier, this is achieved by an amplification of processes that continuously maintain the organization of the tissue in its normal state. Some degree of wounding occurs simply as a result of living. For example, exercise may cause minor damage to muscle fibres – hence the need for periods of rest to allow repair. The same is true of disease, which occurs naturally as part of the living process. Health is not, from this point of view, the absence of damage or disease, but the capacity to respond appropriately through a healing process that maintains satisfactory integrity of the body. Healing need not restore the damaged system precisely to its previous state. All that is required is a restoration of function through an adequate structure. In the case of skin, hairs and sweat glands may fail to be regenerated and scar tissue may be less well organized than normal in terms of collagen fibre orientation and the structure of the ECM.

If the skin remodelling phase does not occur satisfactorily, then the result may be excessive growth and/or disorganization that may lead ultimately to benign or malignant tumours. For example, during wound healing the regenerative phase, initiated by inflammation, may go too far and lead to the formation of superfluous granulation tissue, followed later by fibrosis (primarily, excessive formation of collagen fibres) and contraction which results in malformation, puckering and stiffness, with loss of elasticity. This often happens with burns that have not healed well. Years later, epidermal cancers may occur at these sites. Healing depends upon growth. However, if this is not properly integrated into the other processes of differentiation and remodelling, there is a danger that growth will occur again sooner or later in response to the disorganization, in an attempt to correct it, resulting in cancer.

Summary of Section 3.9

1 The pain that accompanies a skin wound is an indicator of tissue disorganization, signalled by nerves.

2 A sequence of events is then initiated to restore normal organization: inflammation and provisional recovery, with scab formation; regeneration and repair, involving cell division, cell migration and extracellular matrix formation; and finally, remodelling, resulting in a good approximation to the normal state.

3.10 Nutrition and the healing process

Two of the most important conditions required for rapid and effective healing are adequate nutrition and the absence of infection. In the past, good nutrition and dietary rules were among the most important factors that could be used to facilitate healing. With the advent of antibiotics and powerful, selective synthetic drugs, there has unfortunately been something of a decline in the perceived significance of these basic requirements for the restoration of health, except in the treatment of metabolic diseases (i.e. disorders of cell metabolism), where nutrition is the focus of attention. Patients with serious injuries have an elevated metabolic rate, i.e. they use up large quantities of energy and raw materials.

❑ What are the raw materials used in metabolism and in what form is energy made available in cells?

■ Carbohydrates, lipids and proteins; these are broken down to provide both the 'building blocks' and the energy (as ATP) required for the body's synthetic reactions which contribute to growth and repair.

The response to such physical trauma is called the **stress reaction**. The regulatory substances released in this response, e.g. platelet-derived growth factor (PDGF) which stimulates growth in epidermal, dermal, and blood cells, and stress hormones such as adrenalin and noradrenalin, cause increased rates of use of carbohydrates, lipids and proteins. The body's stores of fat and carbohydrate are mobilized in this heightened metabolic state, which may be accompanied by a rise in body temperature (fever).

During the wound repair phase when the fibres and materials of the extracellular matrix are being laid down rapidly and the connective tissue of the dermis is being restored, there is a greatly increased synthesis of proteins. Any serious protein deficiency in the diet inhibits the formation of granulation tissue, resulting in poor healing of skin wounds. Collagen is particularly rich in certain amino acids (the building blocks of proteins), especially glycine, lysine and proline (Table 3.3), so these need to be in plentiful supply, as they are in a well-balanced diet. Other proteins of the extracellular matrix are rich in the amino acids that contain sulphur (methionine and cysteine), adequate amounts of which are therefore also needed in the diet. The formation of collagen fibres, with their elastic and tensile properties, is dependent on a process that requires vitamin C. Thus a deficiency in this vitamin results in poor collagen formation, as well as excessive deposition of other components of the extracellular matrix. The skin is then weak and has a tendency to bleed – hence the characteristic symptoms of severe vitamin C deficiency, scurvy. These consequences of malnutrition also increase the susceptibility to infection, which itself delays healing.

These basic facts about the importance of good nutrition for wound healing and recovery from illness have been known for many years, but they became firmly established during the 1930s when most of the vitamins were discovered and their therapeutic effects were being recognized. For example, the death rate in children with measles in a London fever hospital was reduced from 8.7% to 3.7% simply by a daily supplement of vitamin A from cod liver oil. The rationing system which ensured an adequate distribution of essential foods and vitamins during and in the period following the Second World War was a spectacular success in practically eliminating malnutrition from the UK.

However, the nutritional status of the population is currently getting worse, though the nature of the problem has now changed. One indicator of nutritional imbalance is obesity, which rose from 6% (men) and 8% (women) in 1980 to 8% and 12% in 1987 and to 13% and 16% in 1993. This information is gathered by the Office of Population Censuses and Surveys and used by the Health of the Nation Initiative to monitor trends. Evidently disease related to obesity is becoming ever more common. (Diet and its influence on health are looked at in Book 3, Chapter 7.)

Summary of Section 3.10

1 Successful healing requires good nutrition and the absence of infection.

2 Serious injuries provoke a stress reaction involving the release of hormones that increase metabolic rate throughout the body – often accompanied by fever – and increased growth rates in the damaged tissues.

3 Nutritional deficiencies reduce healing rates and increase susceptibility to infection, which also delays healing.

4 Rationing in the UK during the 1940s and early 50s improved the nutrition of the population; one current and growing nutritional problem is that of obesity.

3.11 Stress and healing

The nutritional needs of the body for successful healing tell us about biological factors involved in reconstructing tissues after damage and restoring a functional whole. However, this is only one of the levels that influence the condition of health. Psychological stress and well-being are equally important as factors in healing. A study by an American research team (Kiecolt-Glaser *et al.*, 1995) presented evidence that carers of dependants with Alzheimer's disease recovered from skin wounds significantly more slowly than a control group that did not have this responsibility. The subjects of the investigation were all healthy women in the 47–81 years age group who had been providing care for several years to either a husband or a parent with Alzheimer's disease. They spent an average of 6.7 hours per day in caring activities. The control group matched the carers in age, income and social class.

There were 13 carers and 13 control subjects who agreed to participate in the experiment. This involved making a small wound to the upper arm by a method routinely used to examine skin cells, a procedure called *biopsy*, which produces a very uniform full-thickness wound (i.e. one that removes both epidermal and dermal tissue). The wounds were dressed, then examined at intervals until they were fully healed. The extent of healing was determined by measuring the size of the wound from photographs, and by the response to hydrogen peroxide, an antiseptic that foams when the wound is not fully healed. To avoid observer bias, these measurements were carried out by a person who did not know which were controls and which were carers. The results are shown in Figure 3.52.

❑ How would you characterize the difference between the two groups?

■ The carers show a delay in healing compared with the controls, though they start to catch up after six weeks.

The investigators also took a blood sample from each individual before the biopsy wound was made. This was used to measure the capacity of the white cells to produce a regulatory protein called interleukin 1, which is important in the formation of the connective tissue matrix and in the migration of fibroblasts. Interleukin 1 levels were found to be significantly reduced in carers compared with controls.

The results of this study, though carried out on a small sample of people, show how psychological stress can affect one of the most basic of the body's repair processes, healing the skin after damage.

❑ Can you think of another indicator of health that is likely to be affected by psychological stress?

■ One of these is vulnerability to infectious disease, which is known to increase in those subjected to stress. (It is also increased by physiological stress, as highly-trained athletes are well aware.)

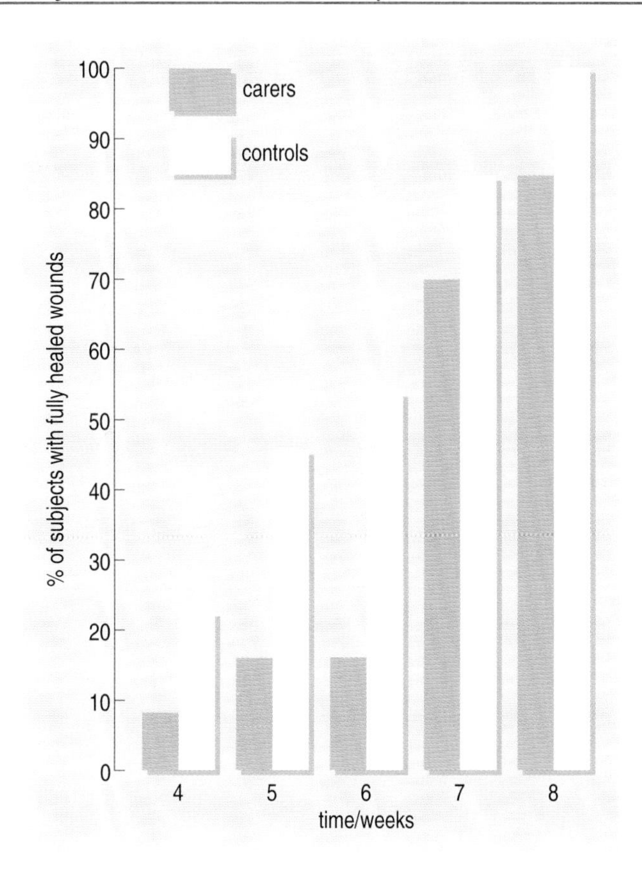

Figure 3.52 Percentage of subjects (carers and controls) whose wounds had healed completely, plotted against time after wounding.

This study points to the significance of what are often called 'states of mind' on the activities of the body, showing that body and mind function as an integrated whole, and that the capacity to heal reflects the condition of the whole person.

Summary of Section 3.11

1. Psychological states have an influence on healing rates, as indicated by a study of biopsy wound healing in carers of people with Alzheimer's disease.
2. Compared with a control group, the carers had a significantly delayed rate of wound healing and reduced levels of a regulatory protein (interleukin) involved in skin regeneration.

3.12 Complementary healing therapies

The unity of body and mind that is illustrated in the previous section is becoming widely accepted as a reality in the health field and the interactions between psychological and physiological states are being intensively explored (as will be described in Book 4). However, not many years ago, the idea that subjective states (emotions, stress, mental images) could affect physiological states was considered either unlikely or impossible by many scientists and doctors, and there are still many sceptics.

Scepticism is a healthy condition to maintain in science so long as it allows possibilities to remain open. There have been many examples of phenomena that have been declared impossible and related therapeutic procedures useless, but where a few years of research and practice have established their worth. Complementary therapies provide many such cases, acupuncture being perhaps the most dramatic instance of a practice that was initially dismissed as unscientific and impossible but is now well established as an effective healing and anaesthetic procedure. Homeopathy is another complementary practice which arouses fierce controversy, being dismissed as scientifically impossible by many professionals. Likewise, there is considerable scepticism over alleged healing by what is called non-contact therapeutic touch (Wirth *et al.*, 1993). Such issues are discussed in TV programme 4, which deals with complementary medicine.

Summary of Section 3.12

Recognition that healing depends on factors such as stress and other influences that are not currently understood within a scientific framework has grown significantly in recent years, and is likely to be extended as a result of the use of complementary therapies and further research on their effects.

Objectives for Chapter 3

After completing this chapter you should be able to:

3.1 Define and use, or recognize definitions and applications of, each of the terms printed in **bold** in the text.

3.2 Understand the basic principles of atomic structure and chemical bonding. (*Question 3.1*)

3.3 Describe the structure and functions of the main molecular constituents of living organisms. (*Question 3.2*)

3.4 Understand the basics of cell metabolism and the role of enzymes. (*Question 3.3*)

3.5 Describe and understand the roles of the various subcellular structures. (*Question 3.4*)

3.6 State the relationship between DNA base sequence, genes and chromosomes and give an account of the process of DNA replication. (*Question 3.5*)

3.7 Appreciate the differences between the structure and function of DNA and RNA. (*Question 3.6*)

3.8 Understand transcription and translation as sequential steps in gene expression and describe the main events in the translation process. (*Question 3.6*)

3.9 Describe the hierarchy of organization of living things, from the molecular components, through organelles, cells, tissues, organs, organ systems up to the level of the whole organism.

3.10 Identify the structural components of normal skin and distinguish epidermis from dermis. (*Question 3.7*)

3.11 Describe how the structure of the epidermis is maintained. (*Question 3.7*)

3.12 Describe the nuclear and chromosomal events that occur during mitosis. (*Question 3.8*)

3.13 Give an account of the stages of wound healing and the basic processes involved in restoring the integrity of the tissue. (*Question 3.9*)

3.14 Explain why good nutrition is necessary for satisfactory wound healing. (*Question 3.9*)

3.15 Describe experiments whose objective is to study the effects of stress on wound healing, and interpret the results. (*Question 3.9*)

Questions for Chapter 3

Question 3.1 (*Objective 3.2*)

Give the structural formula of ammonia, NH_3, and then draw out the structure showing all the bonding electrons, using the convention in Figure 3.7. Would you expect NH_3 to be a polar or a non-polar molecule and why?

Question 3.2 (*Objective 3.3*)

Draw up a table to contrast the structure and role of the macromolecules insulin and glycogen in the body.

Question 3.3 (*Objective 3.4*)

An enzyme early on in a catabolic pathway is found to have a binding site for substrate molecules and another site that binds reversibly to molecules of ATP. Explain the significance of these observations.

Question 3.4 (*Objective 3.5*)

Suggest explanations for the following.

(a) New-born babies have considerable deposits of a particular type of adipose (fat) tissue which appears brownish because of the large numbers of mitochondria present in the cells. New-borns have considerable resistance to low temperatures.

(b) There are epithelial cells lining the intestinal tract which secrete a lubricating mucus – a glycoprotein – and which contain many Golgi sacs and mucus-containing vesicles.

(c) In electron micrographs of liver cells, granules of stored glycogen can be seen which fluctuate in size and number in specimens taken at different times over a 24-hour period.

Question 3.5 (*Objective 3.6*)

In theory, there are two possible ways in which DNA could replicate. The experiment that confirmed that replication is in fact semi-conservative was conducted in 1958 by two American biologists, Matthew Meselson and Franklin Stahl. First of all, they grew bacteria in a solution (usually called a medium) containing ammonium ions (NH_4^+) as the source of nitrogen (required for synthesis of the nucleotide bases), but in which the N atoms were heavier than the normal type (the normal type is shown in Figure 3.7: total number of neutrons + protons = 14; shorthand representation ^{14}N). They used the nitrogen isotope (Section 3.2.1) that has an extra neutron, i.e. ^{15}N for short. *All the bacterial DNA was thus ^{15}N-labelled (heavy) DNA*. Meselson and Stahl then transferred the bacteria to normal ^{14}N medium where they continued to grow and divide.

Using different colours to denote the ^{15}N-labelled (heavy) and ^{14}N (light) DNA strands, draw a simple diagram to show the the outcome of two successive rounds of DNA replication in the ^{14}N medium for (a) conservative and (b) semi-conservative modes.

(*Hint*: start with a pair of DNA strands in your first colour and remember that all newly synthesized strands must be drawn in the second colour. Don't forget to include a key in your diagram.)

Question 3.6 (*Objectives 3.7 and 3.8*)

Figure 3.53 summarizes the process of gene expression. Label the figure by filling the nine boxes, using all the following terms: translation; polypeptide chain; non-coding DNA strand; codon; transcription; mRNA; anticodon; tRNA; coding DNA strand.

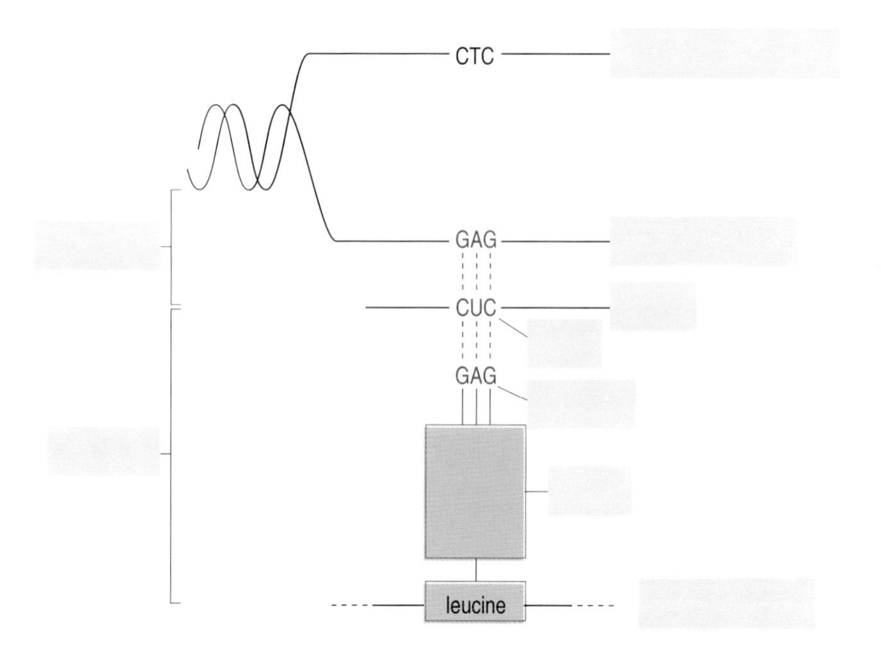

Figure 3.53 For Question 3.6.

Question 3.7 (*Objectives 3.10 and 3.11*)

What is the basic structure of the skin and how is this structure maintained?

Question 3.8 (*Objectives 3.12*)

How does a cell partition complete sets of chromosomes to the two progeny cells which are produced when it divides?

Question 3.9 (*Objectives 3.13–3.15*)

A friend of yours, who spends many hours every day caring for a member of her family with Alzheimer's disease, has a fall and suffers severe grazing on an arm. You observe that the wound is not healing normally, showing persistent tenderness and redness although it is not infected. What would you assume to be happening, and why? What advice would you give to your friend to improve the healing of her wound?

References

Garrow, J. (1994) *British Medical Journal*, **308**, 934.

Kiecolt-Glaser, J. K., Marucha, P. T., Malarkey, W. B., Mercado, A. M. and Glaser, R. (1995) *The Lancet*, **346**, 1194–96.

Wirth, D. P., Richardson, J. T., Eidelman, W. S. and O'Malley, A. C. (1993) *Complementary Therapies in Medicine*, **1**, 127–32.

CHAPTER 4 EARLY DEVELOPMENT

4.1 What is development?

(a)

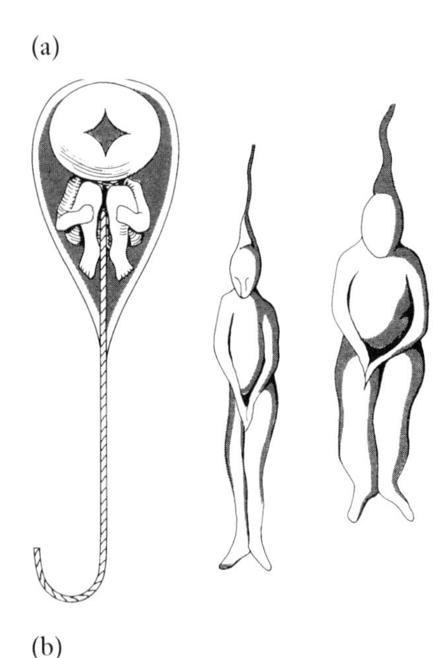

(b)

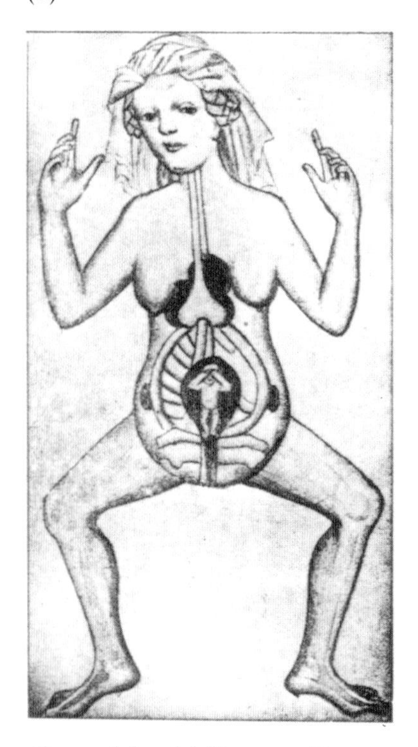

Figure 4.1 (a) Homunculi claimed to be visible in sperm. (b) Pregnancy simplified. (From late 17th century illustrations)

In this chapter we begin to look at the human being in the context of an individual life cycle, examining some of the processes that contribute to the formation of a new person. This is the first time that many of you will have encountered this level of biological detail; we would ask that you take the time to understand it fully at this stage, building on what you have already learned in Chapter 3, as you will require this knowledge not only for this chapter, but throughout the course. We hope to show you that, far from being a dry academic subject, the study of biology allows us to glimpse a dimension of dynamic sophistication and elegance that underpins all the richness and variety of life.

Development can be defined as the collection of processes that produce a whole new individual. From this definition you might surmise that development is a life-long process, as we never stop 'developing'; however, for now we shall confine our discussion to the early steps of the story, beginning with the production of sex cells – eggs and sperm – which have the capacity to fuse together to form a new, unique being. The study of development has a relatively short history. Until the late 17th century it was believed that a new individual was entirely preformed in the sex cells of its male parent. Indeed, some scientists of the day claimed that by using microscopes, invented in the early 17th century, they could actually see tiny preformed people – called homunculi – in sperm cells (Figure 4.1a). The female was thought to act only as an oven in which the baby could grow (Figure 4.1b) until it was large enough to survive birth.

However, the advent of better microscopes did not substantiate these claims, and gradually it became clear that both sperm and an egg, produced by the mother, were needed to make a baby.

Why should we want to make babies? One view is that babies are noisy, disruptive, smelly, exhausting and expensive. They are a large drain on physical resources, especially those of their mothers. Parental input into rearing babies and children lasts a long time – until recent years, with people generally living longer than they used to, parenting was literally a lifetime's work. And yet parents generally find their children rewarding (at least sometimes). They are usually precious to them, and even the most mild-mannered parents will physically fight to protect their child. From an evolutionary point of view, it is important that we *should* keep on reproducing, otherwise our numbers would dwindle and our species might risk extinction. But few people, if any, embark on baby-making with this thought uppermost in their minds. Fortunately, our conscious behaviour is not dictated by this kind of evolutionary imperative: most of us have some freedom of choice, and our decisions are made by taking account of many factors in our lives, some of which may be subconscious, others overwhelmingly practical.

Many babies, of course, just 'come along': the urge for sexual intercourse is very strong, and, in our society at least, has become separated from the urge to reproduce (you will learn more about this in Book 4). But for those of us who have the choice of whether or not to reproduce, a majority – recently quoted as 80% of British adults – chooses to have at least one child. The reasons for this are many.

❑ Can you think of some?

■ People like babies, and want one of their own. Other family members are generally keen for young couples to have families. It is a socially acceptable thing to do. It may help to cement a relationship. People enjoy living in large families.

No doubt you thought of many other reasons.

Make a list of the factors affecting *your* decision whether or not to have one or more children. Ask at least one other person to do the same, and then compare your answers. Note the points of similarity and difference between your answers. Are there any factors which are positive for one person and negative for another? Why do you think this is the case?

The factors listed by you and your friends may have included such items as not wanting to take on added responsibilities, wanting the stability of family life, not being able to afford giving up work, one's (or one's partner's) age, being in an unstable relationship, and doubtless many others. It is likely that many of the reasons given will be social rather than biological: this emphasizes our role as social, highly interactive beings with the ability to adapt our behaviour to suit our environments.

We shall now go on to consider some of these reasons in more detail, and look at *how* people can choose whether to have a child.

4.2 How has the human population grown?

For most of human history there have been relatively few people in the world. Figure 4.2 shows that only over the last 50 years have numbers really shot up, and that we are on target for a world population of six billion or so by the turn of the century.

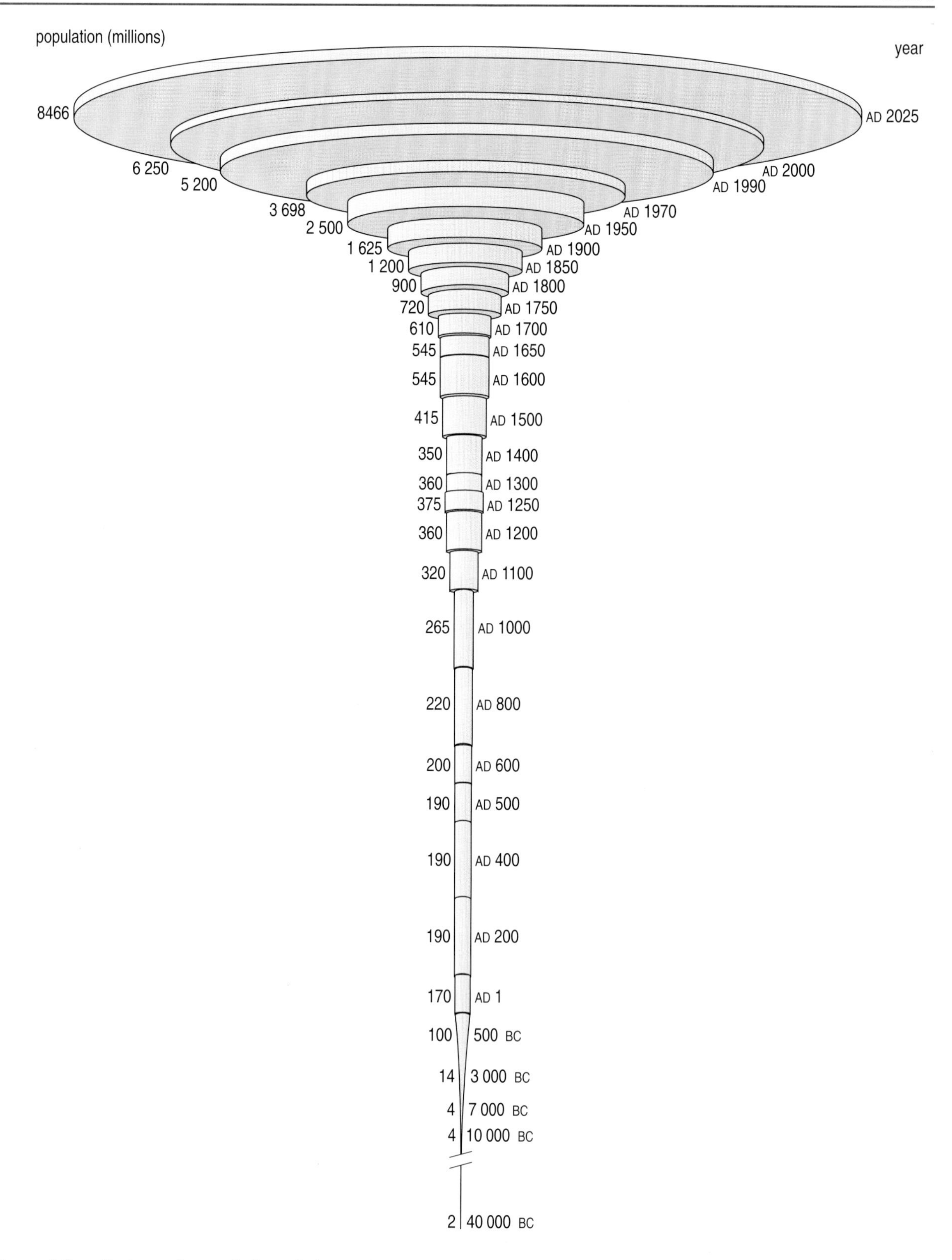

Figure 4.2 Estimated population of the world.

The size of a population depends fundamentally on just two factors: the birth rate and the death rate. When the birth rate exceeds the death rate, there will be net population growth; when the death rate exceeds the birth rate, there will be net shrinkage. When the two rates are balanced, the population remains stable. Common sense dictates that we should be aiming for a stable population which can be sustained by the Earth's limited resources.

❑ Assume that in an ideal world every adult is in a relationship with just one other person. How many children would each couple have in order for the population to be stable?

■ If each couple had two surviving children, they would effectively replace themselves in the population, and overall numbers would remain stable.

Of course, the situation in the real world is a long way removed from this hypothetical ideal. Some people have no offspring; others have one, or two, or several. The United Nations suggests that, based on current trends, the world population will reach between 10.4 billion and 14 billion before levelling off, and eventually declining, due to famine and disease. To put this into a more understandable context, 10 billion people are twice the number who were alive in 1987.

What factors affect the death rate of a population? Broadly speaking, the main factors are the biblical favourites war, disease and famine. These also affect the birth rate: disease and famine generally prevent successful pregnancies (see Section 4.4 below), whereas *after* a war there is generally an increase in birth rate. (This is allegedly due to an increased rate of sexual intercourse by couples who have been separated for long periods.) We in the developed world have been relatively uninfluenced by these factors in recent times. The wars our countries engage in are usually fought at some distance from us; most of us have more than adequate nutrition (at least quantitatively, but see below and Book 3), and modern medical science has significantly reduced the impact of disease for most of us. Yet a glance at the fertility rates of various countries shows that the rate is actually *lowest* among the developed countries (Table 4.1).

❑ Which other country shows a low birth rate?

■ China, where there is a state policy allowing only one child per family.

In view of the generally high standard of health care and nutrition in the developed countries, one might expect the birth rates to be much higher than they are. Assuming that there is no problem of widespread infertility (that is, the biological inability to produce successful pregnancies), a reasonable explanation is that people are *choosing* to limit the size of their families, and have the education and means to be able to select both the

Table 4.1 Some population statistics. This table shows various figures about the populations of some developed and developing countries. Fertility rate is an estimated measure of the rate of *fertilizations*, and includes numbers of live births and numbers of reported abortions; infant mortality rate is infant deaths per 1 000 live births; – means data unavailable.

	Population estimate mid-1993 (millions)	Total fertility rate	Infant mortality rate	Life expectancy at birth (years)	Married women using contraception (%)	People under 15 (%)	Literacy 1990 (male/ female) (%)	GNP per capita 1991 (US$)
UK	58.0	1.8	7.1	76	81	19	>95	16 750
Brazil	152.0	2.6	63.0	67	66	35	83/80	2 920
Egypt	58.3	4.6	56.0	60	48	39	63/34	620
India	897.4	3.9	91.0	59	45	36	62/34	330
China	1 178.5	1.9	53.0	70	83	28	84/62	370
Nigeria	95.1	6.6	84.0	53	6	45	62/40	290
Thailand	57.2	2.4	40.0	68	66	29	95/91	1 580
Poland	38.5	2.0	14.4	71	–	25	–	1 830
USA	258.3	2.0	8.6	75	74	22	>95	22 560

number of their pregnancies and the intervals between them. A woman is fertile for around 40 years, and could in theory have one pregnancy each year. Yet it is unlikely that many women would bear 40 children, even in the absence of contraception, nor would many want to.

Nevertheless, in the developing world it is common for women to have many pregnancies, giving rise to the rapid and unsustainable increase in global population. Large families are highly desirable, and sons are particularly welcome as they will be able to work to support their ageing parents, and will command dowries from their wives' families. In some Hindu communities, sons are also regarded as spiritually 'superior'; only a man's prayers can send souls to heaven. Daughters, on the other hand, are often less welcome, as dowries will have to be found for them unless they are to remain with their parents for their adult lives. This dislike of daughters is at its most extreme in societies that practise female infanticide.

In many developing countries, both food and medical supplies are scarce, which will limit the mother's fertility. Of fertilizations that do occur, it is estimated that more than one-third fail because they are biologically abnormal, or because the mother develops an infection that terminates the pregnancy. Some of these infections can prevent the woman from conceiving again. New-born babies are very vulnerable to injury and disease, and this is another natural means of population control. If the mother is undernourished, it may be some time before she can conceive again, even if she is taking no measures to prevent conception. Breastfeeding can in some cases prevent ovulation – this is known as *lactational amenorrhoea* – so if a mother is still feeding her previous baby she is less likely to start another pregnancy.

Thus, to some extent, family size, and hence population, is limited by the environment. This concept of a natural balance will be picked up again during this course, particularly in Book 5.

The 'natural' ways of controlling family size sound quite draconian to most of us, based as they are on death and famine. In the developed world, and increasingly in developing countries, we have recourse to other methods to limit our families, and it is to these that we turn next.

4.3 Artificial contraception

Humans have separated sex and reproduction: unlike other species, we can enjoy sexual intercourse even (or especially) at times when fertilization is not possible. Many sexual encounters are casual, and in these cases it is often very important to avoid pregnancy. Even during long-term relationships, many couples choose to avoid having children. Thus for many people contraceptive measures are an essential part of their lives. This is not a modern phenomenon: as you will see, birth control has been popular for many centuries. In the absence of other means of family planning, infanticide is said to have been common in times past, and even, in some places, in more recent times, but there are other methods which involve less of a physical and emotional input from the mother. Deliberate attempts to avoid pregnancy probably began with the man withdrawing from the woman before ejaculation. Although this can indeed prevent a sperm from reaching an egg, it is notoriously unreliable, and nowadays people generally prefer to use different and more reliable methods. There are four broad types of contraception available: chemical, mechanical, surgical, and the so-called natural methods. We shall consider them in turn.

4.3.1 Chemical contraceptives

These methods rely to a large extent on an understanding of the physiology of the reproductive process. They are targeted at preventing the production or release of **gametes**, i.e. the sex cells – sperm and eggs – which need to fuse to produce a new individual. To date, most effort in this area has been directed towards preventing a woman from ovulating, i.e. releasing an egg, although more recently trials have begun on 'male pills' which block sperm production.

Ovulation in women generally occurs once every 28 days or so. During part of this time, the lining of the womb thickens, ready to receive a fertilized egg. If fertilization does not take place, the lining is shed as menstrual blood. This gives the 28-day cycle its common name of the **menstrual cycle**. The menstrual cycle is under the influence of two main hormones, **oestrogen** and **progestogen**, produced by the body at varying levels during the menstrual cycle as shown in Figure 4.3. You will learn more of this in Book 4, but for now the important thing to remember is that it is the *relative* levels of the two hormones that determine whether, and when, an egg is released. Chemical contraceptives interfere with this delicate balance and prevent ovulation.

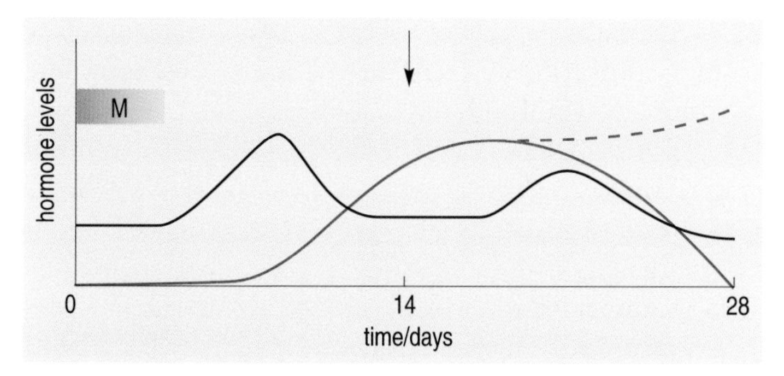

Figure 4.3 Levels of the hormones oestrogen (black line) and progestogen (continuous red line) during the menstrual cycle. M = menstruation. The downward-pointing arrow shows the time of ovulation. (The dashed red line shows what the progestogen level would be if fertilization occurred during the cycle.)

You can see from Figure 4.3 that there are several ways of interfering with the balance between oestrogen and progestogen.

❑ List three ways of doing this.

■ Altering the oestrogen level, altering the progestogen level, or altering the levels of both hormones, but to differing extents.

In practice, the last two methods have been most effective. There are *combined pills*, containing various levels of oestrogen and progestogen, and *progestogen-only pills*. Combined pills give a cocktail of oestrogen and progestogen which alters the natural balance of these hormones, such that the critical balance required for ovulation is never reached. Progestogen-only pills consistently increase the level of progestogen, mimicking the pregnant state (Figure 4.3), so that ovulation and implantation cannot occur. The failure rate (i.e. the number of pregnancies per year per hundred women using a particular method, expressed as a percentage) associated with these pills varies between less than 1% and 4% (that is, between one and four women will become pregnant for every 100 taking the pills for a year), depending on how regularly – at the same time each day – the pills are taken.

❑ From general knowledge, can you suggest why it is important to take the pills regularly?

■ Not only do hormone levels vary throughout a month, but they also vary over each 24-hour period, showing a *diurnal* or circadian rhythm. (If you did not know this, don't worry – you will learn more about the rhythms of hormones elsewhere in this course.)

Thus in order to make sure that the hormone levels do not accidentally reach the critical point for ovulation to occur, it is necessary to take the pills at the same time each day. With the progestogen-only pill, a delay of as little as three hours can result in ovulation and hence pregnancy, although the combined pill is a little more flexible in this respect.

As women go about their daily business, it is not always possible for them to be as regular in their pill taking as is necessary. But there is another option available. It consists of an implanted source of progestogen, which releases the hormone slowly and continuously for a long period. This can be two or three months, from a single injection, to as much as five years from one or more matchstick-sized plastic tubes containing the hormone inserted into the upper arm. As you might expect, the failure rates from this form of contraception are lower than those associated with pill taking.

❑ What would you predict the failure rate to be?

■ You would probably say zero, since pregnancy cannot occur without ovulation.

Progestogen implants have been used extensively in some developing countries, where it can sometimes be very difficult for a woman to keep to a regular timetable, so a lot of information is available about their efficacy. The failure rate is actually as much as 2% over a five-year period. This is because although in some women the administration of this dose of progestogen does indeed prevent ovulation, in some others it does not. In these women, progestogen can still act as a contraceptive, but for other reasons (see below). This variation illustrates an important point: we are all individuals, and although our bodies and physiologies are broadly similar, many differences exist between us, which must be accommodated in any discussion of health and well-being.

The other effects of progestogen are to thicken the cervical mucus (that is, the wet, sticky substance produced by the neck of the womb), making it difficult for sperm to penetrate, and also to alter the lining of the womb, making it difficult for a **conceptus** (newly-formed embryo) to attach there; attachment, or **implantation**, as you will see below, is essential for a successful pregnancy. A look back to Figure 4.3 will remind you that progestogen is indeed normally present at high levels during the third week of the cycle, but drops during the fourth week, when implantation of a conceptus would occur. It is believed that these auxiliary effects of progestogen actually contribute more to its contraceptive effect than does its variable ability to prevent ovulation.

❑ Can you suggest any advantage offered by one of these other effects of progestogen?

■ Thick cervical mucus will hinder the passage not only of sperm, but also of any pathogenic (disease-causing) organisms. Thus, in women, cervical mucus is one of the body's defence barriers.

Of course, no drug is without its side-effects, and contraceptive pill use has been associated with a number of adverse conditions. Perhaps the most serious is the increased risk of cardiovascular disease among women – particularly those who smoke – who have taken the combined pill for a number of years and whose blood pressure is raised. Another potentially serious effect is the apparent increase in some forms of cancer, particularly of the cervix and breast, although the evidence for this is not clear-cut. More commonly, women may experience headaches or nausea. On the positive side, there is some evidence that the pill provides some protection against developing cancers of the ovary and endometrium (the lining of the **uterus**, or womb). Menstrual bleeding is usually lighter and more regular. Users of the progestogen-only pill report different side-effects: their menstrual bleeding tends to be irregular, they have an increased frequency of cysts on their ovaries. They also have an increased risk of ectopic pregnancy, i.e. one where implantation occurs in the wrong place, such as in the **Fallopian tube**, the tube leading from the **ovary**, where the egg is produced, to the uterus. This is an extremely serious condition which can result in infertility or even death.

❑ Why do you think this increase in ectopic pregnancies occurs?

■ Because of the reduced receptiveness of the uterus to implantation.

A lot of research has been carried out into the feasibility of a chemical contraceptive for men. The most popular approach has been to use a cocktail of hormones that interfere with sperm production. The main drawback is that because of the length of time required to produce sperm (more than nine weeks; see below), there is a significant delay before any effect of the hormones is apparent, and furthermore, there is a similar delay when contraception is no longer required, before sperm production is resumed. Moreover, getting the dose right is a problem: even on high doses, some men still produce sperm, and the sperm may be abnormal, which could in theory lead to abnormal pregnancies. Finally, there seems to be some resistance among men to using chemical contraceptives: the term ‘chemical castration’ demonstrates how scared some men are of losing their ability to produce sperm, with all the psychological implications which this might have.

4.3.2 Mechanical methods of contraception

While hormone-containing pills represent a very sophisticated kind of contraceptive, mechanical contraceptives are a straightforward idea: they act by preventing sperm and egg from meeting. Mechanical contraceptives in their simplest form have been around since before Roman times; some are shown in Figure 4.4. The earliest ‘penis protectors’ were allegedly used less for contraception than as protection against disease, and as a badge of rank.

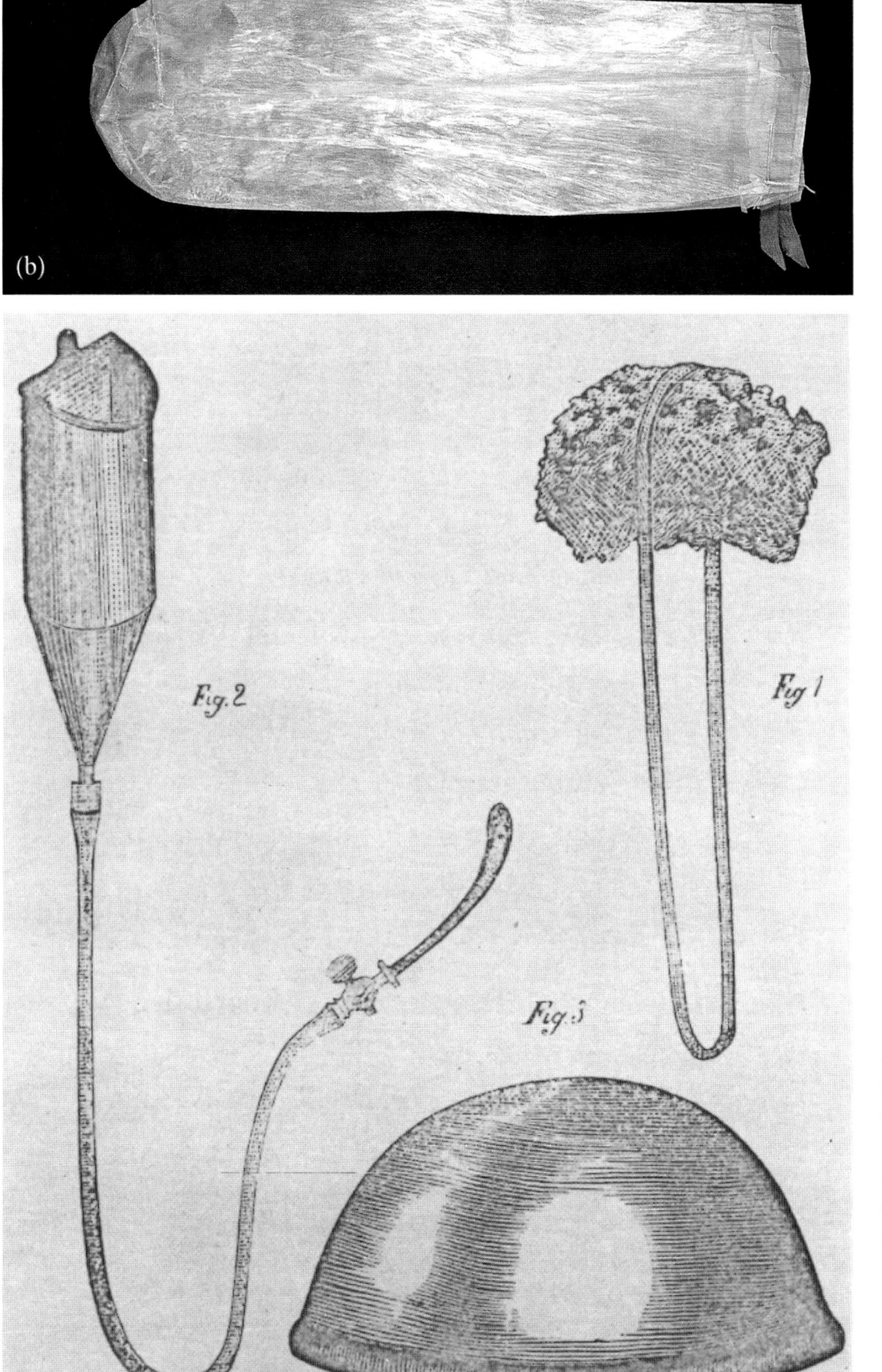

Figure 4.4 (a) Drawing of part of a XIX Dynasty (1350–1200 BC) original, showing an Egyptian wearing something on his penis. (b) Late 18th century English condom made from animal gut and secured by a red silk ribbon. (c) Some 19th century contraceptive devices: Fig 1, a sponge; Fig 2, a syringe (used after intercourse to wash semen out of the vagina); Fig 3, a cap.

Male condoms, placed over the penis to catch the ejaculate and so prevent it from entering the woman, are probably the most widely used form of contraception, particularly as they also provide a barrier to pathogens. Unfortunately, they have very high failure rates: as high as 15% has been reported in some UK studies, although careful use can improve this to 2%.

The high failure rate is partly because they are very thin and split easily during use, but also partly because they are often not put on soon enough, allowing sperm in pre-ejaculatory secretions to enter the vagina. Recently, female condoms have also become widely available. These are designed to fit over the vulva and inside the vagina. They are loose-fitting, and therefore less likely to break, but they are easily dislodged from the vulva, allowing the man's penis to enter the vagina beside, instead of inside, the condom. There are not yet any published failure rates for female condoms, but early results suggest that they are likely to be as effective as the male variety. Condoms have few physical side-effects, apart from occasional allergic reactions to rubber, but they are not popular with all men (or women), as they may be so tight-fitting as to dull sensation.

Although until fairly recently condoms have been the exclusive province of the man, other forms of mechanical contraceptive can be used by the woman. These are all devices that can be inserted into the vagina to prevent ejaculated sperm from passing through the cervix and entering the womb. These mechanical barriers include diaphragms, caps and sponges, early versions of which are shown in Figure 4.4c. These all fit tightly over the cervix, and caps and diaphragms come in different sizes to accommodate women of different shapes. In spite of this, they do not fit perfectly, and need to be used with a spermicide (a substance that kills sperm on contact) to be effective. Caps and diaphragms can be used repeatedly, but sponges are disposable, and are purchased already impregnated with spermicide. Depending upon how correctly they are used, failure rates of between 4% for caps and diaphragms and 25% for sponges are reported. The use of barrier contraceptives with spermicidal properties also has a long history. The ancient Egyptians, in around 1850 BC, were using a variety of pastes inserted into the vagina; honey and crocodile dung seem to have been popular, according to one source (Green, 1971). An alternative method was half a lemon placed over the cervix, which made an effective barrier, while the citric acid in the lemon juice acted as a spermicide.

Figure 4.5 shows the female reproductive tract, highlighting the places where barrier contraceptives are used.

Figure 4.5 The female reproductive tract.

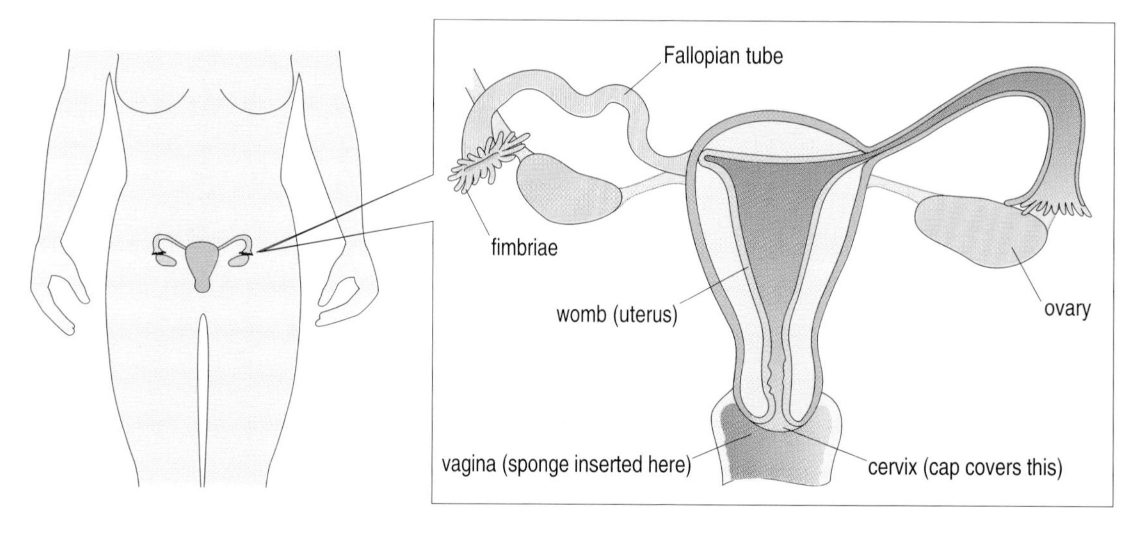

There exists another kind of mechanical device that does allow the meeting of sperm and egg, but nevertheless prevents pregnancy: the intra-uterine device, or IUD. Strictly speaking, this is not a contraceptive as such, but an abortifacient (something that causes an abortion), and for this reason some people have reservations about its use. An IUD is inserted by a doctor into the uterus (womb), where it lies against the inner wall. This appears to set up a reaction within the wall which prevents a conceptus from implanting. Note that once the conceptus *has* implanted it is referred to as an **embryo**.

❑ What other contraceptive already mentioned prevents implantation?

■ This is one of the effects of progestogen.

Some IUDs contain high levels of copper, which is thought to dissolve very slowly in the uterine fluid. The high local levels of copper are thought to disable the sperm in some way, so these IUDs *may*, in fact, be 'proper' contraceptives, preventing fertilization. IUDs have low failure rates, of 1 or 2%, but, unlike barrier methods, they offer no protection against sexually transmitted disease. They can also cause very heavy menstrual bleeding, and users have a higher risk of infections of the uterus and Fallopian tubes, which can lead to infertility.

4.3.3 Surgical methods of contraception

Surgical methods are by and large the most drastic and irreversible ones, ranging from castration to relatively untraumatic tube-tying. Because of the psychological and physiological side-effects, surgical removal of the testes or ovaries is not generally carried out for contraceptive reasons alone, although these operations may be carried out for other reasons, such as the presence of malignant tumours. Any kind of surgical sterilization can be physiologically traumatic for a woman, as it involves cutting or blocking the Fallopian tubes, leading from the ovaries to the uterus, to prevent the eggs from meeting any sperm (Figure 4.6a). The Fallopian tubes are located quite deep within the abdomen, so even 'key-hole' surgery is quite invasive. Surgical sterilization of men (vasectomy), on the other hand, is so straightforward that it can be done in 10 minutes under local anaesthetic, and is often referred to as 'coffee-table' surgery. Here, the tubes leading from the testes, where the sperm are produced, to the penis, are cut before they reach the glands which contribute other components, including fluid, to the ejaculate (Figure 4.6b). Thus, ejaculation takes place normally, but no sperm are released. No hormones are affected by the procedure, so sexual behaviour is unaltered.

You would probably predict that surgical contraceptive methods were very effective, and you would be correct. But once again, failure rates are *not* zero: occasionally, the operation is not performed correctly, the cut tubes will spontaneously repair themselves, and fertility will be restored.

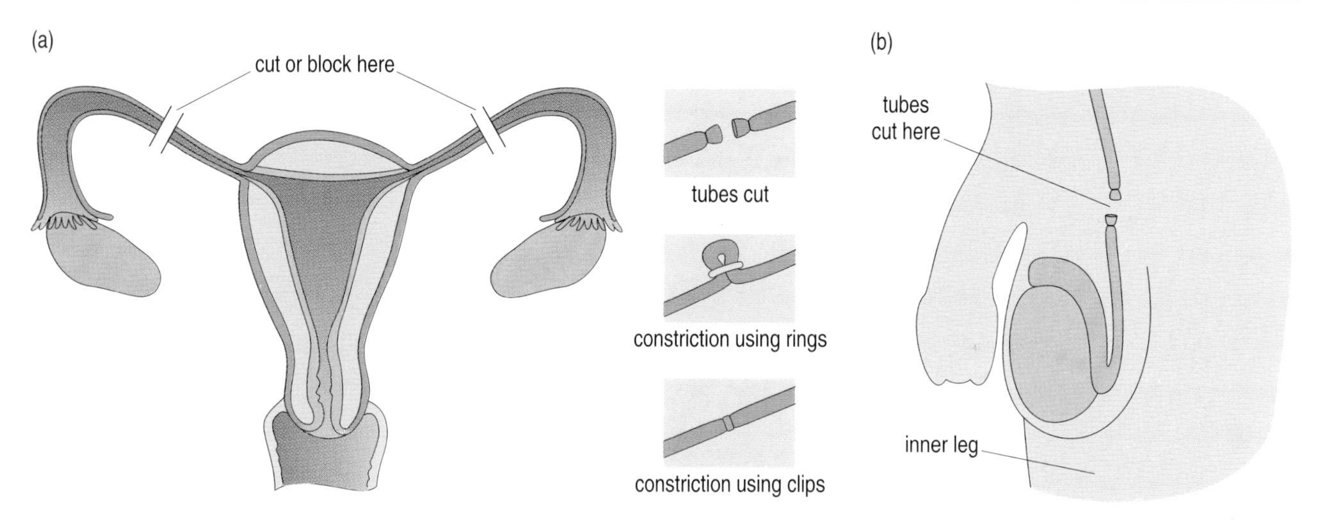

Figure 4.6 Diagram to show where sterilization cuts are made (a) for women and (b) for men. Notice that in (a) the tubes may be blocked rather than cut, using rings or clips.

Reported failure rates are still low, however: about 0.1% is the norm (for both male and female sterilization). One side-effect of male sterilization seems to be an increased risk of developing kidney stones, or prostate or testicular cancer, although this has not been firmly shown. In general, female surgical sterilization seems relatively problem-free, although some women have reported having heavier periods.

Another method of contraception which we shall consider here is surgical abortion. This is, apart from *coitus interruptus* (withdrawal), probably the earliest form of 'contraception', and it is only within the last few decades that it has become a proper surgical process: prior to this time, and in many countries even today, abortion was frequently a 'self-help' method of contraception, and at best was the province of well-intentioned amateurs. Surgical (as opposed to spontaneous) abortion consists of a physical, or occasionally chemical, intervention which causes the growing fetus to detach from the uterus and pass out, with the **placenta** (afterbirth), through the vagina. (**Fetus** is the term used to describe the developing human after the eighth week of gestation, when all the systems of the body have formed.) Surgical abortion can be carried out at any time during pregnancy, but the current legal limit for a surgical abortion in the UK is at 24 weeks of **gestation**. (This is the term used to describe the progress of a pregnancy; babies are generally born after 38 weeks of gestation.) Because modern medical practices can allow *some* fetuses as young as 22 weeks to survive with intensive care, the 'social' limit, i.e. what is generally considered acceptable, is closer to 18 or 20 weeks. Surgical abortion is carried out by inserting instruments into the uterus via the cervix, and either sucking or scraping out the contents. This ensures that all the placenta is removed as well as the fetus: if any part of the placenta is left in the uterus, it may give rise to a serious infection. In times past, it was often sufficient to rupture the membranes surrounding the fetus (hence the popularity of the 'knitting needle' method, in which a knitting needle or other similar object was inserted through the cervix to induce an abortion). Once the membranes are ruptured, the fetus is very prone to infection, which is likely to kill it. A dead fetus is generally rejected by the uterus, which expels it. The success rate of an abortion is very high, although it depends on the precise method used.

Although the majority of abortions are carried out by *physical* removal of the fetus and placenta, *chemical* abortifacients are becoming more widely used. These are commonly referred to as 'morning after' pills, and, although there are several kinds, generally act by administering a cocktail of hormones whose effect is to alter the environment of the uterus so that implantation of the embryo is impossible. These pills can be used for only the first few days after unprotected intercourse, as they will generally not produce an abortion if implantation has already occurred. However, a different hormone, the most widely used form of which is known as RU486, can be used at rather later stages. RU486 is an antiprogesterone, which acts to block the progesterone receptors. It is used in conjunction with a **prostaglandin** to cause contractions. Prostaglandins are a family of hormones made in the body from fats, and are unusual in that they do not need to be transported by the blood to their target organs, but are produced nearby. The effects of prostaglandins are diverse, and some of them are believed to be involved in labour (see Book 4). RU486 acts by inducing delivery of the recently implanted embryo. One drawback of chemical abortifacients is that they need to be administered before a pregnancy has been confirmed, so are probably less suitable for regular use than as an emergency measure, because of the acute side-effects (headaches and nausea) experienced by many women taking them.

Even in the cleanest of modern hospitals, abortion presents a significant risk to the mother. The forcible detachment of a healthy placenta from the wall of the uterus leaves a large number of 'open' blood vessels, and there is a considerable likelihood that haemorrhage (bleeding) will occur, possibly resulting in the mother's death. Although nowadays death from this cause is extremely rare, subsequent infection can render the woman sterile. The introduction of legal abortion in the UK following the 1967 Abortion Act, which allowed abortions to be carried out under cleaner conditions, made a significant difference to the maternal death rate, and this has provided a strong argument for legalizing abortion elsewhere. The argument is as follows: if it is accepted that abortion is going to happen regardless of the law, might it not be better to minimize the risk associated with the operation by allowing it to be carried out properly, rather than under unsanitary, back-street conditions?

The adverse effects of undergoing an abortion are not all physical. Although a woman's initial reaction is often one of relief that she is no longer pregnant, there are often severe long-term emotional effects of the loss of a baby – for, almost always, by the time the abortion is carried out, the woman is all too aware that what is growing inside her is a *baby*, and not just a lump of tissue. This emotional reaction is sometimes overlooked both by the woman herself and by medical staff with whom she comes into contact, and sympathetic counselling should be offered to anyone in this situation. The emotional impact is, of course, particularly severe if the woman actually *wanted* the baby, but underwent an abortion because the fetus was abnormal. There is a large body of literature on the subject of abortion, but space constraints prevent further discussion here.

As mentioned above, abortion is one of the oldest methods of limiting family size. It is believed that in this country one of the major roles of the wise women and healers of olden times was as practitioners of abortion. This is thought to be an important cause of their persecution as witches: the largely male establishment resented anybody who could empower women, particularly in the matter of their own fertility, and they perceived this question of a woman's choice as a threat to their sovereignty. The objection to abortion on the grounds that it takes the fetus' life is a relatively recent development. Even the Roman Catholic Church, until 1869, found early abortion acceptable. This was because of an ongoing debate about when the soul enters the body: if it was not until 'quickening' (that is, when fetal movements can be felt), then abortion before this time did not result in taking a life, and so was permissible. However, in 1869, Pope Pius VI decreed that the soul enters the body at the moment of conception, so abortion at any time is wrong. Whatever the ethical, legal and political debates surrounding abortion, the fact remains that it *is* a widely used and effective means of limiting family sizes throughout the world.

4.3.4 Natural methods of contraception

Many people with particular religious beliefs are fundamentally opposed to the use of artificial methods of contraception. In the developing world, where, as you saw above, the population is frequently increasing at an unsustainable rate, this is a particular problem. For Muslims and Roman Catholics (and others), who may nonetheless wish to limit their families, the preferred option is to use natural family planning methods. The most commonplace method, which involves estimating the 'fertile period' (the time when sperm and egg could meet and result in a pregnancy) from the date of the last menstrual period, only stands a chance of working if the woman has regular menstrual periods. Many women do not, so this method is extremely unreliable. Indeed, in a very short menstrual cycle, it is possible that ovulation may occur immediately after menstruation, and so, because sperm can live for up to five days inside a woman, intercourse within five days of this time – from the start of menstruation – is risky. Other natural methods are based on a close observation of the physical changes that women experience throughout their menstrual cycles, and avoidance of intercourse during the fertile period. In particular, two parameters are most often used: body temperature, and the consistency of mucus secreted from the cervix.

A woman's body temperature rises just after ovulation, and remains elevated for three days. The egg will live for only about two days, so the third day of high temperature marks the end of the fertile period. However, it is not true to say that the beginning of the high-temperature phase corresponds with the start of the fertile period; because of sperm survival, intercourse on any of the five days preceding ovulation might result in pregnancy. Again, because this method cannot pinpoint the start of the fertile period, it is necessary to abstain from intercourse from the first day of menstruation, and resume only after the end of the fertile period. This effectively rules out intercourse for more than half the woman's menstrual cycle. It is also prone to error since other factors, such as illness and some drugs, can also raise body temperature.

The other common natural method of contraception, the cervical mucus method, relies on the fact that the cervical mucus changes in consistency before ovulation. Cervical mucus is one of the body's natural defences, as it helps to prevent infection by presenting a thick, sticky barrier which pathogens cannot easily cross (see Section 4.3.1). For most of the menstrual cycle the mucus is present in only small amounts, but is very viscous. However, about five days before ovulation, more is produced, and it is 'wetter' and more slippery.

❑ Can you suggest what the advantage of this might be?

■ It will be easier for sperm to penetrate into the uterus and Fallopian tubes at a time when an egg might be there to be fertilized.

The change in cervical mucus therefore defines the start of the fertile period. However, once again the method is not foolproof, and many women find it difficult to distinguish between 'wet' and 'dry' days.

In contrast to the artificial contraceptive methods mentioned above, which are all fairly easy to use, natural methods require very careful observation, counselling and instruction by a trained practitioner, and are generally not very suitable for many women who have irregular cycles or naturally variable cervical mucus. Failure rates are high: as many as 20% of women using these methods will become pregnant each year. There are no side-effects caused by physical intervention, although the requirement to be aware of one's body temperature can sometimes take away the spontaneity of a sexual encounter.

However, new techniques are on the horizon. If you look at Figure 4.7, you will see that another hormone, luteinizing hormone (LH) has a production pattern that varies throughout the menstrual cycle. (LH is made by the pituitary gland, a part of the brain.) The interesting thing about this hormone is the peak it shows around the time of ovulation. Recently a test has been produced to monitor LH levels in urine, and give a colour change when levels are rising or high. By avoiding intercourse on the 'high LH' days, it is believed that fertilization can be avoided. There is not yet any evidence to suggest how effective this test is, although the technology it uses – which space prevents us from discussing here – suggests that it will be very sensitive.

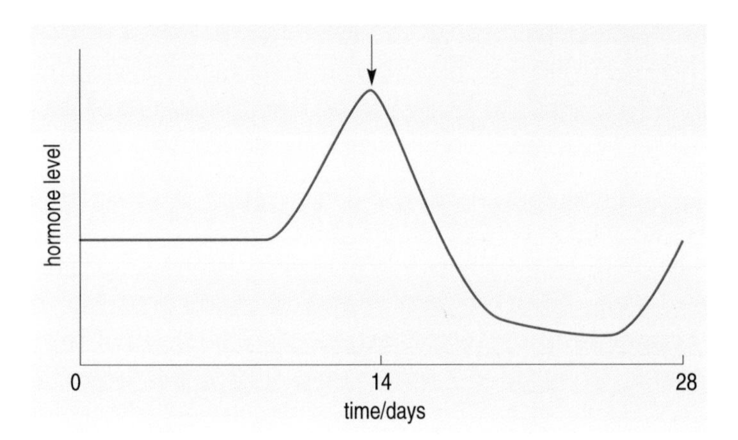

Figure 4.7 Pattern of production of luteinizing hormone (LH) through the menstrual cycle. The downward-pointing arrow denotes the time of ovulation.

The choice of whether or not to practise contraception, and if so, which method to adopt, is a complex one. There are physical and emotional advantages and drawbacks to all the methods; religious, economic and cultural pressures also play a very important part in the decision. Most individuals will have to make some kind of choice about the matter at some time in their lives, however, so it is important that the relevant information be freely available to all who need it. In this country there is widespread availability of family planning clinics, which are a good source of advice about suitable methods for particular individuals.

Summary of Section 4.3

1. Many people wish to limit the number of their offspring, and so resort to contraceptive measures.
2. Chemical contraceptives interrupt the production of gametes, or prevent implantation.
3. Mechanical or barrier contraceptives prevent egg and sperm from meeting and, in the case of IUDs, prevent implantation.

4 Surgical methods of contraception involve physical alteration of the reproductive tract so as to prevent eggs and sperm from meeting.

5 Abortion is a traditional and widely used form of birth control, but it sparks ethical and legal debate.

6 Natural methods of contraception are the only means ethically acceptable to some people, but they restrict intercourse during much of the menstrual cycle, and are less reliable than other methods.

7 A reductionist understanding of the reproductive process has allowed the development of better contraceptives.

4.4 Preparing for conception

Now let us go on with our story and assume that we have decided the time is right to have a baby. The primary requirement for conception is that healthy gametes should be produced. We shall therefore look first at how gametes are made, and then examine some of the factors affecting their quality. But we must start with an explanation of what gametes are, and what sets them apart from other kinds of cell. In other words, what makes gametes special? Gametes are the cells that fuse to form a new individual, as you saw above, but why would the fusion of *any* two cells not do the job just as well? In fact, experiments have been done in which pairs of body cells have been fused together artificially, but in no case has an embryo, far less a living person, ever been produced. Perhaps we need to think about different sorts of cells.

4.4.1 Why are cells different?

All the cells in your body are 'your cells' but, as you learned in Chapter 3, if you could look at them using a microscope you would see that there are many obvious differences in their appearance. The shape of the cell, position of the nucleus, the number of mitochondria and the presence or absence of various cytoplasmic structures are all clues to the cell's function. Some such differences are visible by microscopy, as shown in Chapter 3, Figure 3.46, but there are many others which are apparent only at the level of the molecules within the cells.

The molecules within cells can be classified broadly as carbohydrates, lipids or proteins. As you learnt in Chapter 3, many of the characteristic differences shown by different types of cell are due to the different proteins found in these cells. This is an important concept, which you should remember.

❑ From general knowledge, and from Chapter 3, can you name any specific proteins found in the body?

- ■ Well-known proteins include collagen, which enhances the stretchiness of skin, keratin, the protein that forms hair and nails, and insulin, the hormone involved in controlling the level of blood glucose. Perhaps you thought of others: many are listed on the labels of cosmetics or food items, or in cookery books.

Every cell in the body contains thousands of proteins, but different types of cell contain different proteins. For example, collagen is made by several different types of cell, including skin cells, keratin is made by skin cells only, and insulin is made by a particular group of cells in an organ called the pancreas (see Book 2). All cells make proteins by assembling amino acids, which are obtained from food. There are only 20 amino acids used in the body, so differences between proteins (each of which can be hundreds of amino acids long) arise from differences in the order in which the amino acids are joined together.

- ❑ How does a cell 'know' how to assemble amino acids in a particular order?

- ■ From the genetic code – the instructions carried by the DNA making up the chromosomes in the cell nucleus (Chapter 3).

The chromosomes contain all the information needed to produce the molecules out of which an individual is made: not just about *which* proteins out of the whole lot to make, but also *when* to make them and *where* (in which cells of the developing embryo) they are made.

- ❑ If all cells of an individual carry the same chromosomes, and therefore the same instructions, how can different *types* of cell make different proteins?

- ■ They use different subsets of information from the chromosomes (i.e. different sets of genes).

Because the chromosomes carry *all* the information needed for the molecular composition of a new individual, they are very important structures.

- ❑ Where do a cell's chromosomes come from?

- ■ Ultimately, all an individual's cells are derived from the one original cell that resulted from the fusion of an egg with a sperm: the fertilized egg. So the cell's chromosomes came from the gametes' chromosomes.

- ❑ Each cell of a particular species contains the same number of chromosomes (23 pairs, i.e. 46, in humans). What would you predict would be the number present in a cell resulting from the fusion of two others?

■ It would have twice as many. 46 chromosomes in each of the fusing cells would give a fertilized egg containing 92.

But since each cell contains the *same* number (46), this cannot be what happens. We can resolve this paradox by considering our original question: what makes gametes special? The answer (at long last!) is that human gametes contain only *half* the normal number of chromosomes, i.e. 23 – only one member of each pair. How this is accomplished we shall see below.

4.4.2 The reduction of chromosome number: meiosis

If you look at the chromosomes shown in Figure 4.8 you will see that they have been lined up in pairs. The members of each pair are of similar shape and size, and unlike the members of other pairs. At a molecular level these distinctions are maintained: the order of the bases in the DNA is very similar in both members of a pair, but is quite different from that found in other pairs. By 'very similar' we mean that the order of the particular genes on each chromosome of the pair is the same, but the exact order of the *bases* within those genes may be slightly different.

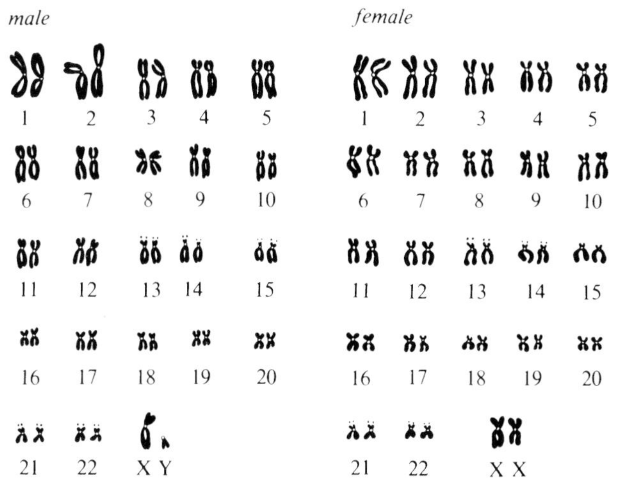

Figure 4.8 Karyotypes of male and female human chromosomes. (X and Y are the sex chromosomes – see later.) To produce pictures like these, chromosomes from a dividing cell are photographed down a microscope. Each individual chromosome's photograph is cut from the resulting print, and the chromosomes are lined up and matched into pairs based on size and the position of the constriction which is found somewhere along each chromosome's length. (This constriction is called the **centromere**, and has structural properties which enable it to attach to the mitotic spindle during cell division – see Chapter 3, Figure 3.51.) The resulting matched pairs of chromosomes are photographed again, to give a karyotype, as shown. Note that such pictures are obtained *only* for dividing cells, in which the chromosomes are replicated – each one consisting of two strands, which are held together at the centromere. In fact, each strand is itself a chromosome, but while still joined to its partner like this, is called a **chromatid**.

❑ If the order of bases in two copies of a particular gene is slightly different, what does this imply about the protein that the gene codes for?

■ The amino acids *may* also be slightly different, if the base change caused a change in the code (see Chapter 3).

It is these subtle differences in the amino acid sequences of our proteins that make us individuals. For example, we all have genes for hair colour, but we do not all have the same colour hair! These different forms of the same gene are called **alleles**.

The fact that the members of a pair of chromosomes carry the same genes, albeit slightly different forms of them, means that each body cell is actually carrying two sets of genetic information. One set is derived from the mother, the other from the father. Each gamete, however, carries only *one* set of chromosomes; and each chromosome is a mixed combination of alleles from the chromosome pair that the mother inherited from *her* father and mother. The same is true of each sperm cell from the father. (The cell division process by which this gene mixing comes about is called **meiosis**, and is discussed below.) Thus, when fertilization occurs the fertilized egg will contain the two sets of chromosomes characteristic of the body cells. The individual chromosomes, however, will be different (i.e. have different combinations of alleles) from those carried by either the mother or the father – this is why we are not identical to our parents or to our siblings (unless we have an identical twin, who would be derived from the same conceptus). What happens is that during meiosis a process occurs which allows alleles to swap between members of the pair. This *recombination* of alleles is of vital importance to us, both as individuals (as we have seen, this process is what *makes* us individuals), and as a species, because it gives rise to the genetic variations that allows our species to be flexible, make compromises and survive (see Chapter 1).

Because each pair of chromosomes is qualitatively different from all the other pairs, a mechanism is needed to make sure that each egg and sperm cell has a complete *set* of chromosomes, and not just any 23. This is achieved by the process of meiosis, which is shown diagrammatically in Figure 4.9. The key points to remember about meiosis are: (a) that it halves the number of chromosomes per cell, and (b) that it gives rise to new gene combinations.

You already know quite a lot about mitotic cell division from Chapter 3. As with mitosis, by the time they become visible each meiotic chromosome has replicated, i.e. it has been accurately copied along its entire length (Figure 4.9a). The pairs of chromosomes move so that they come to lie side by side (Figure 4.9b), and the pairs migrate towards the equator of the cell (Figure 4.9d). During this process, the arms of the chromosomes make contact with each other (Figure 4.9c), and members of a pair can cross over and exchange material by two chromatids breaking at the corresponding point, then rejoining with the 'wrong' chromatid.

Figure 4.9 A generalized diagram of meiosis, showing the sequence of events for just two pairs of chromosomes. (For clarity, the meiotic spindle is not shown.) Although this diagram illustrates the basic process, note that there are many species, including humans, in which the precise details vary.

DIVISION I OF MEIOSIS

(a) Chromosomes appear, divided into chromatids. Red chromosomes have been inherited from the mother, blue chromosomes from the father.

(b) Chromosomes pair up: the two chromosomes A form one pair, the two chromosomes B form another.

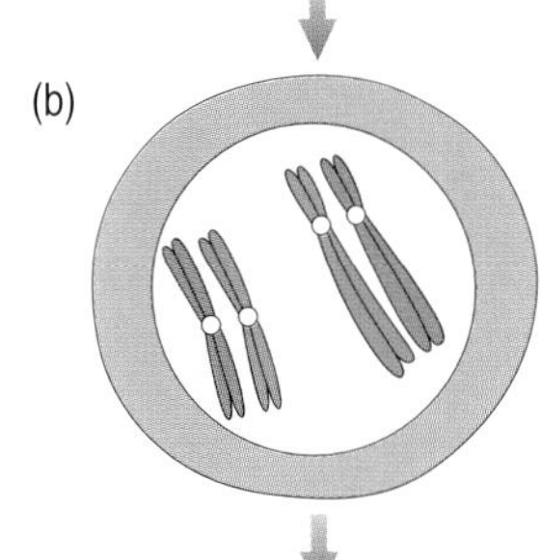

(c) Chromatids cross over and exchange material.

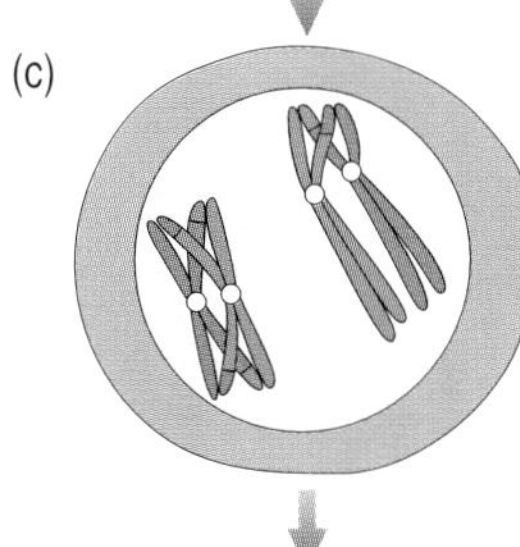

(d) Nuclear envelope disappears and chromosomes line up at equator of cell.

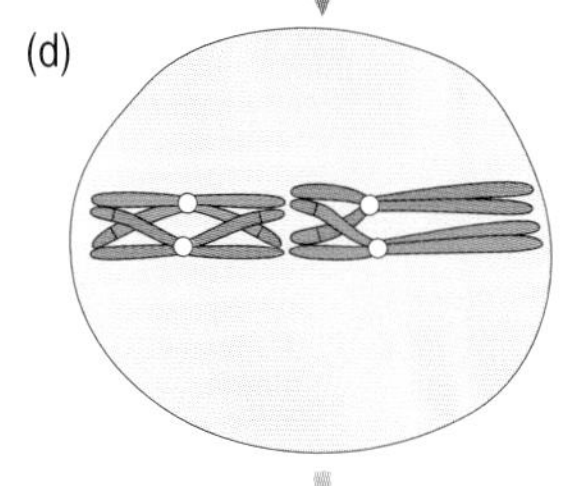

(e) Chromosomes pull away and move to opposite poles of the cell.

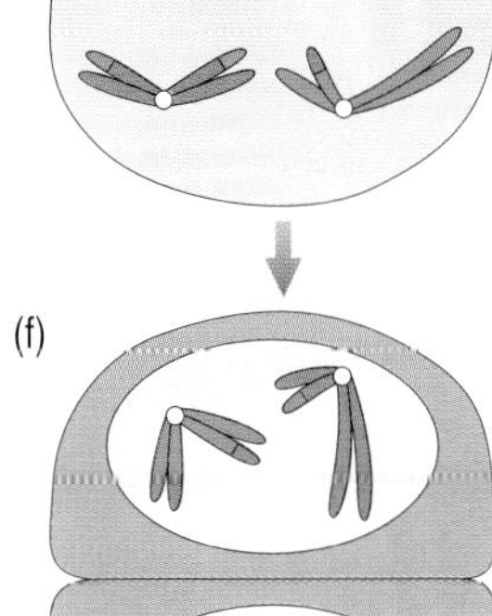

(f) Nuclear envelopes develop around both sets of chromosomes and cell divides into two.

DIVISION II OF MEIOSIS

(g) Beginning of division II. (Only one of the two cells of stage (f) is shown here – the upper one.)

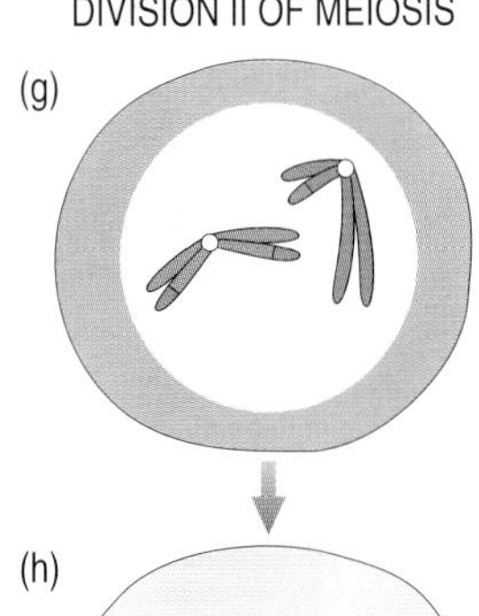

(h) Nuclear envelope breaks down and chromosomes align themselves at equator.

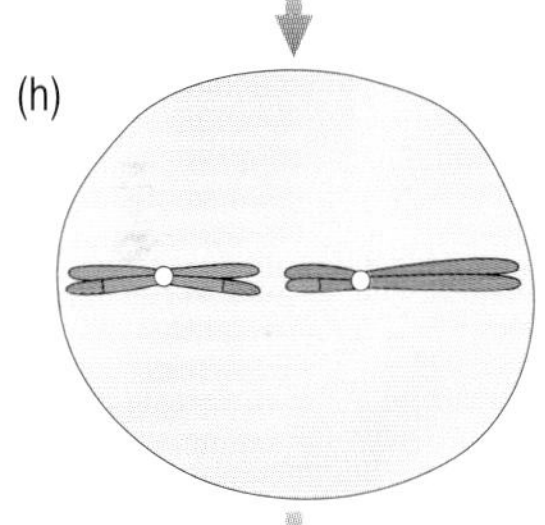

(i) Centromeres divide and chromatids separate and move to opposite poles.

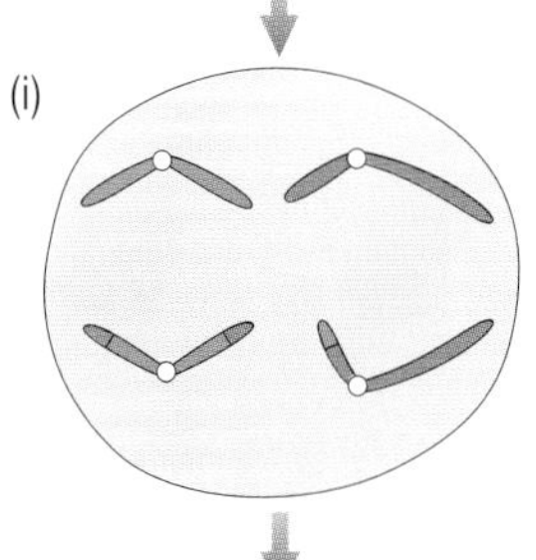

(j) New nuclear envelopes enclose the two sets of chromatids (i.e. chromosomes) and cell divides into two.

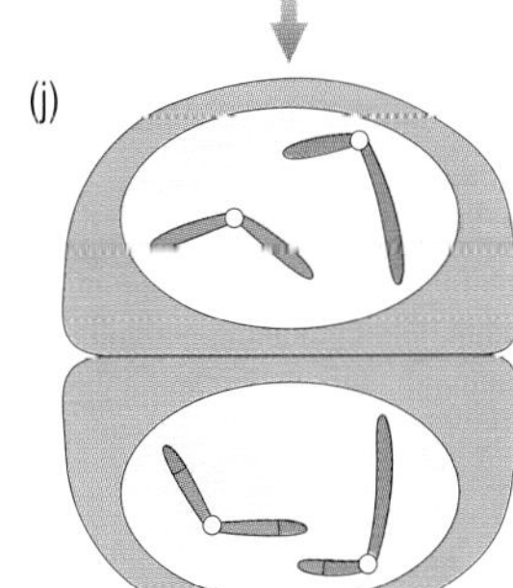

Because all the chromosomes have replicated and become chromatid pairs, there are actually sets of *four* chromatids lined up together at stage (c), allowing a large number of possible exchanges. Because of the way the pairs of chromosomes were aligned at the equator of the cell, one member of each pair (i.e. one set of two chromatids) moves to each pole (Figure 4.9e). Note that it is *not* the case that all the chromosomes originating from this individual's mother goes towards one pole, while all those originating from its father goes to the other: the pairs are *randomly assorted*, so that at each pole there is a mixture of maternally-derived and paternally-derived chromosomes, although there will be a complete set of 23 at each pole (Figure 4.9f). The cell membrane now pinches in, producing two new cells.

Although each new cell now holds 23 chromosomes, there is still twice the amount of genetic material as is needed for a gamete.

❑ Why is this so?

■ Because each chromosome has replicated, giving two copies of each.

The two copies (i.e. the pair of chromatids) are held together at the centromere, the constriction lying part of the way along each chromosome that is the point of attachment to the meiotic spindle (not shown in Figure 4.9).

This stage is illustrated in Figure 4.9g. So meiosis needs to proceed further, to reduce the genetic material to one copy per gamete. The 23 chromosomes in each new cell assemble at the equator once more (Figure 4.9h), and one chromatid is pulled towards each pole via the centromere (Figure 4.9i). Once again, the cell divides (Figure 4.9j), yielding from the original one a total of four cells, each of which contains *one* copy of *each* of 23 different chromosomes. These four cells are now ready to be gametes.

❑ Think back to mitosis, the type of cell division you learned about in Chapter 3. What is the most important difference between meiosis and mitosis?

■ In mitosis, there is one round of DNA replication and one cell division. In meiosis there is also one round of DNA replication, but there are two cell divisions. This ensures that each of the four resulting cells contains only one copy of the genetic information.

Also, whereas in mitosis the chromosomes in the two progeny cells are identical with those of the parent cell, in meiosis the gametes' chromosomes are subtly different because of random assortment of the chromosomes and the exchange of material between similar strands that has taken place during the first cell division.

But although the cells now have the right number of chromosomes to be gametes, they need to undergo various steps of maturation before they can take part in fertilization. We shall now look at these steps in men and women.

4.4.3 Gamete production in men

A sexually mature man is producing sperm all the time at a rate of around 300–600 per gram of testis per second. This provides the 500 million or so which are released at each ejaculation. But the formation of an individual sperm takes about nine weeks (64 days). Sperm are produced in the testes, and production is most efficient at a temperature several degrees lower than the normal body temperature of 37 °C. For this reason the testes (plural of testis) are suspended outside the body cavity in the scrotum, and if the testes fail to descend from their original position near the kidneys to this location during development, sperm production will not occur.

Figure 4.10b shows the structure of the adult testis. It consists of a large number of coiled tubes, called seminiferous tubules, which empty into a collecting area called the rete testis. From here larger tubes, the vasa efferentia, carry the mature sperm to the epididymis, where various substances are added to improve sperm 'quality' (this will be explained below). From the epididymis leads the muscular-walled vas deferens, contractions of which help ejaculation. The vas deferens empties into the urethra at the base of the penis (Figure 4.10a).

Sperm production takes place in the seminiferous tubules. There are three stages to the process: a mitotic phase, in which several rounds of division mean that cell numbers are increased; a meiotic division which reduces the chromosome number in each cell and provides individual allele combinations; and a maturation and packaging phase, in which the chromosomes are readied for transport in sperm outside the man's body.

Figure 4.10 (a) The male reproductive system. (b) A section of testis; for simplicity, the figure shows only one seminiferous tubule per lobe; in reality, there are thousands. One of the tubules is shown extended

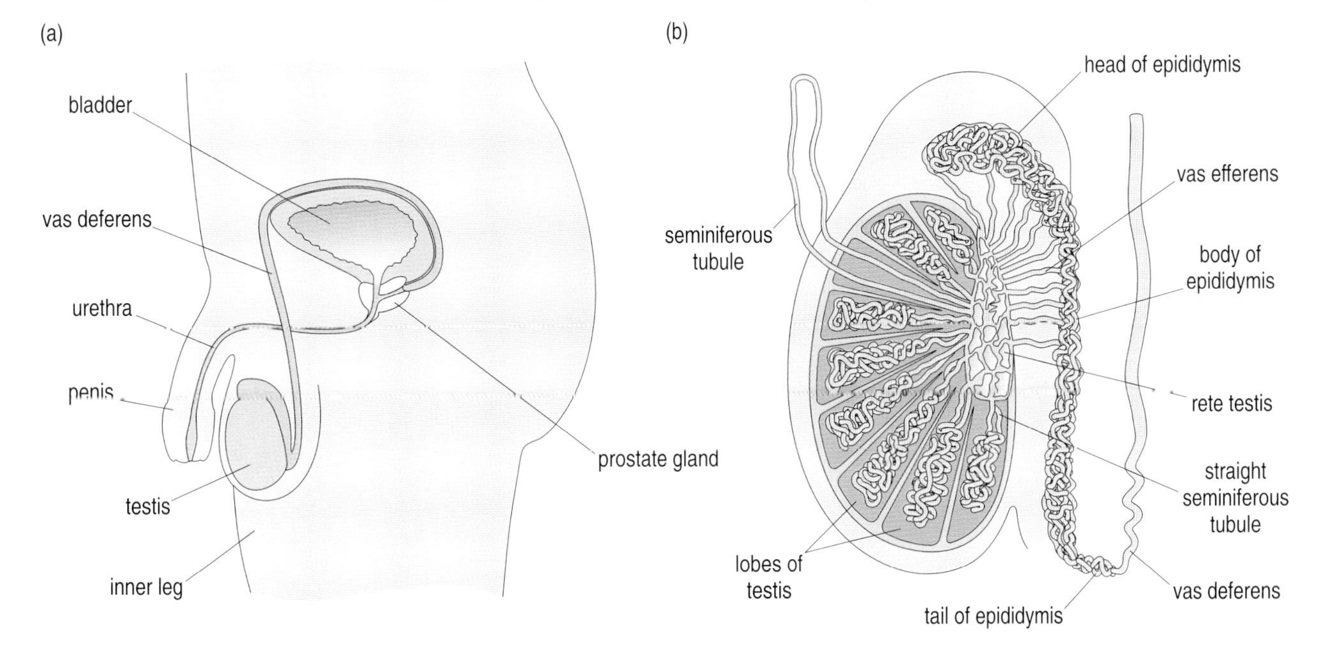

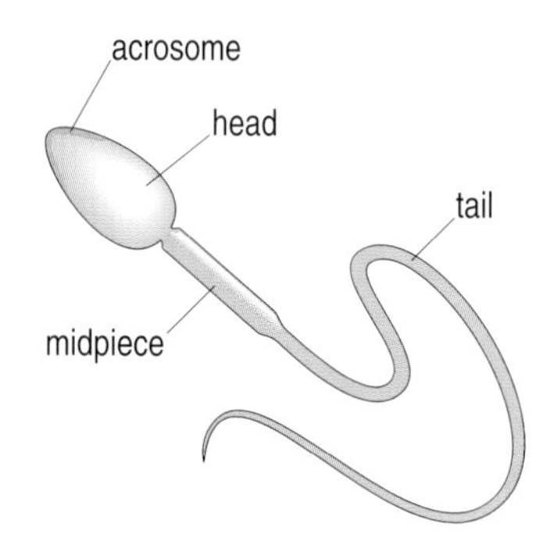

Figure 4.11 Structure of a mature human sperm.

The cells destined to become sperm are known as *spermatogonia*, and they undergo several rounds of mitosis until a population of cells known as *spermatocytes* is produced. Some of these enter meiosis, and after this a population of *spermatids* results, which are ready to be packaged into sperm. Figure 4.11 shows a mature sperm: it has a head, a midpiece and a tail. There is also an attachment to the head, an enzyme-containing vesicle called the *acrosome*, which is vital for fertilization (see below).

The head contains the chromosomes, tightly packed with protein so that they can fit into the small space available. The midpiece is the power supply of the sperm, as it contains the mitochondria which provide the energy needed for the sperm to swim. Swimming is accomplished by means of the tail, which can beat rapidly, although does not do so until ejaculation.

Sperm formation is quite a complicated sequence of events, and of the spermatogonia that begin the process, only a proportion will successfully complete it, the rest dying at different stages along the way. During the whole process, the developing cells move from the outer layer of the seminiferous tubule inwards towards the lumen (the hollow part) of the tube. This means that a cross-section taken across the seminiferous tubule will show a characteristic array of cells at different stages of sperm development, as shown in Figure 4.12. By the time the cells reach the lumen they are complete sperm, but they are not yet able to fertilize an egg. For this they need to be bathed in secretions from various sources. Just what these various secretions do is not entirely clear, but some which are sugars seem to act as a source of energy for the swimming process. One source of secretion is the epididymis (Figure 4.10b); another essential source is the prostate gland (shown in Figure 4.10a). The whole mixture of sperm cells and the secretions they are bathed in is called **semen**, and this is what is ejaculated.

These, then, are the mechanics of the process, but how is it controlled? We shall now take a brief look at the control mechanisms.

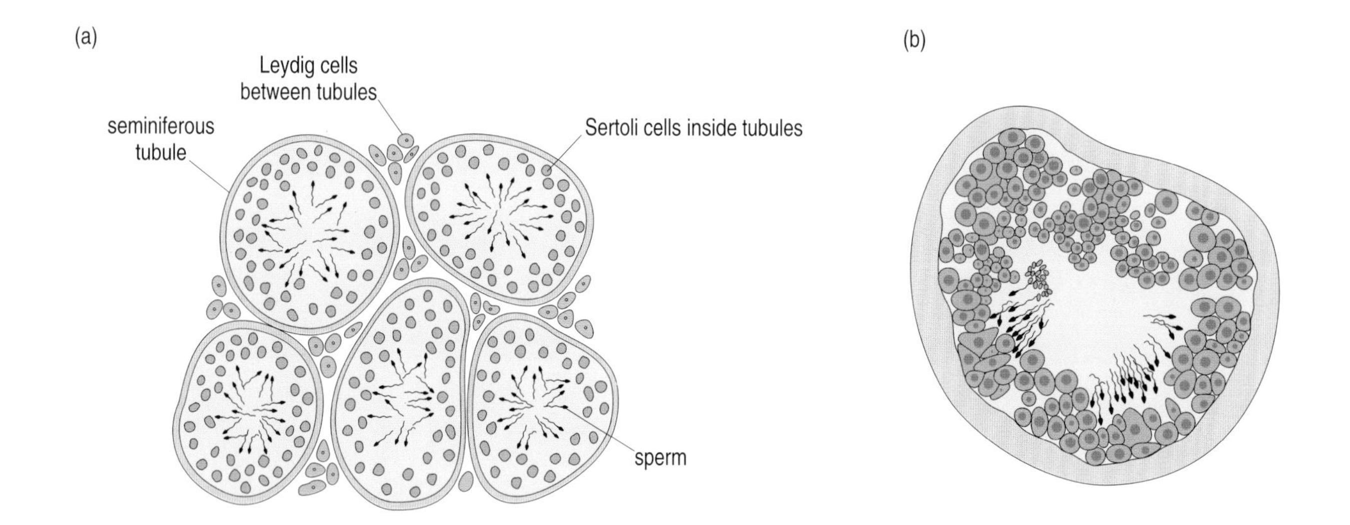

Figure 4.12 (a) Cross-section through a set of seminiferous tubules. (b) Cross-section through a single seminiferous tubule, showing 'wedges' of tubule at different stages of sperm production.

4.4.4 Hormonal control of sperm production

The most important hormone involved in controlling sperm production is a steroid called **testosterone**. This is produced in the testis itself, by the Leydig cells (see Figure 4.12a). The testosterone is released from the Leydig cells between the tubules, and taken up by the neighbouring Sertoli cells. The Leydig cells are stimulated to make testosterone by two other hormones, luteinizing hormone (LH) and follicle-stimulating hormone (FSH), which are both produced by the pituitary gland and reach the testis via the blood. (Although subsequently found to be present in males, these hormones were originally named in relation to their role in females.) Once the Sertoli cells have taken up the testosterone, they convert it to a much more effective substance, dihydrotestosterone (DHT), and this is what stimulates the spermatogonia to enter the sperm production pathway, first by undergoing several rounds of mitosis, then one meiosis. FSH and LH also play a role in maintaining the pathway, and in producing high *local* levels of DHT within the testes.

❑ Why do you think it might be important to have local differences in hormone levels?

■ So that only *some* of the spermatogonia embark on the sperm production pathway.

Sperm do not live for more than a few days once formed, so cannot be stored for long. The production of sperm to ensure a constant supply is illustrated in Figure 4.12b, where different 'wedges' of the tubule are at different stages of development. This shows that different areas were started off at different times.

You will learn more about the role of hormones in sexual behaviour in Book 4. However, to summarize for now, the production of good-quality, fertile sperm requires first that the testis be at the correct temperature. Spermatogonia are triggered to enter the developmental pathway by a cocktail of hormones, in particular DHT, the production of which is controlled by the brain. Sperm production involves three phases: mitosis, meiosis, and packaging. The completed sperm then have to be mixed with a variety of substances before they are ejaculated. Only sperm that contain a correct set of chromosomes, can swim powerfully, and have an intact acrosome, are candidates for fertilization.

4.4.5 Gamete production in women

It is time now to turn to the question of how female gametes – eggs – are made. There are substantial differences between sperm and eggs, and consequently their production pathways are very different.

❑ Can you give one *similarity* in the production of eggs and sperm?

■ Both are gametes, so must undergo meiosis to reduce the number of chromosomes by half.

In contrast to sperm, which contribute, along with a set of chromosomes, only a very small quantity of cytoplasm to the fertilized egg, the egg itself contributes a large amount of cytoplasm. Indeed, the egg is among the largest of human cells, being in the range of 90–120 μm (i.e. around one-tenth of a millimetre) in diameter – practically visible with the naked eye. As you will learn later, the growing embryo lives without the benefit of a placenta for the first two weeks of its life, so the egg's cytoplasm must supply many of its needs during this time (the rest can be obtained from the liquid in which it is bathed – see below). It is quite demanding for a cell to produce so much cytoplasm – making all those proteins, for example, is expensive in terms of energy and raw materials – so the developing egg is nourished by a group of 'ladies-in-waiting' called *granulosa cells*, which develop alongside the egg, and remain in close contact with it.

Egg production takes place in the ovary, which, unlike the testis, remains in the abdomen at 37 °C. As with sperm production, there are three phases involved, but they are not sequential. Mitosis and meiosis occur as before, but the packaging part of the maturation phase is replaced by a growth phase which increases the amount of cytoplasm, and which occurs concurrently with a large part of the meiotic phase. Only one egg is ovulated each month, so that you might expect the whole development phase to take around a month to complete. You may therefore be surprised to learn that the process *can* take more than 50 years! This is because egg development begins before birth, in the first month of gestation in fact. The cells destined to become eggs begin their mitotic cycles then, and *all* the cells begin meiosis before birth. This means that at birth, a girl will have in her ovaries all the oocytes (pronounced 'oh-oh-sites' – partially matured eggs) that she will *ever* have. If these are destroyed, for example by exposure to radiation, the woman will be infertile. Unlike the situation with sperm, which die a few days after their completion, at this stage of development the oocytes remain alive, but enter a kind of suspended animation known as *arrest*. All will stay arrested until puberty, after which time some will be stimulated to carry on through meiosis. Although there are more than two million oocytes present at birth, only a few hundred will ever proceed to maturity (and fewer still will be fertilized). Nevertheless, after puberty a few are reactivated every day, resulting in a steady trickle of maturing oocytes, one of which will be released each month.

Each oocyte is enclosed in a follicle (small sac), whose cells, as we mentioned above, encourage its maturation and growth (see Figure 4.13). The oocytes are arrested at a very early stage of meiosis, at the time when the four chromatids (i.e. the two pairs of duplicated chromosomes) are in intimate contact with each other (see Figure 4.9c). They resume meiosis, but in an unusual way: at the end of the first division, half the chromosomes, but *almost all the cytoplasm*, go to one cell. The remaining chromosomes are discarded in a small bag on the outside of the large cell, called the first polar body, and are subsequently destroyed. The 'big' oocyte resumes meiosis, its chromosomes lining up at the equator for the second time, but the process is then arrested again. The oocyte is ovulated in this state. Meiosis is not reactivated until 2–3 hours after fertilization when,

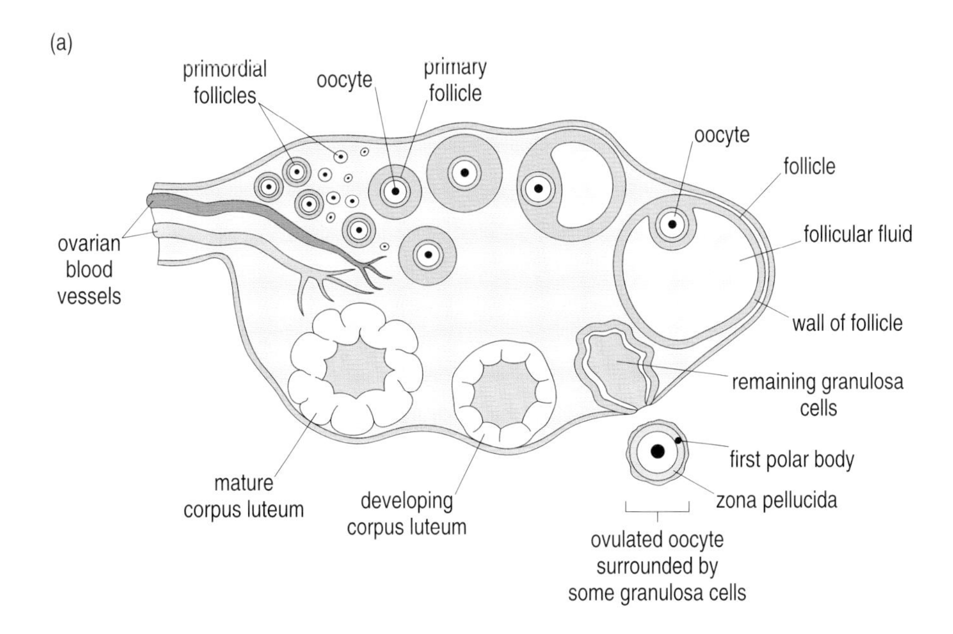

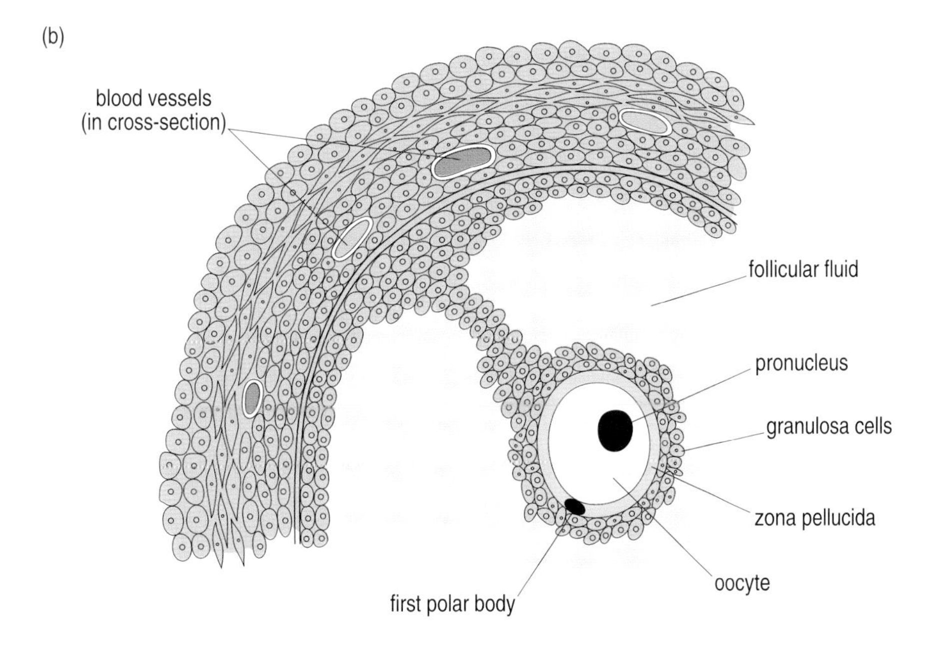

Figure 4.13 (a) Diagram of ovary to show the development of a follicle and the production of a mature oocyte. Maturation proceeds clockwise from the top left. One of the primordial (undeveloped) follicles is stimulated, and starts to grow. (b) Enlarged diagram showing the follicle containing the oocyte, a lot of fluid, and the granulosa cells which nourish the oocyte. Note that after ovulation the follicle itself develops into a structure called the corpus luteum, which plays a role in pregnancy.

once again, one set of chromosomes is expelled in a polar body (the second) and lost. Thus, in contrast to male meiosis which results in four gametes, in women there is only one, which is relatively very large; Figure 4.14 shows the relative sizes of the egg and the polar bodies.

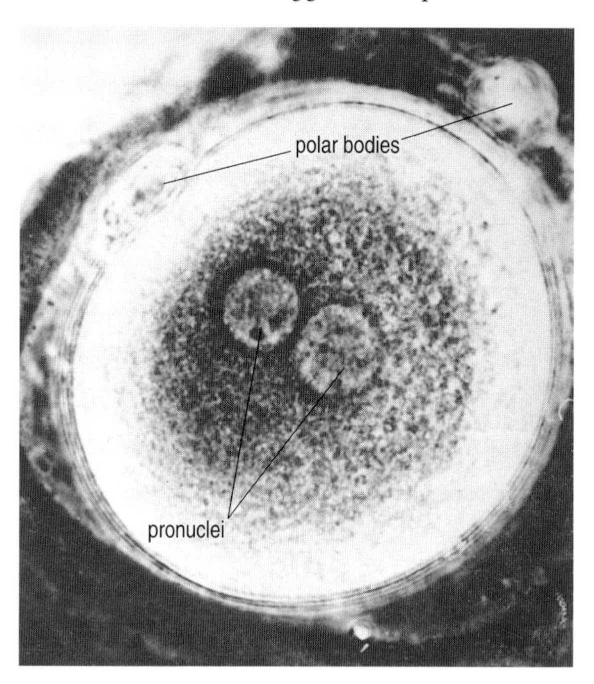

Figure 4.14 Photograph of fertilized egg with two polar bodies. This photograph also shows the two sets of chromosomes present in a fertilized egg, one set derived from the mother and the other derived from the father. Because the sets of chromosomes do not constitute a proper nucleus, it is more accurate to refer to each as a *pronucleus* (plural: pronuclei). The polar bodies contain the other sets of chromosomes derived from the two cell divisions of the meiosis that produced the egg: they will eventually detach and degenerate.

The maturation of the oocyte is often called the follicular phase, as it occurs while the oocyte is within the follicle. It is characterized by a massive amount of protein synthesis in the oocyte, which loads up the cytoplasm with all the 'goodies' that a newly fertilized egg will need. The surrounding follicle cells also undergo mitosis to form a thick layer around the oocyte. They make a protective jelly coat, called the *zona pellucida*, between themselves and the oocyte, but maintain contact with it by means of long strands of cytoplasm which pass through the zona. All this happens spontaneously, without any hormonal influences from the brain or elsewhere. However, towards the end of this first step, some of the follicle cells develop the ability to respond to FSH and LH, and this ability is crucial for the process to go further.

Because of the slow trickle of oocytes into the first phase of maturation, at any point in time there are always a few follicles which have reached the end of this first step. There are two possible fates for these follicles: either they will degenerate and die, or, if sufficient levels of LH and FSH are present, and the cells can respond to them, another round of mitosis will occur in the follicle cells, making the follicle bigger. At this stage, the oocyte itself does not increase in size, but the follicle enlarges even more because its cells begin to secrete a fluid which surrounds and cushions the oocyte (see Figure 4.13b). The follicle cells also start to make certain sex hormones, including testosterone and oestrogen. As with the testis, the ovarian follicle does this in response to hormonal cues from the brain. The oestrogen is important: it causes even more of the follicle cells to become sensitive to LH, and this, in its turn, is vital for further maturation. Once again, the increase in sensitivity must coincide with an increase in levels of LH, otherwise the follicle will degenerate. If the timing is good, then the oocyte can enter into the final maturation phase, the pre-ovulatory phase. The pre-ovulatory phase coincides with the progress of the oocyte through the first meiotic division, and it results in the oocyte being ovulated, i.e. released from the surface of the follicle, together with the first polar body (Figure 4.13a). Once this occurs, the follicle continues to secrete hormones, but different ones, and in different amounts: it becomes a *corpus luteum*, making progestogen, not oestrogen, in preparation for a possible pregnancy.

Although both the resumption of meiosis and cytoplasmic maturation are stimulated by an increase in local levels of LH, the oocyte itself is not sensitive to LH: the effect comes via the LH-sensitive follicle cells. By the time of ovulation the oocyte is at the surface of the follicle, separated from the outside world by only a thin layer of cells. Increased fluid pressure within the follicle causes it to pop, and the egg, still surrounded by some granulosa cells but no longer firmly attached to them, is ejected. It enters the Fallopian tube (see Figure 4.5), where it is wafted gently downwards by the fimbriae lining the tube. Fertilization can take place in the Fallopian tube if sperm are present there.

❑ Within how many days must intercourse have occurred for there to be fertile sperm in the Fallopian tubes?

■ Within five days (Section 4.3).

4.4.6 Hormonal control of egg production

As you can see from the preceeding section, hormones play a crucial role in the maturation of the oocyte. Figure 4.3 showed you how levels of oestogen and progestogen vary throughout the menstrual cycle, and suggested that hormone balance is important for a woman's fertility, but you can now see how subtle the control really is. Cells have to develop sensitivity to hormones at the times when the hormones are likely to be present, otherwise the entire operation will fail.

❑ At which stages of egg production is this important, and which hormones are involved?

■ FSH and LH at the end of the first stage of maturation; LH alone just prior to ovulation.

Once again, then, there is a significant contribution to the entire process by the pituitary gland.

4.4.7 Factors affecting fertilization

It is useful at this stage to summarize the main factors involved in a successful fertilization. First and foremost, fertile gametes must be made. This depends fundamentally on the health of the prospective parents. If they are diseased or undernourished, or have been exposed to high levels of radiation, then not only will they not produce healthy gametes, but they will probably not want to engage in the kinds of activity that might bring their gametes together.

DNA replication and protein synthesis are metabolically expensive processes. Our metabolism only runs if we supply it with a source of energy and the correct raw materials, and these are obtained from our food. It follows therefore that if we cannot eat enough of the 'right' food, i.e. that containing sufficient energy and building blocks, our metabolism will run down. Ultimately this will result in death, but before that time the body will find ways to maximize its resources by cutting down on non-essential activities, including reproduction. It is well documented that undernourished women do not have menstrual cycles, and though there is nothing as easily measured in men, it is clear that sperm production suffers too in times of hardship. Indeed, there is some concern about the observation that over the past century in the UK there has been a drop in men's average sperm count. The reason(s) for this remains a mystery, and so far there has been no adverse effect on fertility, but there is obviously a possibility that, unchecked, it could result in an increasing proportion of infertile men.

Any process that interrupts the correct production of hormones is likely, as you have seen, to have an adverse effect on gamete production. Many sex hormones are derived from the lipid *cholesterol* which, although it generally receives a very bad press on account of its propensity for blocking arteries, is a vital constituent of the healthy body. It, or precursors required to synthesize it, must be present in the diet. As you know from Chapter 3, cholesterol is a component of cell membranes, and is also vital for the production of several hormones, the steroid hormones. Figure 4.15, which you need not memorize, and which is included only for interest, shows the chemical structures of the main sex hormones, and how these structures are related to the structure of cholesterol. (Recall from Chapter 3 that chemical formulae can be written out to show the relative positions of the atoms in a molecule. This gives an idea of the *shape* of the molecule, which, as you know, is very important for its function.) Although we are not suggesting that everyone should adopt a high-cholesterol diet, it is clear that some at least is necessary for reproductive success.

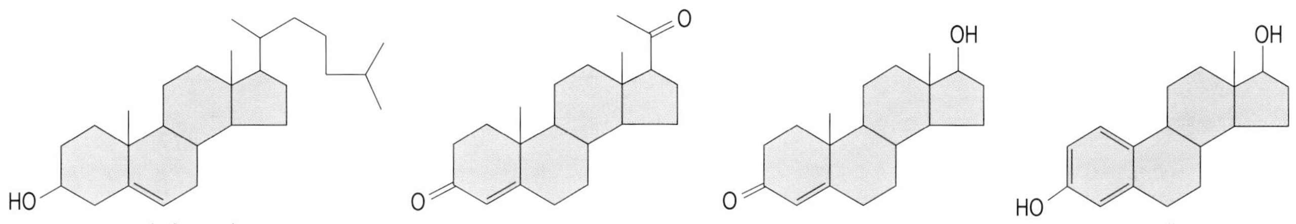

HO cholesterol O progesterone O OH testosterone OH HO oestradiol

Figure 4.15 Structures of cholesterol, progesterone (a progestogen), testosterone and oestradiol (an oestrogen). The cholesterol fused ring structure is shown in colour. (*Note*: these are *skeleton* representations – none of the carbon atoms and directly attached hydrogen atoms are shown. Each line denotes a single C to C bond, with the C atoms at the ends of the lines and H atoms attached to them; so a 'free end' represents a CH_3 group, a 'bend' is a CH_2 group and where three lines converge it means there is a CH group.)

We have looked at some of the factors affecting the production of healthy gametes, but there remains a further important criterion for fertilization.

❑ What is it?

■ Biologically successful sexual intercourse.

Biologically successful intercourse is defined as intercourse in which the man can maintain an erection for long enough to allow sperm to enter the woman's vagina. This usually means that ejaculation has to occur, but some leakage of sperm may take place before this. The maintenance of an erection depends not just on physical factors but, very importantly, on psychological ones too. There is no requirement for female orgasm to occur, and there is certainly no substance to the old wives' tale that 'you can't get pregnant unless you come'; interestingly, though, there *is* some evidence that a woman's emotional state may affect her ability to conceive. (Unfortunately, not wanting a baby is not an effective contraceptive!). Of course, it is difficult to conduct experiments to clarify this, and for the time being this evidence remains rather circumstantial. Nevertheless, it seems clear that, in humans, the process of reproduction is more than just a physical one. You will learn much more about this in Book 4, where the interactions between mind and body will be addressed.

Summary of Section 4.4

1. Gametes are special cells because they contain only one set of chromosomes instead of the more usual two sets.
2. The chromosome number is halved by meiosis.
3. The crossing over and random assortment of chromosomes in meiosis produces a unique set of genes in every gamete, and thus in every individual (except for identical twins, who are derived from the same conceptus).
4. Sperm production involves many rounds of mitosis, one meiosis (involving two cell divisions), and a maturation and packaging phase. Millions of sperm can be produced every day.
5. Egg production also involves mitosis followed by meiosis, but the meiosis is arrested until the oocyte is stimulated to re-enter division. Only one egg is produced per menstrual cycle, and only one gamete is produced as a result of each meiosis.

6 Gamete production in both men and women is influenced by hormones.

7 Good health is necessary for the production of healthy gametes.

4.5 Fertilization

Now that we have dealt with the basic biology, we can resume and give more detail to our story, and return to where we left it: fully mature, strongly swimming sperm have been deposited in the vagina, and will begin their race to the newly ovulated egg.

❑ Where do the sperm have to go to fertilize an egg?

■ Through the cervix, up through the uterus, and into the Fallopian tubes (see Figure 4.5).

Immediately following ejaculation, the semen coagulates, which reduces loss from the vagina due to gravity. However, it soon liquefies again, and most of it leaks from the vagina, but within a minute or two of ejaculation some sperm have already swum through the cervix and entered the uterus. They may be helped on their way both by the wafting motion of cilia lining the entrance to the cervix, and also by muscular contractions of the woman's reproductive tract if she has experienced orgasm, but these are not required. It used to be thought that sperm do not swim purposefully up the reproductive tract: that their direction of swimming is completely random, and, indeed, changes frequently, and that sperm only get to the 'right' place by luck. However, it has recently been reported that sperm have a net tendency to swim up the Fallopian tube containing the recently ovulated egg, rather than the other tube. It is not clear how the sperm accomplish this, but it is possible that they are responding to subtle chemical signals. Whatever the mechanism turns out to be, the first hurdle is getting through the cervix.

❑ Why is this difficult?

■ The cervix is plugged with mucus. This is normally very viscous, but becomes thinner around the time of ovulation.

Of the many sperm ejaculated, only about 100 get through to the uterus. These then undergo a process called *capacitation*, the precise nature of which is not known, but is necessary to give them the *capacity* to fertilize an egg. The process can only occur in the uterus when it has been 'primed' by oestrogen (that is, at a particular point in the menstrual cycle), and takes several hours. Even after capacitation, sperm are still not fully able to fertilize: they require one final change, called *activation*. Activation involves changes to the membranes surrounding the sperm, including the one surrounding the acrosome. This develops holes, releasing from inside the acrosome an enzyme called hyaluronidase. Another change at activation is in the swimming properties of the sperm. Instead of the regular, wave-like beats used up to this stage, the tail starts to beat in a periodic, whiplash movement, propelling the sperm along in lurches. Activation must take place very close to the egg, as once it has occurred the sperm will not survive for long.

The sperm become activated in the Fallopian tube, where they will meet the moving egg, still surrounded by some follicle cells. The follicle cells are removed by the action of the hyaluronidase from the acrosome (by digestion of the polysaccharide material holding the cells together) and the sperm stick to the outside of the zona pellucida, as shown in Figure 4.16. Other enzymes from the acrosome (e.g. protein-digesting ones) produce a path through the zona, and the whiplash movement of the tails propels the sperm along their paths, leaving them in close proximity to the egg. The membrane of *one* of them fuses with that of the egg. This sperm stops all swimming movements immediately, and a change takes place in the egg membrane which effectively stops any other sperm from fusing with it. Enzymes are released from the egg's surface which alter the structure of the zona and prevent any further sperm from penetrating it.

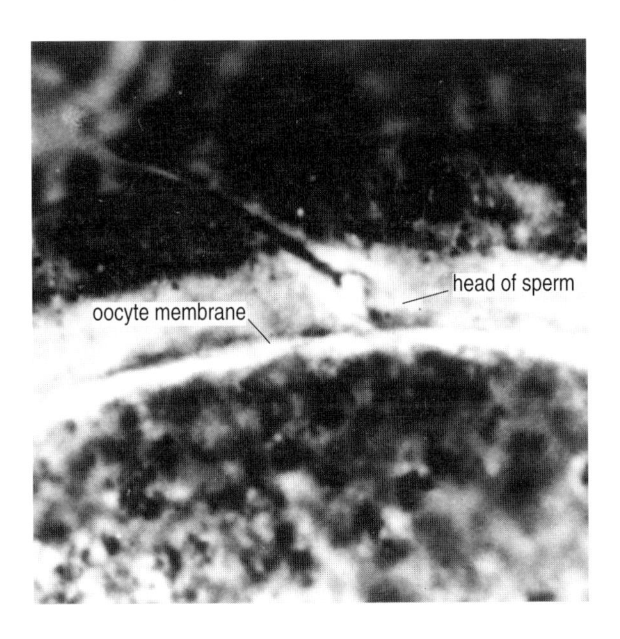

Figure 4.16 Photo showing a sperm stuck to an egg.

These events take place within a few minutes of fusion, and the chromosomes of the successful sperm begin to move into the egg cytoplasm. The final part of fertilization, taking longer but absolutely vital, is the egg's resumption of meiosis to get rid of the extra set of chromosomes. Two to three hours after fusion the second polar body is extruded, and the fertilized egg is left with one set of maternal chromosomes to complement the paternal set from the sperm. Although the two chromosome sets are independent to begin with, and so are called pronuclei (these can be seen in Figure 4.14), by the time of the first division of the fertilized egg the chromosomes have mingled and will all come to lie in the same equatorial plane. This would appear to mark the time at which the embryo becomes autonomous, but as you will see below, this is not necessarily the case. However, it is certainly a prerequisite for a new individual.

If the egg is old, i.e. was ovulated several hours before fertilization, the processes described above may not occur accurately.

- ❑ Can you predict what sort of errors might occur?
- ■ If more than one sperm fused with the egg, or if the second polar body was not extruded, there would be too many sets of chromosomes in the fertilized egg. This would produce abnormal embryos.

It is estimated by most practitioners that more than 50% of all conceptions are genetically abnormal, and it has been suggested that this results in part from the fertilizing of old eggs. Of course, good maternal health and adequate nutrition are also important for the production of good eggs. To maximize the chance of healthy offspring, perhaps intercourse should be timed so that sperm are present as soon as the egg is ovulated. But, as you have seen for practitioners of natural methods of contraception, timing ovulation may be more easily said than done, and even when it can be pinpointed, it may be socially inconvenient to have intercourse at that time. Many abnormal embryos are in fact aborted naturally, and persistent abortion of this sort *may* be one cause of infertility.

- ❑ Can you suggest other biological reasons for infertility?
- ■ There are a vast number of points at which the fertilization process might go wrong. But broad categories are failure of sperm production, failure of egg production, and failure of the *interactions* between them.

It is sometimes impossible for structural reasons for egg and sperm to meet. This may be because sperm are not ejaculated properly, or because there is some kind of anatomical blockage in the woman's reproductive tract. In these cases *in vitro* fertilization (IVF) may help. This involves removing eggs from the woman's ovaries just before they would normally ovulate, and mixing them with sperm that has either been ejaculated or, if ejaculation is not possible, with sperm removed from the vas deferens. The woman can be treated with hormones so that several eggs ripen at the same time; the sperm can be concentrated if necessary. Either eggs or sperm (or both) can be taken from a donor, rather than from the couple themselves. The gametes are mixed in a dish (*in vitro* means 'in a glass') and observed under a microscope. Enzymes necessary for the whole process, such as hyaluronidase, can be added. Once fertilization has occurred, and the embryos appear to be progressing normally (see Section 4.6), they can be placed in the woman's uterus, where they may implant and grow normally. Although several embryos may be produced by this method, usually only two or three are given the opportunity to implant (the rest are frozen for future use). This is to reduce the risk of multiple pregnancies, which are not desirable for the welfare of the mother or the babies. A variation of this is GIFT, gamete intra-Fallopian transfer, where the gametes (sperm and eggs), taken from whatever source, are mixed together, then immediately placed in the Fallopian tubes. This ensures that the conditions for fertilization are as 'natural' as possible. In both cases, it is important that the woman should be at the correct point in her cycle so that her uterus is able to receive an implanting embryo. This may require her to have hormone treatment.

IVF and GIFT are valuable techniques for overcoming blockages, and their use has made many couples happy, although in general success rates are low: of the order of 10% per menstrual cycle, varying to some extent depending on the practitioner involved. But sometimes couples are infertile for no obvious physical reason: both produce healthy gametes, and have no physical abnormalities. It turns out that some women unwittingly produce antibodies that interfere with their partner's sperm, and prevent their survival.

Antibodies are very large protein molecules, and as you will learn later in the course (Book 2), they play an important part in the immune system, the body's machinery to fight infectious disease. As with most proteins, the key to an antibody's functions lies in its shape.

❑ Why is the shape of a protein important?

■ It allows that protein to interact with (bind to) other specific molecules.

Antibodies have the ability to distinguish between molecules that 'belong' in the body, that is, are 'self', and other, non-self, molecules. Antibodies patrol the whole body, being carried around in the blood and lymph, but also penetrating the tissues to maintain maximum surveillance. Their response to meeting a non-self molecule is to attach to it, and form an insoluble complex which can be removed by other parts of the immune system. Usually this response is limited to invading microbes, but occasionally antibodies are produced against inappropriate targets – in this case, components of the sperm's membrane. The bodies of women in whom this happens recognize the sperm as non-self, and the immune system responds by making antibodies against them. When the antibodies attach to the sperm's membrane, they prevent it from fertilizing an egg, and infertility results. This type of infertility is not the result of a disease but of an over-enthusiastic defence mechanism. The important point to note from this is that even a normal body mechanism, if deployed inappropriately, can result in abnormal functioning – *dis*-ease. The treatment for this kind of infertility is to 'de-sensitize' the woman, that is, inject her with very low levels of the offending sperm membranes, in an attempt to fool her body into thinking that these are just another few self molecules. Perhaps surprisingly, this treatment often works remarkably well.

Summary of Section 4.5

1 After ejaculation some sperm penetrate the cervical mucus, and on arriving in the uterus become capacitated.

2 A few sperm swim up the Fallopian tube containing the recently ovulated egg.

3 In the tube the sperm become activated. This involves changes to the membranes and a change in the swimming pattern.

4 Enzymes from the acrosome allow the sperm to get next to the egg, by removing follicle cells and digesting a path through the zona pellucida.

5 The membranes of the egg and one sperm fuse, allowing chromosomes from the sperm to enter the egg.

6 The egg resumes meiosis.

7 In cases where 'natural' fertilization cannot take place, IVF or GIFT may allow conception.

4.6 Small beginnings – the first two weeks

Let us now return to the Fallopian tube, where a fertilized egg is assembling its chromosomes prior to commencing a series of mitotic divisions which will eventually give rise to the millions of cells that make up the human body. Obviously these millions of cells do not just exist as an amorphous mass: they are *differentiated* into many different types of cell, and they are organized into recognizable, discrete structures: tissues and organs. This is accomplished by a coordinated sequence of complicated events, yet the fascinating choreography of development is based, like ballet, on a few relatively straightforward steps. In the very first days of embryonic development the stage is set for the full repertoire of movements, so understanding the early days allows us insights into the later processes, not yet fully understood, that lead to the development of a new-born baby.

Figure 4.17 shows the stages of development of the human conceptus (early embryo) during the first week or so. These stages are called *pre-implantation*, because during this time the embryo lies free within the mother's reproductive tract. From the time of fertilization until the early **blastocyst** stage, roughly 5 days, the conceptus is in the Fallopian tube. A blastocyst is the first structure that looks different from the blob of cells that is the conceptus. A blastocyst consists of two distinct types of cell, and is swollen by fluid which is produced by one of these types. We shall discuss this further in a moment. By the time the blastocyst arrives in the uterus, the hormonal cycle has progressed, so that the uterus will be able to accommodate the embryo, and allow it to attach to the thickened uterine wall. Prior to this, implantation is impossible.

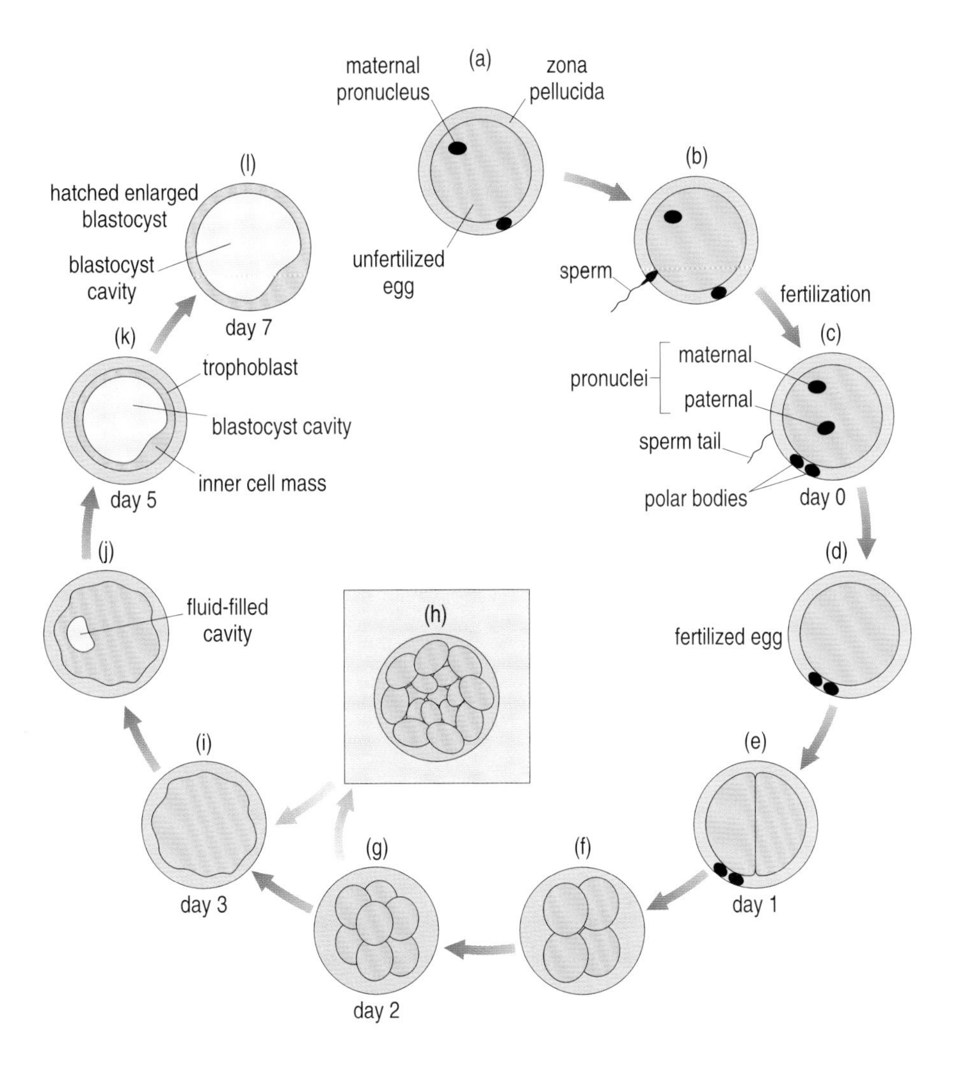

Figure 4.17 The stages of early human development. (*Note*: from the fertilized egg onwards each cell contains two sets of chromosomes in its nucleus – for simplicity, the cell nuclei are not shown beyond stage (c).)

❑ Which hormone is involved in preventing implantation?

■ Progestogen (Section 4.3.1).

The divisions resulting in the stages shown in Figure 4.17 are called *cleavage* divisions. This means that the divisions take place without any net cell growth, resulting in the cells roughly halving in size at each division. In this way, the large amount of cytoplasm in the egg is reduced to more manageable amounts. You will recall that the cytoplasm in the egg provides the conceptus with many of the nutrients it needs to get it started; as the conceptus becomes metabolically active, synthesizing its own nucleic acids and proteins, it becomes less dependent on components in the cytoplasm, and more reliant upon external supplies of nutrients from the surrounding fluid.

Three waves of cell division result in an 8-cell conceptus whose cells are all in contact with the 'outside', i.e. the uterine fluid, through the zona pellucida (Figure 4.17d–g). But the fourth division is different, and, as you will see, vital for further development. The division from 8 to 16 cells is *asymmetric*, that is, the division does not occur across the centre of the cell, but is closer to one end than the other.

❑ What is the consequence of this asymmetry?

■ The resulting cells will be of two different sizes (Figure 4.17h).

At the 16-cell stage, the first differentiative event has occurred: there are two populations of cells: eight large cells on the outside, and eight smaller cells on the inside. Size is not the only difference between them; although cells of both populations have a nucleus, cytoplasm and a cell membrane, the constituents of the cytoplasm and the cell membranes are different. In particular, one of the differences in the membranes means that the smaller, inner cells are more adhesive than the larger, outer cells, and form a tight ball of cells lying within, though still connected to, the outer shell. You would not expect the *constituents* of the nucleus to be different, as the same DNA is present in all cells of this individual, but different proteins are made in the two populations (in particular, different membrane proteins, giving rise to the differential adhesiveness).

❑ Think back to Chapter 3. What does the synthesis of different sets of protein imply is happening at the level of the DNA?

■ Different genes are being transcribed.

Around the time of the 8- to 16-cell division, the conceptus undergoes a morphological (shape) change, called compaction, in which the cells flatten on each other, and the outlines of individual cells become hard to distinguish. This stage, sometimes referred to as a *morula*, from the Greek word for mulberry, is shown in Figure 4.17i. At this stage it is hard to see individual cells; in fact, unless the cells are separated by various laboratory treatments, it is not possible to see the two different cell types present in an intact embryo at the 16-cell stage. (Figure 4.17h is included for clarity even though it is not a stage normally seen.) Once compaction has occurred, it is impossible to see subsequent cell divisions, although they do occur at frequent intervals. By the time the fluid-filled cavity of the early blastocyst appears (Figure 4.17j), there are about 64 cells, and a mature, expanded blastocyst can contain more than 250 cells (Figure 4.17k).

As mentioned above, the blastocyst consists of two cell types which are derived from the asymmetric division at the 8-cell stage. The larger cells resulting from that division form the **trophoblast** cells, which surround the fluid-filled cavity, and indeed are responsible for making the fluid. These cells will form the embryo's contribution to the placenta, the organ which will supply nutrients to and remove waste products from the embryo throughout the rest of gestation. The smaller cells from the asymmetric division form the **inner cell mass**. These are the cells that will give rise to the embryo proper, and to some of the membranes surrounding it. The trophoblast cells do not undergo much further differentiation, but the inner cell mass cells have a long developmental road ahead of them.

How interchangeable are the two cell populations? Is differentiation into two different types of cell really a permanent step, or can it be reversed? The answer seems fairly clear-cut: once a cell has differentiated, there is no turning back. If a trophoblast cell is put into the middle of a group of inner cell mass cells it will *not* contribute to the inner cell mass, but will be pushed to the outside of the clump, and discarded. This happens because trophoblast cells have different protein molecules on their surfaces from those found on the surfaces of inner cell mass cells.

❑ Can you remember a property of some of the surface molecules which might account for this?

■ Adhesiveness. Adhesion between cells is mediated by surface proteins whose shapes allow them to interact with each other.

In fact, as you learnt in Chapter 3, many surface proteins – the glycoproteins – carry short chains of carbohydrate on them and it is the carbohydrate part that is responsible for the stickiness. Adhesion of one cell to another can be regarded as an early form of communication of the 'Is there something there – Yes or No' variety. But it seems that cells are able to recognize similarities and differences between themselves purely on the basis of their adhesive properties. This differential adhesion is what pushes the trophoblast cell out of the clump of inner cell mass cells. The inner cell mass cells adhere more tightly to each other than they do to the trophoblast cell, with the result that it is gradually excluded from the group. The concept of differential adhesion is an important one, and you will meet it again.

It seems clear that by the time the inner cell mass and the trophoblast are visibly distinct, they have differentiated to such an extent that they cannot substitute for each other. But what about earlier stages? It has been known for many years that in some animals the fate of early embryonic cells has not been determined: cells from 2- and 4-cell embryos can substitute for each other and can even, if separated from their fellows, give rise to complete new embryos. It is ethically difficult to do this kind of experiment with human embryos, but what little experimental evidence there is seems to accord with what has been found in other mammals: until the 8-cell stage, embryonic cells are developmentally 'plastic', that is, their fate is not yet sealed. By the 8-cell stage, however, preparations are clearly under way for the asymmetric division that marks the first differentiation, and the cells cannot safely be tampered with.

Summary of Section 4.6

1. For the first week after fertilization, the conceptus (early embryo) floats freely in the female reproductive tract, obtaining some of its nutrients from the fluid in which it is bathed.
2. The fertilized egg begins a series of divisions to give 2, then 4, then 8 undifferentiated cells. There is no net cell growth, so each generation of cells is smaller than the last.
3. The 8- to 16-cell division is different, and important. Each cell divides asymmetrically, to yield one large, outside cell and one small, inside cell.
4. The two populations of cells are different from each other, and cannot substitute for each other in development. They are differentially adhesive, and express different subsets of genes.
5. The large, outside cells give rise to the trophoblast, which will make the placenta. The small, inside cells make the inner cell mass, which will form the embryo and associated membranes.

4.7 A new life

There is a common belief that life begins at the moment of conception, i.e. when a sperm fuses with an egg. This is a step forward from past years, when life was alleged to start at the time of 'quickening', i.e. when a woman could feel her fetus moving inside her. However, both these opinions suffer from an underlying falsehood: that life 'begins' at all. Life is a continuum; gametes are produced by living parents, and fuse to produce new living individuals, but unfused gametes are nonetheless alive and capable of metabolic activities. So the question is not 'when does life begin?', but rather 'When does a new individual come into being?' The answer to this highly controversial question could, in theory, be at any time from the moment of penetration of the egg membrane by the sperm up to the time of birth, and, indeed, the Roman Catholic Church debated for centuries the time when the soul enters the new being (see above). As with many contentious issues, people's opinions will be drawn not just from biology, but from psychology and sociology too. However, for now we shall concentrate on a biological answer to the question.

Common (female) experience suggests that a new individual is alive, and certainly kicking, from around 17 weeks of gestation. Heartbeats and other spontaneous movements can be detected by ultrasound scans several weeks earlier than this. And, if life is defined in terms of metabolic activity, a new individual is alive by two weeks of gestation, when pregnancy tests can detect a substance, chorionic gonadotropin, made by the trophoblast cells as part of placental activities, so the conceptus is metabolically active at this stage at least.

So let us now address the question of the development of individuality.

❑ What characteristics distinguish one individual from another?

■ Almost any answer is correct here! You might have mentioned eye colour, fingerprints, or handedness, for example. But a more molecule-orientated answer might have mentioned enzymes, or genes (an individual's inheritance).

As you saw above, a new individual, resulting from the fusion of two gametes, is characterized by having two sets of genes, and therefore two alleles of each gene. Some of the alleles will have come from the mother, and others from the father.

❑ In view of this, how could you distinguish a new individual from its mother?

■ By identifying an allele that had been inherited only from the father.

It is technically quite tricky to carry out this kind of test in humans. It is not possible to start with laboratory breeding stocks with well characterized alleles as geneticists can do in some other species! So recourse has been made to studying traits that could only ever have come from the father. One such trait is the sex of the embryo. Sex, as you may know, is determined by the chromosomes (although you will learn more of this in Book 4): among the 46 human chromosomes are two that are called sex chromosomes. There are two sorts of sex chromosomes: X chromosomes, which are among the largest, and Y chromosomes, which are among the smallest. Every female has two X chromosomes in each of her cells; every male has one X chromosome and one Y chromosome. The sex chromosomes behave as a pair during meiosis, lying together at the equator, then separating to opposite poles of the dividing cell, even though they are physically ill-matched in terms of size (see Figure 4.8, left). You will see the reason for this in Book 4.

❑ Which sex chromosomes will be present in the gametes of men and women?

■ Eggs will always contain one X chromosome among their 23. Sperm contain *either* an X chromosome *or* a Y chromosome. (If you are at all unsure about this, go back to the diagram of meiosis, Figure 4.9.)

This means that an X-containing egg may be fertilized by either an X-bearing sperm or a Y-bearing sperm. In the former case the embryo will have two X chromosomes and be female; in the latter case the embryo will have an X and a Y and be male.

The sex chromosomes do not only carry genes determining sex; most of their length is taken up with genes concerned with quite different characteristics such as blood group. In fact, the only part of the Y chromosome that is *essential* for maleness is a tiny area called *Sry* (formerly *tdf*, testis-determining factor). Another gene on the Y chromosome codes for a protein that is a normal part of the (male) cell membrane, and it is possible to distinguish the presence of this protein on the surface of pre-implantation embryos.

❑ What does this tell us?

■ If the protein is there, it *must* have been made by the embryo's cells, as the mother cannot make it – she has no Y chromosome. So this means that the embryo's own genes are active, and it is beginning to become a male individual.

The earliest that this protein has been detected on embryos is the 4-cell stage. This and other evidence tells us that the embryo's genes have probably been 'switched on' by the 4-cell stage, i.e. about 2 days after fertilization. This is long before most mothers have any inkling that they might be pregnant. As we have mentioned before, more than 50% of conceptions will fail, and the majority of these failures are at very early stages. A significant proportion is probably due to an inability of the embryo's genes to become functional. In these cases, implantation will not occur, and menstruation will take place at the normal time, washing the failed embryo out, and leaving the woman with no sign that she was ever pregnant.

Strictly speaking, then, the question we asked at the beginning of this section – When does a new individual come into being? – has been answered from a biological perspective. The embryo *begins* to function independently at the 4-cell stage, but it is certainly not capable of independent existence at this stage. It still uses cytoplasmic components from the mother, although these are progressively metabolized and by the blastocyst stage, few, if any, remain. Once the embryonic genes have been switched on, early development seems to proceed automatically, with little reference to the outside world; this is why early embryos can be successfully grown in the laboratory. The embryo, moving slowly down the Fallopian tube into the womb, is constantly bathed in various secretions, but all that it appears to need is a suitable temperature (37 °C), an appropriate level of acidity, an energy source, raw materials for synthesis as well as the right mixture of ions. Large glycoprotein molecules, present in the tubal secretions, seem to help stabilize the cell membranes, which are fragile because the cells are so big.

A few days later, however, the picture is very different. The embryo begins to burrow into the wall of the womb, and a placenta is formed. The placenta is a large and important organ, and it is made jointly by the embryo and its mother. You will learn the details about its formation and functions in Chapter 5. By the time the placenta begins to form, about two weeks have passed since fertilization. Only if a woman has very regular 28-day menstrual cycles, and her period has not appeared on time, will she have any inkling that she may be pregnant. We shall go on to look at this aspect of reproduction now.

4.8 How does a woman know she's pregnant?

Our description of the developing embryo has, so to speak, detached it from its mother. But we should remember that on the other end of the placenta is a woman whose reaction to her pregnancy may lie almost anywhere in the scale of human emotion, and whose behaviour during her pregnancy will have an enormous effect on its outcome. This section attempts to look at the pregnancy from the mother's point of view. Of course, it cannot possibly be applicable to *all* women in *all* pregnancies – no text could ever do that – but it will attempt to draw out some general facts.

As you have seen above, much has been said throughout the ages about the beginning of pregnancy and the formation of a new individual. For most women, this is frankly irrelevant: fertilization and the first cell divisions pass unnoticed. The first sign of pregnancy for many women is a missed menstrual period, but this is by no means diagnostic. Women who have infrequent or irregular cycles may not be aware that they have missed one; even in women with regular cycles there are several reasons why a period might be missed.

❑ Give some of these reasons.

■ The woman may be undernourished, or ill, or breastfeeding, or be taking a lot of exercise.

Some women who are eager to become pregnant, and who are highly aware of their bodies, claim to 'feel different' only a few days after biologically successful intercourse. Just what these feelings consist of is difficult to ascertain, nor do they seem to be common to all women, but there is no doubt that they are real enough to the women concerned. Symptoms such as sore breasts and feeling bloated are certainly common in pregnancies, but do not usually become apparent until a couple of weeks after fertilization, a time when some women experience these symptoms anyway pre-menstrually. Conversely, women who either have no idea that they might be pregnant, or those who do not want to be, and unconsciously deny the possibility, can go for several months before finally realizing that they are in fact pregnant. Even symptoms such as nausea and vomiting, which some women find overwhelming, are often shrugged off as 'a tummy bug'. So there is a significant emotional component in the recognition of pregnancy.

Whether or not a woman will welcome a pregnancy, it is often very important for her to *know* as early as possible whether a pregnancy has occurred. Pregnancy testing must surely have had a long history; it is related that some wise women could tell, apparently just by looking, whether a woman was pregnant. Indeed, even today there are mystics who claim that a woman's 'aura' changes when she is pregnant. If this is true, it might be related to the 'different feelings' that some women experience. However, these 'tests' cannot be substantiated by a scientific explanation, and other, more reproducible tests have been sought. One early favourite was the observation that if urine from a pregnant woman was rubbed on the back of a female toad, the toad would be induced to lay eggs. Urine from a non-pregnant woman would not do this. This was the pregnancy test of choice for many years, but it was not very accurate.

❑ Can you suggest why not?

■ If the toad was old or ill it might not be able to lay eggs. (On the other hand, if it was ready to spawn anyway, it might do so regardless of this external stimulus.)

The reason why pregnant women's urine can cause a toad to spawn is that it contains hormones which can penetrate the toad's skin and have the appropriate physiological effect in the toad's body. This led to the search for more accurate tests, and modern ones have got it down to a fine art.

Although many of the hormones in a woman's urine are present whether or not she is pregnant, there is one hormone that is absolutely diagnostic of pregnancy. It is called **chorionic gonadotropin**, and is made by trophoblast cells in the placenta, as you will see later. Currently available tests detect a very small amount of chorionic gonadotropin in urine, and so can be used very early in pregnancy – in fact, as soon as the placenta is formed, at around two weeks after fertilization. This time corresponds with the time a period would be expected if the woman were not pregnant, so can be used reliably as soon as the period is missed.

If a woman does find herself pregnant, what can she expect? Pregnancy is a time of enormous physical and emotional changes, and these are often difficult to cope with. To begin with, the physical effects of early pregnancy can be extremely unpleasant. The nausea and vomiting of morning sickness can be very severe, and although in many women the symptoms abate after a while, in others they persist right through the pregnancy. Sickness is thought to be due to the high levels of progestogen circulating in the blood.

❑ Where is the progestogen made?

■ It is made initially by the corpus luteum.

At later stages, progestogen is made by the placenta, and it is necessary to maintain the pregnancy. Women differ in the degree of morning sickness that they suffer – some do not suffer at all, and in many women the amount they suffer varies with each pregnancy. It is not known whether this is because women produce different amounts of progestogen, or whether they can tolerate different levels of it (or both). Whatever the reason, this does not alter the fact that the woman may feel very ill. During this time she may nonetheless have to function as efficiently as always, while perhaps concealing her condition – for it will not at this stage be obvious to anybody (except perhaps mystics) that she is pregnant.

Other 'normal' symptoms of early pregnancy are generally easier to cope with. Sore breasts, again resulting from exuberant hormones, are often no more painful than what the woman may experience every month during her menstrual cycle. An increased appetite, something which, rather perversely, often accompanies morning sickness, is usually easily dealt with. Many symptoms associated with pregnancy, such as raised blood pressure or fluid retention, do not usually occur until later on, when the fetus has grown bigger.

One thing that must be emphasized is that all the changes associated with pregnancy are *for the baby's benefit*. A normal, healthy woman is well adapted to her diet and lifestyle, so it is likely that she will find any body changes unpleasant. Added to the physical discomfort is often severe emotional upheaval, particularly if it is a first pregnancy. The prospect of fundamental lifestyle changes are sometimes difficult for people to accept. Even when a pregnancy is very much wanted, it is not unusual for women to suffer periods of depression and worry about the future. Part of the worry surrounds practical issues like jobs, housing and money, but most women also worry about the pregnancy itself, and whether the baby will be healthy. As you know, many pregnancies are unsuccessful for the very reason that the embryos are *not* healthy, and, perhaps mercifully, many abnormal fetuses are spontaneously aborted. However, a small number of babies are born with some kind of abnormality, and this knowledge cannot fail to be a source of worry, especially if there is a family history of disease or if the woman is in her late 30s or older.

In our society, pregnancy is generally very popular with the non-pregnant, particularly in the 'correct' social circumstances. But even the most universally welcomed pregnancy is likely to appeal rather more to by-standers than to the mother-to-be herself. The combination of physical effects and deep emotional disturbances can cause uncharacteristic mood swings which may be hard to live with. All this makes pregnancy seem like a case for intensive care (and perhaps contributed to the pervasive view that pregnancy is an illness), but even the worst of pregnancies is only temporary, and this knowledge gives many women cause for cheer, especially if they are looking forward to having their baby. Many women differentiate in their own minds between 'expecting a baby' and 'being pregnant', and the feelings associated with each can be diametrically opposed. Most women, however, achieve some balance that they can live with, so that although reproduction may not be an entirely pleasurable experience, it is usually not all bad, either. The anticipation, the challenge, the excitement of feeling movements, and generally feeling special are all aspects of pregnancy that most people thoroughly enjoy.

4.9 Selecting the sex of a child

Once a pregnancy has been established, many couples are anxious to know the sex of their unborn baby. The reasons for this are many, ranging from the prosaic (will the baby be able to use its brother's or sister's old clothes) to the deeply religious (as described for Hindus in Section 4.2 above). In many communities there is so much social pressure on mothers to produce the 'right' sex (usually male) that infanticide of the 'wrong' sex is widely practised. Because this is illegal in most societies, figures to substantiate this claim are not available. However, it is estimated that between 50 000 and 80 000 female fetuses are *aborted* each year in India simply because of their gender. Although cultures differ in their regard for individual human life, it is probably true to say that no mother, whatever her beliefs, can be completely unmoved by the death of her baby. Although this matter is complicated by cultural factors, a lot of pain might be avoided if couples could choose the sex of their child.

Conduct a straw poll among your acquaintances, particularly any who are expecting a baby. Do they, or would they, want to know the sex of their unborn baby? What are their reasons for feeling this way? How many would like to be able to choose in advance the sex of their child?

You probably found that many people do not mind what sex their baby is as long as it is healthy. However, some people have very strong feelings about this, one way or the other. This may be for cultural reasons, as we mentioned above. But it may also be because they have a family history of a genetic disease which affects only one sex, and therefore want a child of the other sex, which will be unaffected. Or it might simply be because they already have one or more children of a particular sex, and want some variety.

❑ When does the sex of a baby become apparent?

■ As you saw in Section 4.7, one of the first 'markers' of the new male individual is carried on the Y chromosome. This is first expressed at the 4-cell stage, so this is when embryos of different sexes can be *functionally* distinguished.

The sex of a fetus can be determined using *amniocentesis*. This technique involves sampling the amniotic fluid (the liquid that surrounds the fetus): a hollow needle is inserted through the mother's abdomen and some of the amniotic fluid is withdrawn. In the fluid are found some fetal cells which have been shed during development. These can be tested for biochemical or chromosomal abnormalities, if the family history suggests that this is a possibility. The appearance of the chromosomes will also reveal the sex of the fetus. Amniocentesis is not routinely used, however, as it brings a risk of miscarriage.

As far as telling the sex of a fetus simply by looking, it is, of course, necessary to wait until the genitals are visible. Although the genitals are quite well developed by about 12 weeks' gestation, they are very small, and hard to distinguish. In practice, it is often possible to tell the baby's sex by 17–18 weeks, at the time of a routine ultrasound scan. However, unless there is a good reason for telling the mother (such as to give her information about whether the baby is likely to suffer from an inherited disease affecting only one sex), many hospitals withhold this information on the grounds that some people would procure an abortion if the baby were found to be of the wrong sex. We leave you to think about the ethics of this.

How easy might it be to choose an embryo's sex? There are two approaches to this that are technically possible at present. One involves looking at the chromosomes of the fertilized egg or conceptus, to see if a Y chromosome is present or not. This is clearly possible only when the embryo is outside the mother's body at this stage.

❑ When might this be the case?

■ During an *in vitro* fertilization procedure (IVF).

In most IVF procedures, more embryos result than can realistically be replaced into the mother. The general rule is that no more than three embryos should be put back.

❑ Why do you think this is?

■ If all three should implant, this is a major physical load for a woman to cope with. Since there is only a certain amount of space in the uterus, too many embryos will mean that resources will be limited, and the babies will be small.

Excess embryos are usually frozen or discarded. However, since all the embryos are cultured in the laboratory for a short while to ensure that they are developing normally, it is a simple matter to use one cell of each to check the chromosomes. If embryos are sexed in this way, only those of the chosen sex can be replaced in the mother, ensuring that the baby will be the right sex. Loss of one cell is not a problem for the health of the embryo; as you saw above, at early stages the embryos are developmentally plastic, and one cell can easily replace another.

Embryo sexing is all very well in cases where the mother is undergoing IVF, but is clearly impractical for most cases.

❑ Can you suggest another, easier technique for ensuring an embryo of a particular sex?

■ Fertilizing an egg with either an X-bearing or a Y-bearing sperm. (Section 4.7).

This is fine in theory, but how can the two types of sperm be reliably separated? They are produced together, and the ejaculate contains a mixture of roughly equal amounts of each type. They look the same, are the same size, have the same proteins on their surface, and swim with the same random path.

❑ Look back at Figure 4.8. Can you suggest a difference between X-bearing and Y-bearing sperm?

■ The Y chromosome is smaller than the X chromosome. Therefore Y-bearing sperm contain less DNA than X-bearing sperm.

Because DNA accounts for a significant part of the weight of a sperm (remember there is very little cytoplasm), the difference in amount of DNA means that Y-bearing sperm are a little bit lighter than X-bearing sperm. Thus, you might expect them to be able to swim a little further or a little faster than X-bearing sperm, assuming the energy supply within the sperm is constant. This is difficult to demonstrate in practice. Because sperm swim in a completely random way, it needs sophisticated microscopes and measuring equipment to trace the paths taken by individual sperm after ejaculation. Furthermore, manipulating sperm so as to be able to measure them can interfere with their ability to fertilize.

There is one approach that has yielded some promising results. Although sperm move in a random way, it seems that they have a *tendency* to swim up a gradient of concentration (that is, from a dilute solution to a more concentrated solution) of some sugars. If ejaculated sperm are placed in a tube, on top of a carefully layered gradient of sugar, with a dilute solution at the top and a more concentrated solution at the bottom, then after a while the sperm can be shown to have swum down the tube (see Figure 4.18). It is claimed that those sperm that have swum the furthest are the lighter ones, that is, the Y-bearing sperm. The heavier, X-bearing sperm do not swim so far along the tube. In theory, taking sperm from the top or bottom of the tube would be expected to give pure populations of X- and Y-bearing sperm respectively. Based on this approach, many sperm-sexing centres have been established, which claim to be able to produce babies of the desired sex by artificial insemination with such 'pure' populations of sperm. Unfortunately, the success rate achieved by such methods is little different from the 50% that would be achieved with any pregnancy. This is due to the random paths taken by the sperm: although they have a *net* tendency to swim up a gradient, the tortuous path they take can negate any separating effect of the weight difference (which is less than 1%). So the most that can be obtained is two populations of sperm *enriched* in X- or Y-bearing members.

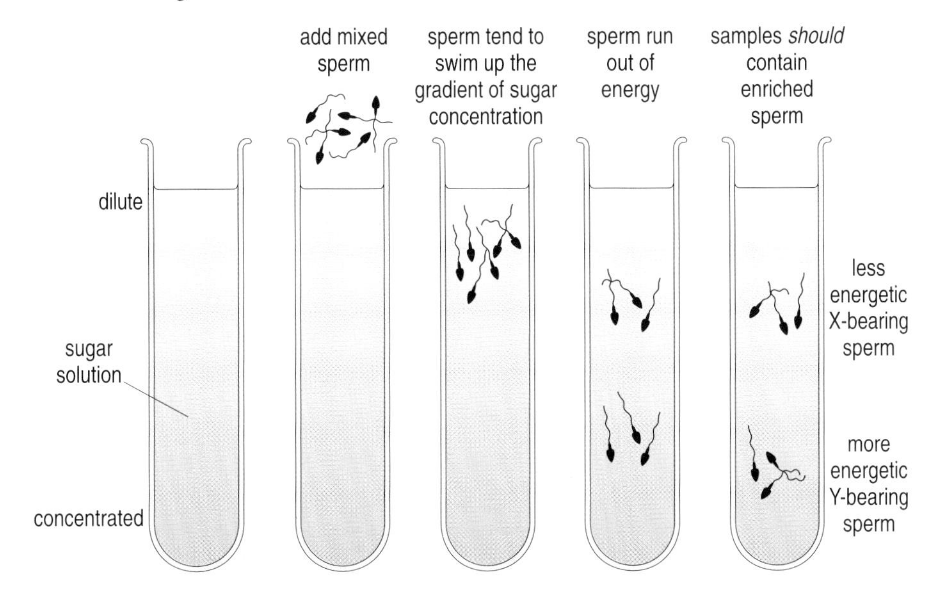

Figure 4.18 Gradient separation of X- and Y-bearing sperm.

In spite of the lack of certainty offered by sperm selection techniques, they enjoy a certain amount of popularity, particularly among people whose cultures strongly favour one sex of child. It seems clear that many people find this kind of intervention in their reproductive processes preferable to abortion or infanticide.

4.10 Review

In this chapter we have looked at some of the factors which influence a couple's decision to have a child. For people who choose not to do so, at least for some part of their lives, a range of contraceptive practices is available. We have looked at the biological processes involved in producing gametes, and at the precise sequence of events necessary to produce a healthy, fertilized egg. We have seen how this egg develops, first as a pre-programmed, free-living embryo, then becomes buried within its mother's tissues. We have looked at some of the effects that this has on the mother herself, and have glimpsed some of the physiological sacrifices that a mother must make for the sake of her baby. Finally, we put reproduction into a more social context, and looked at the ways in which cultural preferences for babies of a particular sex can be fulfilled to some extent by modern practices.

Objectives for Chapter 4

After completing this chapter you should be able to:

4.1 Define and use, or recognize definitions and applications of, each of the terms printed in **bold** in the text.

4.2 Explain the scientific basis for the main methods of contraception. (*Question 4.1*)

4.3 List the factors affecting fertilization. (*Questions 4.2 and 4.5*)

4.4 Describe with the help of diagrams the early stages of embryonic development. (*Question 4.3*)

4.5 Describe the main developmental forces at work during early embryonic development. (*Questions 4.3 and 4.4*)

4.6 Discuss social attitudes towards fertility, and suggest how these can change with culture (including religion) and education. (*Questions 4.5 and 4.6*)

Questions for Chapter 4

Question 4.1 *(Objective 4.2)*

Figure 4.19 is a graph showing how the viscosity (thickness and stickiness) of a woman's cervical mucus changes with time. Day 0 is the start of her menstrual period.

(a) Can you suggest when ovulation might be occurring?

(b) What would be the effect on mucus viscosity if the woman took a daily dose of progestogen?

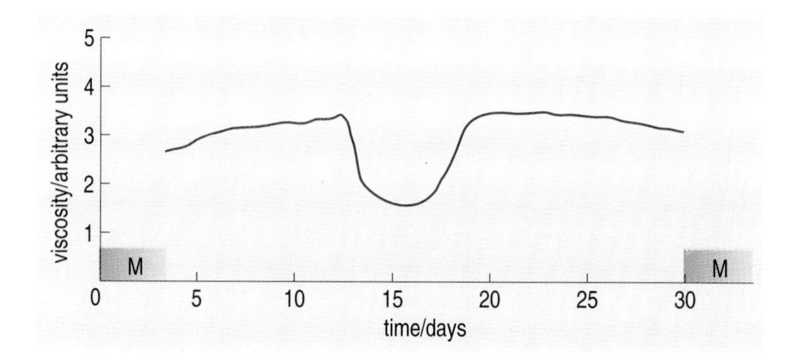

Figure 4.19 Relationship between viscosity of cervical mucus and time over the course of one menstrual cycle. M = menstruation.

Question 4.2 *(Objective 4.3)*

State two main differences in the meiotic divisions that take place in males and females.

Question 4.3 *(Objectives 4.4 and 4.5)*

Describe the first four divisions undergone by a fertilized egg. In what way are the first three different from the fourth?

Question 4.4 *(Objective 4.5)*

A sponge is a simple animal consisting of only two cell types, one of which encloses the other. If the cells are separated by passing the sponge through a sieve, and thoroughly mixed, in time they will reaggregate and the two cell types will resume their correct respective positions. Suggest how this may happen.

Question 4.5 *(Objectives 4.3 and 4.6)*

Is it correct to say that an ejaculate consists of two populations of sperm? Is this important?

Question 4.6 *(Objective 4.6)*

The Roman Catholic Church is currently strongly opposed to abortion, yet 200 years ago it was not. Why has this change come about?

Reference

Green, S. (1971) *The Curious History of Contraception*, Ebury press, London.

CHAPTER 5 THE DEVELOPING EMBRYO

You are advised to watch Video programme 2, *The human embryo: a developmental journey*, before you begin your study of this chapter. You may also wish to view it again afterwards, when the terms used in the programme will be more familiar to you.

5.1 Introduction

The development of a complex human body from a single cell – the fertilized egg – is one of the most remarkable and moving spectacles in the whole of biology, though one that is not usually on display. To reconstruct this complex process in detail requires patient and painstaking accumulation of evidence obtained by a diversity of means. We are still far from understanding all the coordinated changes of shape that systematically transform the embryo from an initial disc of cells into the intricate organization of the human form, with all its tissues and organs interacting cooperatively to produce a coherent, dynamic, creative being. Of course, equivalent miracles of coordinated transformation occur in the development of any type of multicellular organism, whether it be a sea-urchin, a buttercup, or a fruit-fly. And the detailed study of the developmental processes that occur in such species has contributed a lot to our understanding of the factors involved in guiding the sequence of shape changes that define the normal pathway of development in any species, including the human. These factors include the activity of genes as well as environmental influences, acting at particular times during development to produce characteristic effects. We shall be looking at a number of these effects in humans, for diseases and disabilities may originate or strike at these times. Such insights can lead to ways of attempting to prevent or correct them. However, the primary focus of this chapter is to understand the basic characteristics of human development.

Because in embryonic development we are dealing with a process in which three-dimensional structures are emerging and transforming in time, it is essential to use a lot of pictures to visualize what is going on. This chapter makes extensive use of diagrams, so you are encouraged to spend time looking at the pictures and viewing the associated video programme, *The human embryo: a developmental journey*, which will allow you to see the processes that transform a single cell into a complex human being. You will also meet new words describing the various structures that are generated during development. These structures are given their technical names, but you need remember only the terms that are printed in bold.

5.2 The first two weeks of development

5.2.1 The beginnings of differentiation

Chapter 4 included a description of the earliest stages of development of the conceptus, from the fertilized egg to the blastocyst that is just beginning to implant into the uterus (shown in Figure 5.1). During these stages the conceptus derives some of its nutrients from the fluid in the Fallopian tube. As you learnt in Chapter 4, the first events after fertilization of an egg by a sperm are a series of cell divisions in which the initially large egg subdivides into two, then four, eight, and 16 cells, producing a spherical ball of cells. At this stage a process begins which initiates the emergence of an embryo and the structures that envelop and protect it.

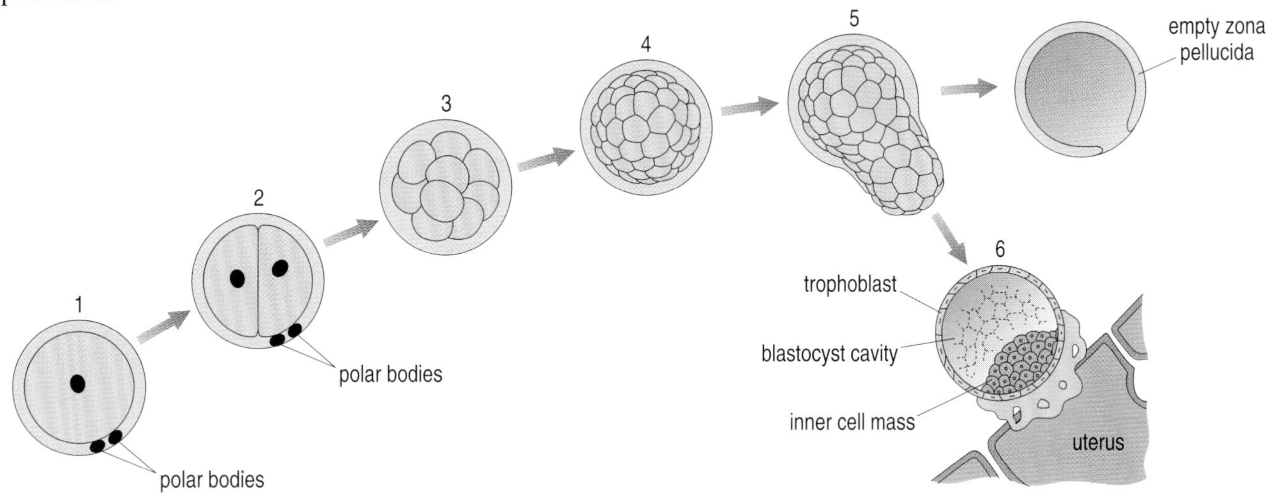

Figure 5.1 The development of the human conceptus: (1) fertilized egg; (2) two-cell stage; (3) 16-cell stage; (4) blastocyst; (5) hatching blastocyst; (6) blastocyst implanting into the uterus on day 7.

❑ Can you remember what this first step is which starts the development of the embryo?

■ The uniform, spherical ball of cells separates into an inner group of cells called the inner cell mass and an outer layer called the trophoblast.

This separation is a result of differential adhesion – the inner cells are stickier than the outer ones, so they form a compact aggregate. How does this process of cell differentiation occur? What is happening here is typical of how complexity arises out of initial simplicity in developing embryos. An undifferentiated group of cells produces a spatial pattern which creates slightly different environments, to which certain of these cells respond by making different proteins. As you learned in Chapter 3, these proteins are products of genes. So some genes in particular cells get switched on or amplified in their activity and others are shut off or decrease their activity (i.e. there is differential gene expression), resulting in cell differentiation. We saw this type of process in the epidermis (Chapter 3, Section 3.8).

In the spherical ball of Figure 5.1, the cells are dividing rapidly but they are getting smaller since there is no overall growth of the conceptus until it reaches the uterus. The cells that find themselves in the interior of the sphere experience a slightly different environment from those on the surface, since they are completely surrounded by other cells and so make more intimate contact with them. They become more adhesive, while the cells at the surface begin to spread and flatten against the outer jelly coat, the zona pellucida. These changes are accompanied by modified gene activities in the cells. This is the beginning of the differential gene expression between the surface cells which become the trophoblast, from which the placenta will form, and the inner cell mass, from which the human being will develop.

5.2.2 Formation of the placenta and the early embryo

Human conceptuses will continue to develop independently, even *in vitro*, so long as they remain within their protective coat, the zona pellucida. However once the blastocyst has emerged from this chamber, development will stop unless contact is made with the mother to begin formation of a placenta, a metabolically very important organ which is made by the concerted action of both the embryo and the mother (once the blastocyst has implanted, the conceptus is called an embryo). The growing embryo must be given a rich supply of raw materials and oxygen and, equally important, its waste products, such as carbon dioxide, must be removed before they poison it.

❑ What are some of the raw materials?

■ Carbohydrates, fats and proteins, which provide both the building blocks and the energy for the synthesis required for growth.

Because of the eventual size of the fetus, it is not possible for these components to reach it simply by bathing its surface: they must be actively supplied by a large blood circulation. This is the role of the placenta – a specialized structure where fetal and maternal circulatory systems are in very close proximity. The two circulations remain separate, however, for reasons which will become clear later.

By the time the blastocyst is hatching from the zona pellucida, it has reached the uterus, as shown in Figure 5.1. The menstrual cycle has moved on by a week since ovulation and the hormones described in Chapter 4 (see Section 4.4 and Figure 4.3) are now such that the wall of the uterus is ready to accommodate a pregnancy. The outer surface of the trophoblast cell layer makes contact with and then attaches to the uterus.

❑ How do you think this is achieved?

■ By specific adhesion between the cells.

Only those trophoblast cells in the vicinity of the inner cell mass can adhere to the uterine wall, as illustrated in Figure 5.1 (stage 6). In the majority of cases, implantation takes place in the upper half of the uterus. Several blastocysts can implant into a uterus, though this is unusual and it imposes a severe strain on the mother later in the pregnancy.

The trophoblast cells proliferate and invade the wall of the uterus (Figures 5.2 and 5.3). Some of the uterine tissue becomes eroded and eventually destroyed by the trophoblast cells, releasing its contents which include massive amounts of raw materials. These are taken up by the trophoblast cells and transferred to the embryo. The spaces (or lacunae) left by the death of uterine cells fill with blood from the mother, further increasing the supply of nutrients for the embryo (see Figure 5.4).

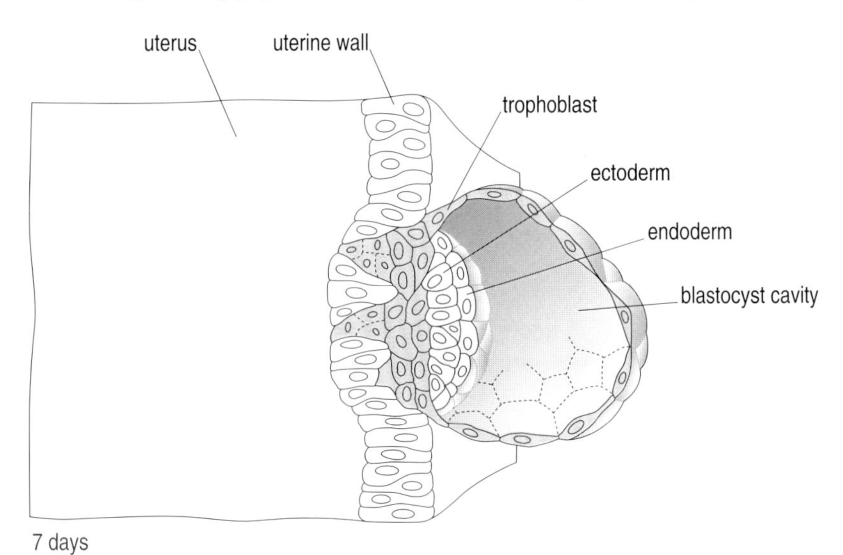

Figure 5.2 The newly implanted embryo (day 7). The trophoblast (surface) cells of the blastocyst adjacent to the inner cell mass adhere to the uterine wall and proliferate into the uterine tissue while the inner cell mass forms two layers of cells: the endoderm next to the blastocyst cavity and the ectoderm beneath the endoderm.

A large area of the uterus becomes engorged with fluid, and changes occur in the extracellular matrix which accompany growth of new blood vessels. Later the trophoblast cells become arranged in finger-like projections, called **villi**, which protrude into the uterine wall (shown in Figure 5.8). Within the embryonic villi lie blood vessels. Because the villi have very thin walls, the fetal blood is in close proximity to the maternal blood lying in the tissue between them.

The uterus can participate in this complex embedding process of the blastocyst only if it is suitably primed with the correct cocktail of hormones, in particular a high level of progestogen followed by a pulse of oestrogen. Other hormone regimes, such as those of the earlier and later parts of the menstrual cycle, produce a uterus that is hostile to embryos and will actually kill them, though just how this happens is not known. One hypothesis is that the correct hormones sensitize the uterus so that it becomes receptive to signals from the blastocyst. Once established, the pregnancy must be maintained, which also requires a continuous, high level of progestogen. This is achieved initially by the persistence of the corpus luteum (Chapter 4, Section 4.4.5), which normally disintegrates during the latter half of the menstrual cycle but which is able to survive if it receives the right signals from the developing embryo. One of these signals is the hormone chorionic gonadotropin, which is produced by the trophoblast cells. Later on, progestogen is made by the placenta itself (Chapter 4, Section 4.8), which develops from the trophoblast. The whole picture, then, is one of a 'conversation' between the embryo and the mother. Now we need to look at the changes that are occurring to the inner cell mass during the process of implantation.

The layer of cells that covers the inner cell mass on the surface next to the blastocyst cavity becomes distinct from the cell layer beneath it. These two layers are given the names **endoderm** and **ectoderm**, respectively, both of which are distinct from the trophoblast cells that grow into the uterine tissue (Figure 5.2). Now something happens that is essentially a repeat of the process that separated the inner cell mass from the trophoblast, creating a space between them. As the ectoderm cells grow and divide, a cavity appears that grows larger to produce the amniotic cavity which fills with amniotic fluid (Figures 5.3 and 5.4). At the same time a process occurs that is characteristic of all developing animal embryos: the cells of the endoderm, which are also growing, spread over the inner surface of the trophoblast (arrows in Figure 5.4) . We shall find this type of spreading movement occurring repeatedly at different stages of development.

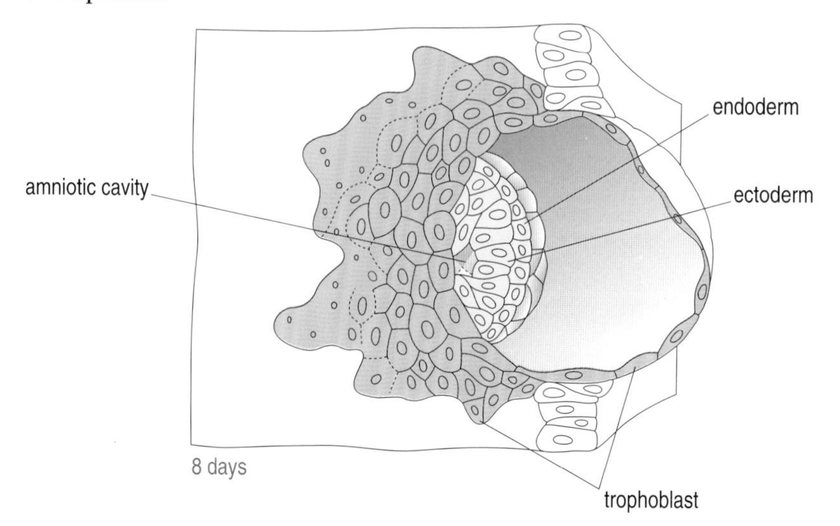

Figure 5.3 As trophoblast cells proliferate and begin to form the placenta on day 8, a cavity appears in the ectodermal layer, which is the beginning of the amniotic cavity.

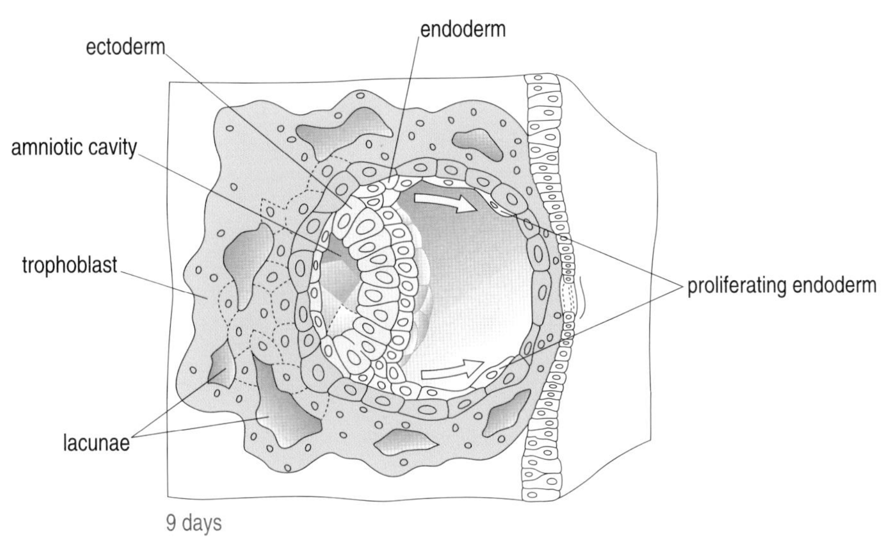

Figure 5.4 The embryo at 9 days has an expanded amniotic cavity within the ectoderm, and the endoderm is proliferating over the inner surface of the trophoblast (arrows). Implantation is complete, the uterine wall now covering the implantation site. Spaces or lacunae are forming in the trophoblast tissue around the embryo that will soon be filled with maternal blood from capillaries in the uterine tissue.

❑ What aspect of wound healing in the skin, described in Chapter 3, Section 3.8.2, depends upon cells spreading over a surface?

■ The epidermal cells spread over the reconstituted dermis, closing the gap made by a wound.

The whole structure is now fully implanted into the lining of the uterus, from which it is drawing nutrients.

Next it is the turn of cells at the outer margin between ectoderm and endoderm to perform the same movement as the endoderm: these cells proliferate and spread over the inner surface of the trophoblast, between it and the newly formed layer of endoderm cells (Figure 5.5, arrows). The inner cavity is now called the yolk sac. The region between the new cell layers is called the *chorionic cavity* which enlarges to generate a space into which the future embryo will grow (Figures 5.6 and 5.7). The cellular membrane surrounding this cavity is called the chorion, while that surrounding the amniotic cavity is the amnion. As shown schematically in Figure 5.8, at the end of the second week of development the embryo is no more than two hollow and somewhat flattened spheres within a fluid-filled cavity, connected by a stalk to a rapidly growing envelope surrounding the cavity. These elaborate preparations create the space and the conditions that will allow the human being to develop from the tiny **germ disc** which is the junction between the two little flattened spheres.

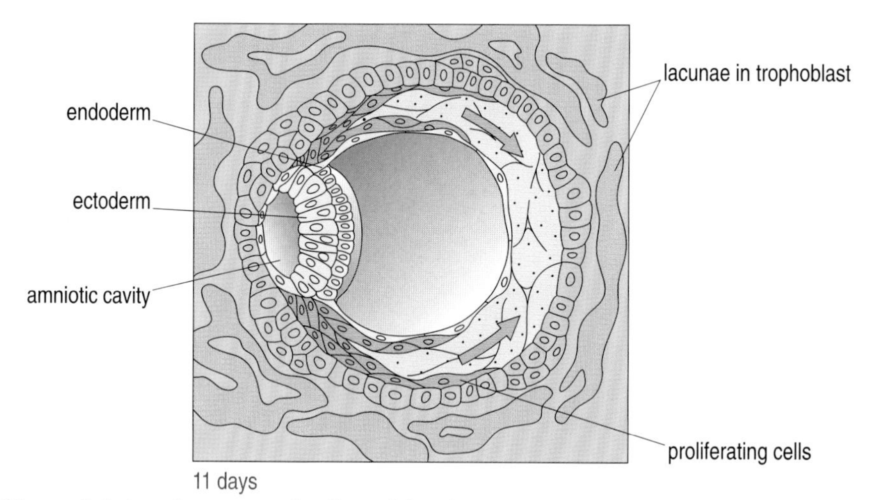

Figure 5.5 Another wave of cell proliferation and migration from the margins of the embryo (shown by arrows) occurs at 11 days, cells spreading over the surfaces between the surrounding trophoblast and the newly formed layer of endoderm cells.

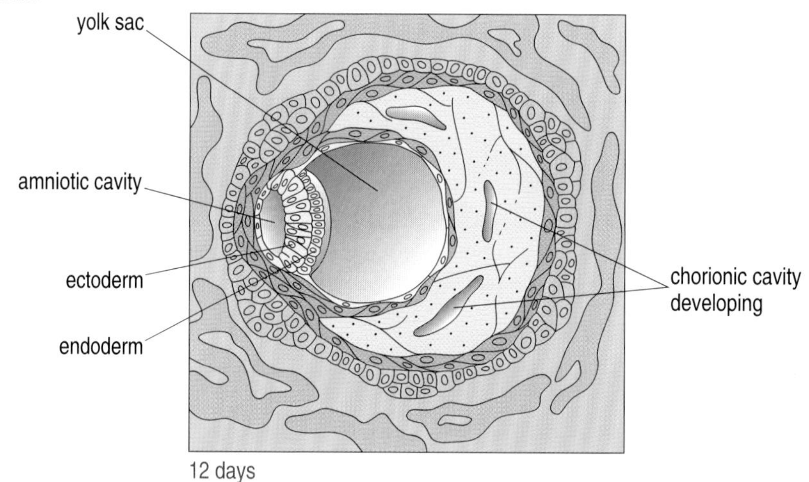

Figure 5.6 The inner chamber next to the endoderm is called the yolk sac, while the non-cellular material between the new cell layers begins to disintegrate to form the chorionic cavity. The embryo is now 12 days old.

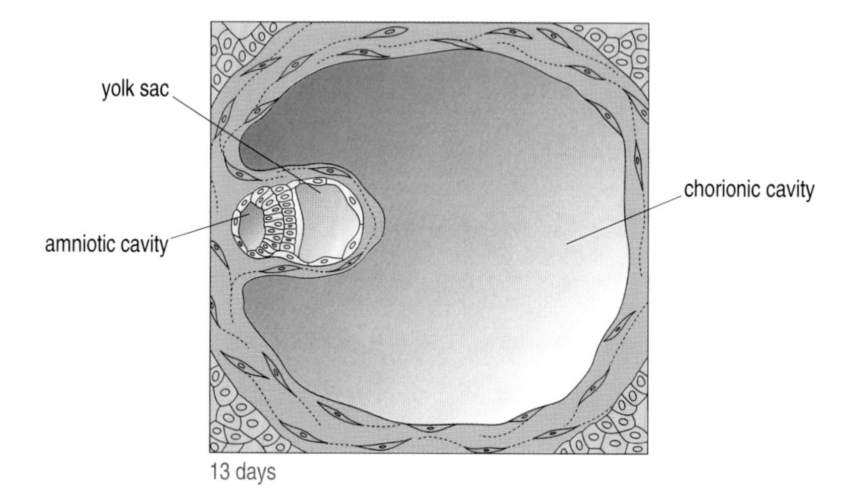

Figure 5.7 By 13 days the chorionic cavity has enlarged greatly relative to the amniotic cavity and the yolk sac which lie on opposite sides of the cell layers that will produce the embryo.

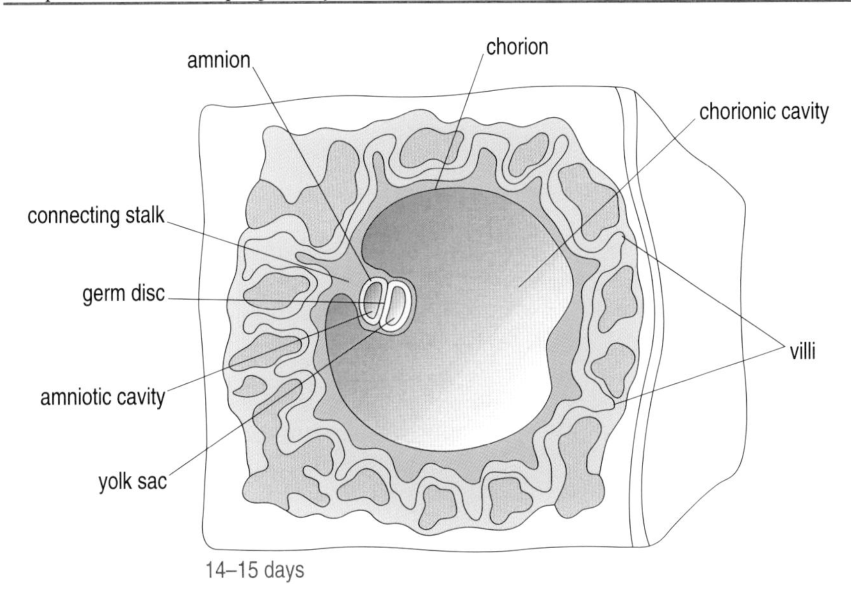

Figure 5.8 By the end of the second week the villi of the placenta are developing rapidly among the trophoblast lacunae, which are filled with maternal blood. The chorionic cavity, surrounded by the chorion, is very large relative to the yolk sac and the amniotic cavity. Between these cavities is the germ disc consisting of two cell layers: ectoderm next to the amniotic cavity and endoderm next to the yolk sac.

Summary of Section 5.2

Human development begins with the transformation of a single fertilized egg into the chambers, envelopes, and structures which establish the conditions for the emergence of a human being during the second week of development. The processes involved may appear complex and intricate, and the array of descriptive terms is formidable; however, the main point to grasp is that these processes depend upon the repetition of a few basic cellular activities: cell division, cell adhesion, separation of cell sheets to form cavities, and spreading of cells over surfaces. Separation of the inner cell mass from the outer layer of cells in the initial spherical ball of cells to form the blastocyst cavity is the first of these. Then the continued division of cells in the inner cell mass results in a new cavity within the mass (amniotic cavity) while the cells lining the mass spread over the inner surface of the trophoblast – the blastocyst cavity now becoming the yolk sac. This growth and spreading continues, with another cavity forming between the cell sheets (the chorionic cavity), while the original inner mass is enveloped by an extension of the spreading cell layer. The result is the structure shown in Figure 5.8. These movements are all shown in Video programme 2.

The placenta forms from cells of the trophoblast as implantation occurs. The uterus must be in the right state to participate in implantation and to continue the pregnancy, which depends upon hormones such as oestrogen and progestogen. These are maintained by a 'conversation' between the placenta, the corpus luteum, and other maternal hormone responses. The uterine cells disintegrate in response to trophoblast proliferation, releasing nutrients that are used by the developing embryo. A massive blood system develops in the placenta in which maternal and fetal blood remain separate but in intimate contact, allowing transfer of nutrients and oxygen to the embryo, and wastes to the mother.

5.3 Weeks three and four of development

So far there is nothing that looks remotely like an organism in the developing embryo, let alone a human being. However, during the next few weeks of development the tiny, flat germ disc between the two flattened spheres, no more than a fraction of a millimetre in size, will transform into a recognizable organism with head, tail, the beginnings of limbs and eyes and all its internal organs, as shown in Figure 5.9. At five weeks the human embryo, now 8 mm in length, could be mistaken for an embryonic newt, chick, or sheep, so similar is the basic plan of the body in these different species. Not until another three weeks of development have passed (week 8) is the unique form of the human being clear and distinct in gross structure from its evolutionary relatives.

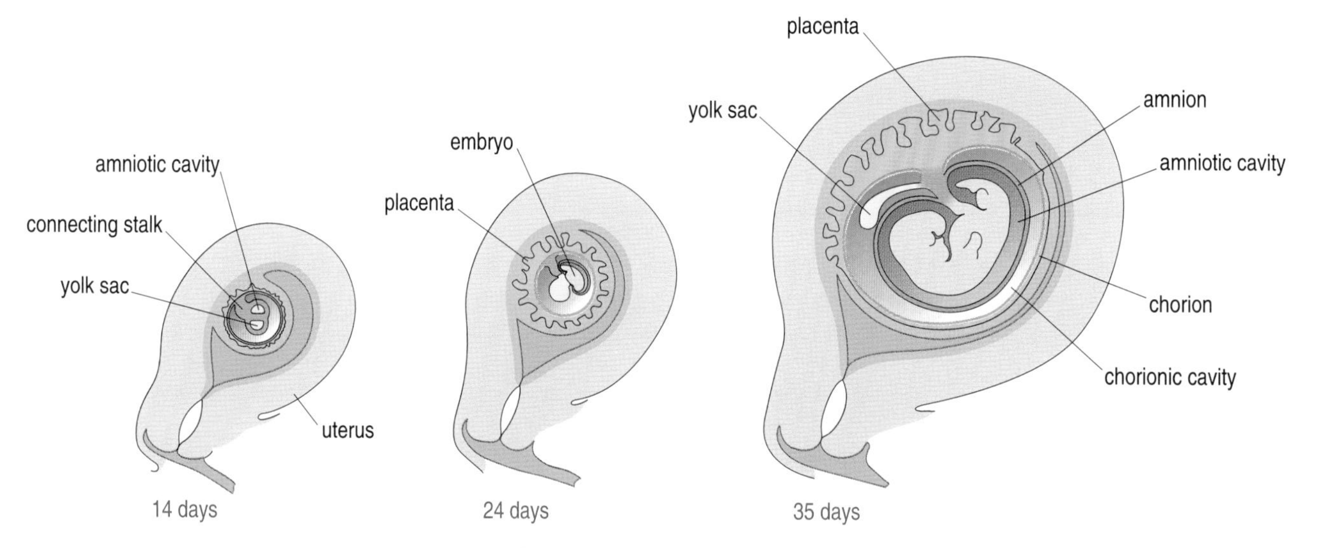

Figure 5.9 The gross changes that occur in the form of the embryo from weeks 2 to 5 are shown schematically, illustrating the growth of the placenta, the development of the body from the germ disc, the membranes surrounding the embryo (amnion and chorion) and the cavities they enclose, and the uterus.

There is no need to examine in great detail the processes that transform the germ disc into a human being. However, it is important to get a feeling for the main types of cell movement that occur in producing the different structures of the human body. The reason for this is twofold.

First, as already mentioned, the embryo repeats the same types of activity over and over again during development, but as the context changes so do the consequences. Once you see these repetitions or iterations, the whole process whereby progressive complexity of form arises becomes much easier to grasp. It is rather like oregami, the Japanese art of paper-folding, whereby a complex form can be generated by repeating similar types of simple folding operation. Slight changes in the pattern of folding also produce a variety of different forms, just as different species of vertebrate arise by initially small differences which get exaggerated as development proceeds. Embryos are experts at the folding art, to which they add the capacity of cells to flow over surfaces and to migrate individually.

The second reason for understanding basic human embryology is that it provides important insights into fundamental questions about human health and disturbances to it. Why do environmental influences affect developing embryos in particular ways at particular stages of pregnancy? How do genes act during development, and can their effects be altered? How are twins produced and what can we learn about human heredity from the study of twins? So, although human development is intrinsically fascinating in its own right, our main reason for looking more deeply into this process is to gain insight into the nature of the processes that have produced us in the first place, in order to see how an individual life is an unbroken whole from conception to death, each stage of which is of equal importance for expressing the unique attributes of a person and realizing the fulness of a life, a fundamental aspect of which is health.

On about day 15 of development, a faint groove called the *primitive streak* appears along the midline of the germ disc (Figure 5.10). This is the first sign of any structure within the embryo itself. It defines the central axis of the future human being, where the backbone will form.

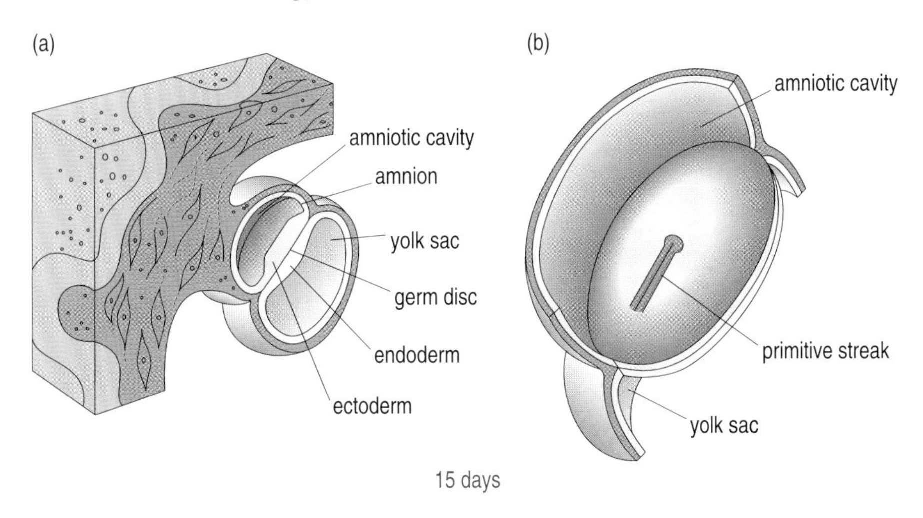

Figure 5.10 The first detailed embryonic structure to form in the germ disc is the primitive streak, appearing on day 15. (a) A section through the embryo showing the location of the germ disc between the amniotic cavity and the yolk sac. (b) A more detailed view of the germ disc showing the primitive streak forming in the centre.

The two layers of the germ disc, the ectoderm on top and the endoderm beneath, are doing something familiar: they are separating in the middle, forming a cavity which causes the disc to bulge (Figure 5.11). Cells move from the centre of the bulge into this cavity, so producing the primitive streak. As they move into the cavity, these cells spread out, moving between ectoderm and endoderm, as shown in Figures 5.11b. This crucial process is called **gastrulation**, and creates a third cell layer in the embryo – the **mesoderm**. The three **germ layers** (ectoderm, mesoderm and endoderm), as they are called, now start to interact with one another by signals that pass between them, causing further changes of structure and more complex interactions. The basic processes involved are like those we encountered in the skin, in Chapter 3, where cells organize themselves into layers and 'talk' to each other by chemical signals, inducing their neighbours to make particular proteins and to assume particular shapes – cylindrical or cuboidal or flattened – with either a tendency to migrate or to stick together, depending on their adhesiveness. Genes in cells get switched on or off according to the signals received from their neighbours, and the products of gene activity influence the properties and behaviour of the cells. Let's now look at the way this cascade of developmental events proceeds in the production of a human body.

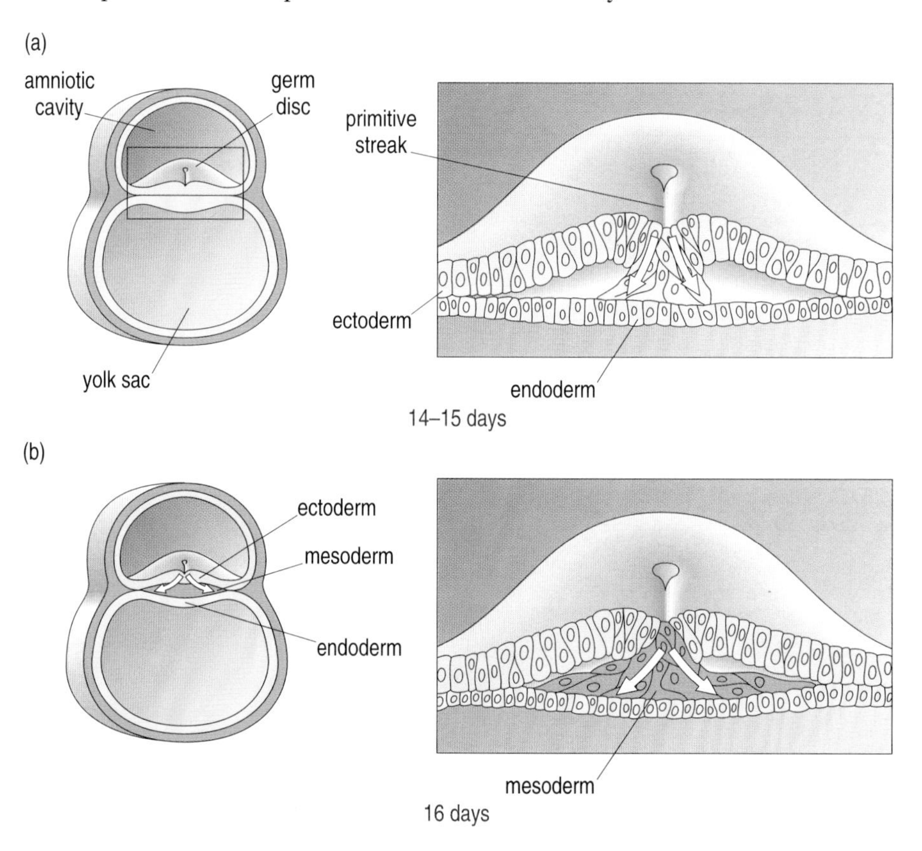

Figure 5.11 The primitive streak is formed by the inward movement of ectoderm cells into a cavity that forms in the middle of the germ disc, between ectoderm and endoderm, the cells spreading out in this cavity (see arrows) to form a new germ layer in the embryo, the mesoderm. This process, which begins on about day 15, is called gastrulation.

5.3.1 Formation of the body axis and the nervous system

Referring again to Figure 5.11, cells flow from the ectoderm into the cavity along the midline of the germ disc and spread out, forming a new layer of cells, the mesoderm. The cells that occupy the central axis of the new layer under what was the elongating groove stick tightly together due to increased adhesion, separating from their neighbours and producing a rod-shaped structure called the **notochord**, which is the beginning of the backbone (Figure 5.12: day 17). Then the cells on either side of the notochord also become adhesive and stick together, but the process doesn't go so far as to result in complete separation of groups of cells from their neighbours. They round up into mesodermal aggregates but remain in contact (Figure 5.13: day 18). As this occurs, the ectodermal cells on top begin another characteristic movement: under the influence of the underlying mesoderm, they bulge up on either side of the midline and begin the process that will produce the nervous system.

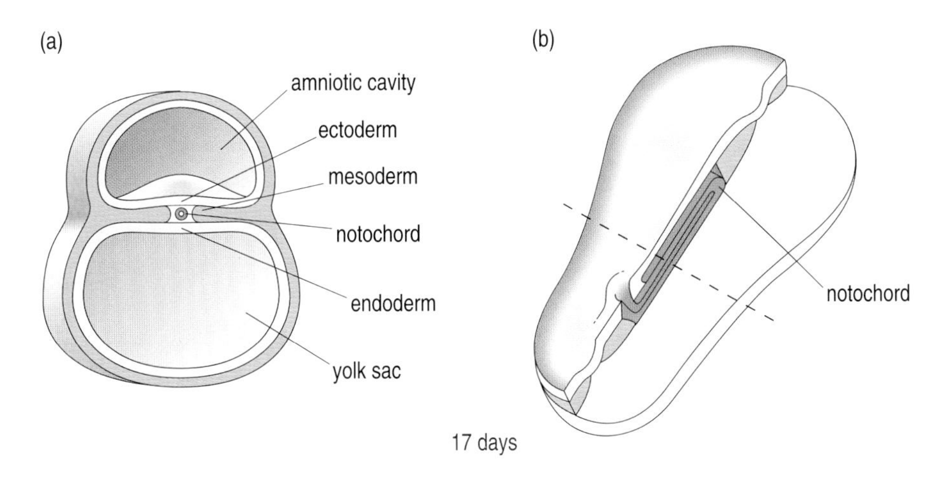

Figure 5.12 On day 17, the notochord forms as a rod produced by highly adhesive mesoderm cells along the central axis of the developing embryo, and is the precursor of the backbone. (a) A section through the amniotic cavity (above) and the yolk sac (below), with the notochord seen as a ring in the centre of the germ disc. In (b) the germ disc is shown in perspective, with the rod-shaped notochord running down the middle. The dashed line shows where the section is made to give the view of the notochord shown in (a).

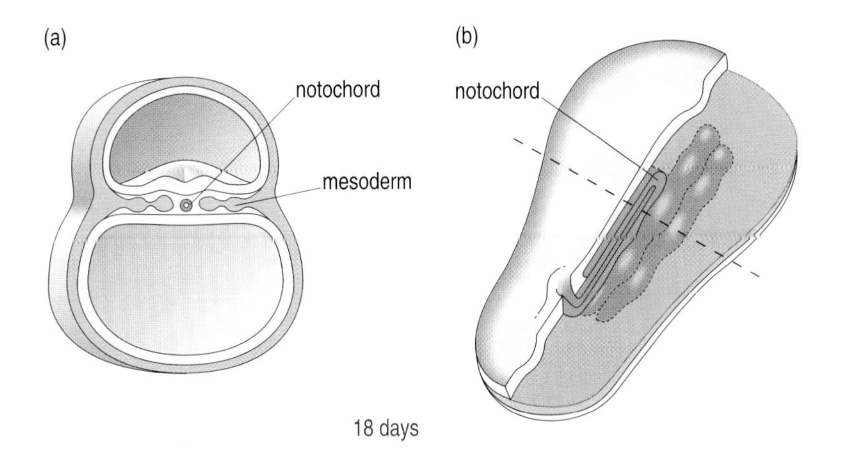

Figure 5.13 On either side of the notochord, mesoderm cells form aggregates. Above them, the ectoderm begins to buckle into the neural folds or ridges, beginning the process that produces the nervous system. As in Figure 5.12, the dashed line in (b) shows where the section is made to give the view shown in (a).

The next stage in this process is shown in Figure 5.14: by day 21, the mesodermal aggregates have developed into a series of bead-like structures (**somites**), which lie on either side of the notochord, forming two rows down the elongating axis. These will develop into the muscles that lie on either side of the backbone, resulting in a repeating or *segmented* structure that is also shared by the nervous system (Book 2, Chapter 3). The overlying sheet of ectodermal cells folds into a distinct central groove, called the neural groove (neural means relating to the nervous system), shown in the cross-section in Figure 5.14a. Here you can also see the hollow central region of the bead-like somites. The upper part of the neural folds come together and fuse along the midline of the embryo, producing the **neural tube** (or neural canal) from which the brain and spinal cord develop As shown in the sequence of three-dimensional images in Figure 5.15, the neural folds meet in the midline of the embryo and fuse, first in the middle of the embryo, which is where the upper neck will be located. Fusion spreads forwards to produce the beginnings of the head and backwards to produce the spinal cord. The back muscles or somites become prominent as bulges on either side of the neural tube as closure occurs. A side view of the embryo at 26 days (Figure 5.16) shows something that is recognizably an animal, with a head and a tail, though it is not yet so clear that this is going to become a human being.

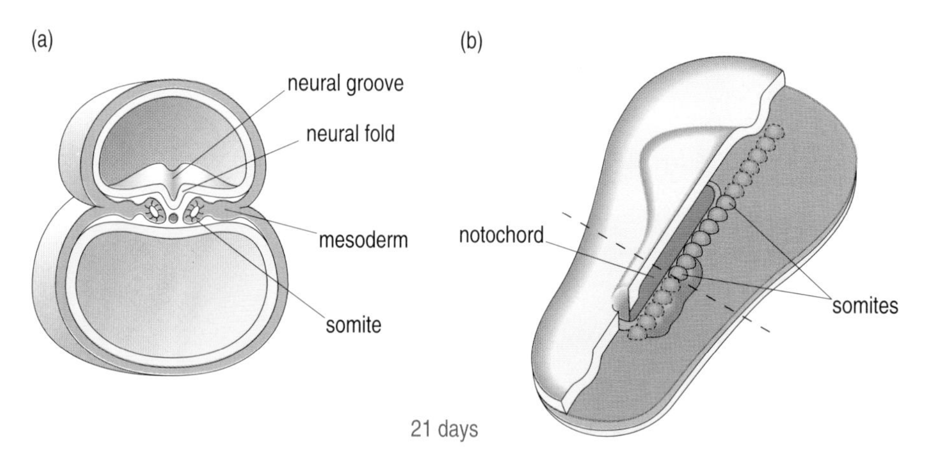

Figure 5.14 The mesodermal aggregates on either side of the notochord round up, forming rows of bead-like structures called somites which develop into the muscles associated with the backbone. As they form, the overlying ectoderm deforms further to produce a central neural groove flanked by neural folds, precursor of the nervous system. Again, the dashed line in (b) shows the plane of section of (a).

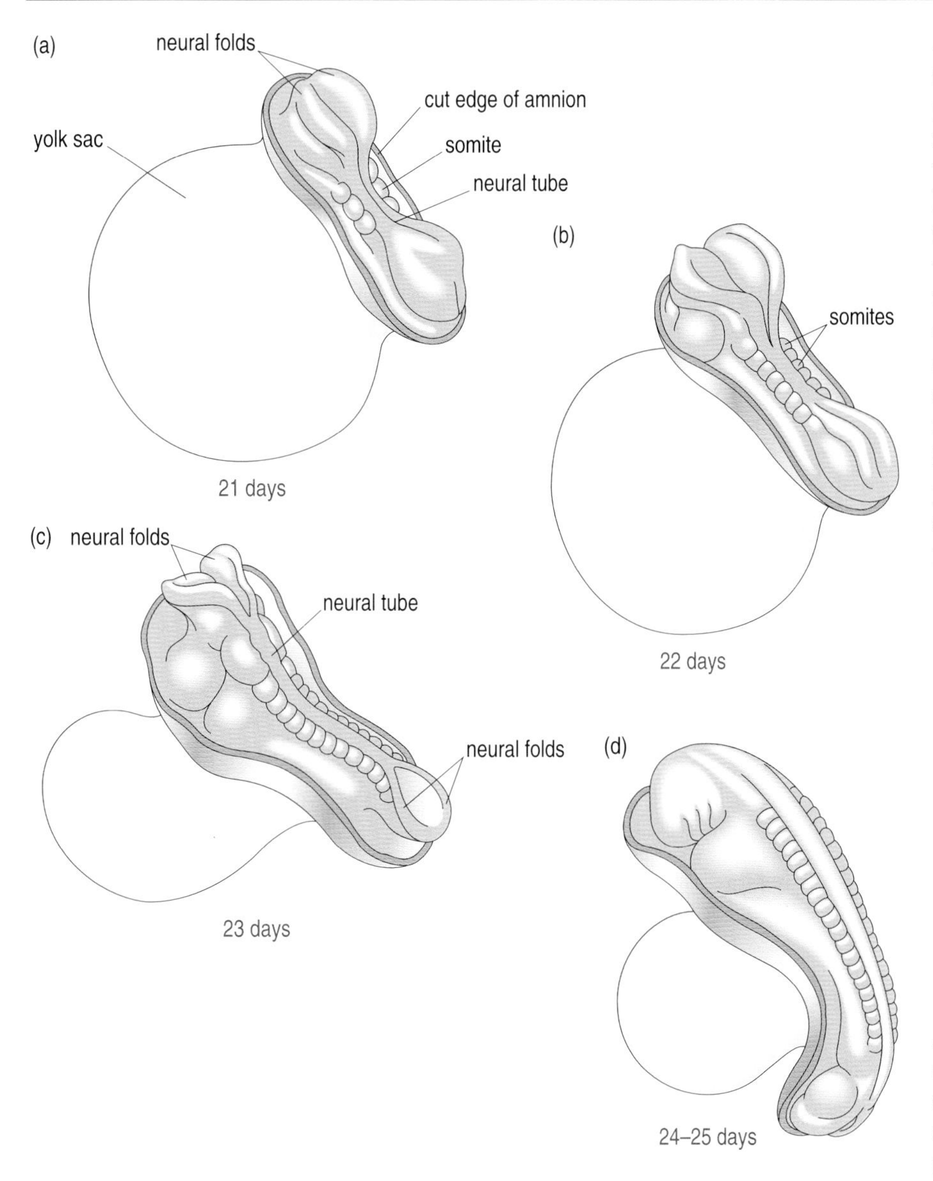

Figure 5.15 The neural folds first meet and fuse on day 21 to produce the neural tube in the middle of the embryo (a), which is where the upper neck will be located. Fusion then proceeds towards the top (anterior) end, where the brain will form, and towards the bottom (posterior) end, where the spinal cord develops (b and c), with growth and elongation resulting in the formation of a curving tail, shown in (d). The somites develop as paired bead-like structures on either side of the neural tube. The yolk sac lies beneath the embryo.

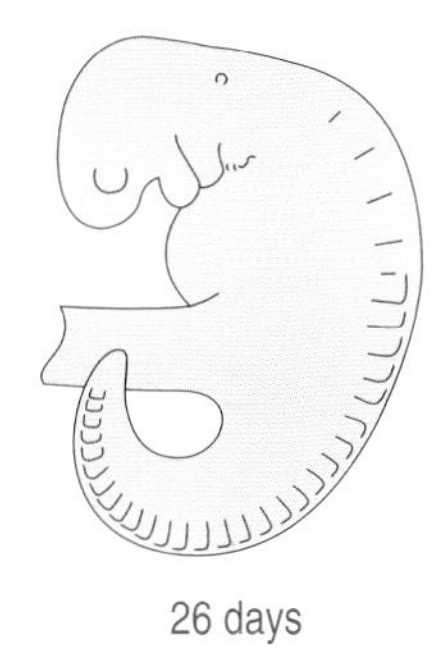

Figure 5.16 The external form of the human embryo at 26 days showing head, back, tail, and the somites running alongside the neural tube. (actual size $\times 6$)

5.3.2 Genetic and environmental influences on axial development

The events just described, occurring during the third and fourth weeks, are crucial for development of the human embryo, especially of the nervous system and the basic structure of the body – the backbone, muscles, and limbs (to be described later). The interactions which occur between the cell layers during the formative movements of gastrulation are essential for normal **morphogenesis** (the emergence of form or structure). This is therefore a critical, sensitive period when either defective genes or adverse environmental influences can cause the developing embryo to go off course, resulting in a spectrum of disturbances from minor to major. Some of these are readily corrected while others result in more serious defects and can be lethal.

One of the processes that is sensitive to genetic and environmental influences is the closure of the neural tube at either the head (anterior) or the 'tail' (posterior) end at the stage shown in Figure 5.15c. Failure of posterior closure leads to a condition called **spina bifida** (divided spine). Mild cases can be corrected by surgery to the baby after birth, involving skin grafting to cover the exposed tissue, but more severe cases can be fatal. The term used to refer to disturbances of neural tube formation, whether of the brain or the spinal cord, is **neural tube defects** (NTDs). As long ago as 1965 there was a report that a good diet and adequate vitamins, particularly folic acid, are significant in preventing these defects. It was known that a woman who had already borne a child with a neural tube defect had a 5% chance of bearing another, whereas the incidence in the population as a whole is about 1 in 500 (0.2%).

❑ What are the possible factors that could account for the higher rate of occurrence of NTDs in certain women?

■ It could be diet, as already mentioned; it could be a genetic factor; or it could be some combination of the two.

A trial was conducted on a group of 111 women who had borne a child with an NTD and who intended to become pregnant again (Laurence *et al.*, 1981). They were divided into two groups. Those in one group received folic acid in pills which they took from the day they ceased contraceptive precautions (60 women), while those in the other group received pills that contained no folic acid (a placebo). The women did not know to which group they belonged, nor what vitamin was being tested. This means that a group of women did not receive the potential, although uncertain, benefit of folic acid, though they had been consulted before and had agreed to participate in the trial.

❑ What is your view of such trials, which are designed according to recognized scientific procedures to eliminate any psychological influence from the results?

■ You might take the view that if there is some evidence that a vitamin such as folic acid can reduce the incidence of NTDs then it should be made available to *all* pregnant women and that scientific tests that may put some children at risk are unethical. On the other hand, you may believe that obtaining accurate scientific evidence is essential for establishing reliable medical treatments and that this justifies experimental procedures that may deny some children potential benefits, as long as the mother understands the situation and agrees to the procedure.

The results of the trial strongly supported the hypothesis that folic acid significantly reduces the recurrence of NTDs. None of the women who complied with the procedure of taking folic acid bore a baby with an NTD, whereas in the other group there were six recurrences. A larger trial reinforced this result, showing that folic acid reduces the risk of recurrence from 5% to between 0.28 and 0.17%. A good diet with plenty of green leafy vegetables, an excellent source of folic acid, is normally quite adequate to provide the nutrition required to largely prevent NTDs, providing this is taken prior to and during pregnancy. But to be sure, folic acid supplement is now recommended. However, there appears to remain a residual genetic effect in a small proportion of the population that cannot be eliminated entirely by diet.

This example shows that good nutrition can reduce the appearance of defects that may involve genetic tendencies. Human disorders result from some mixture of genetic and environmental influences, and it is estimated that less than 2% of human diseases can be attributed primarily to genetic causes. However, the rare instances in which defects can be traced to direct genetic factors, in humans or in other species, can be very informative about basic causes of the disturbance. Here is one.

There is a mutation in mice called *curly-tail* which mimics closely spina bifida in humans. The condition is due to delayed closure of the neural tube. It was discovered in the late 1980s by Andrew Copp and colleagues, working in Oxford, that this delay is associated with changes in the relative rates of cell division in the three germ layers of the embryo. The mutant has a reduced rate of cell division in the lower layers (endoderm and mesoderm), whereas the ectoderm cells divide at the normal rate. Rate of growth and elongation of the tissues along the head-to-tail axis is primarily dependent on cell division rate.

❑ What would you expect to be the result of this imbalance in the elongation of ectoderm in the mutant (given that the layers are fused together)?

■ The more rapidly dividing ectoderm cells, producing more rapid elongation of the upper layer of cells, would tend to cause the ectoderm to curve around the more slowly elongating lower layers, producing a curly tail.

A further result is that the abnormal curvature at the posterior end of the embryo tends to pull the neural folds apart, resulting in delayed closure of the neural tube and the occurrence of spina bifida in the more extreme cases. If this is also a cause of the condition in humans, then it becomes clearer why many different factors, genetic and environmental, can contribute to the condition. Cell division is a very basic process that is influenced by a host of disturbing factors so that no single cause can be identified.

Summary of Section 5.3

The first sign of a body to emerge in the germ disc is a central groove which elongates. This results from the inflow of cells from the surface to form a new cell layer, the mesoderm. Down the centre of this new layer, under the

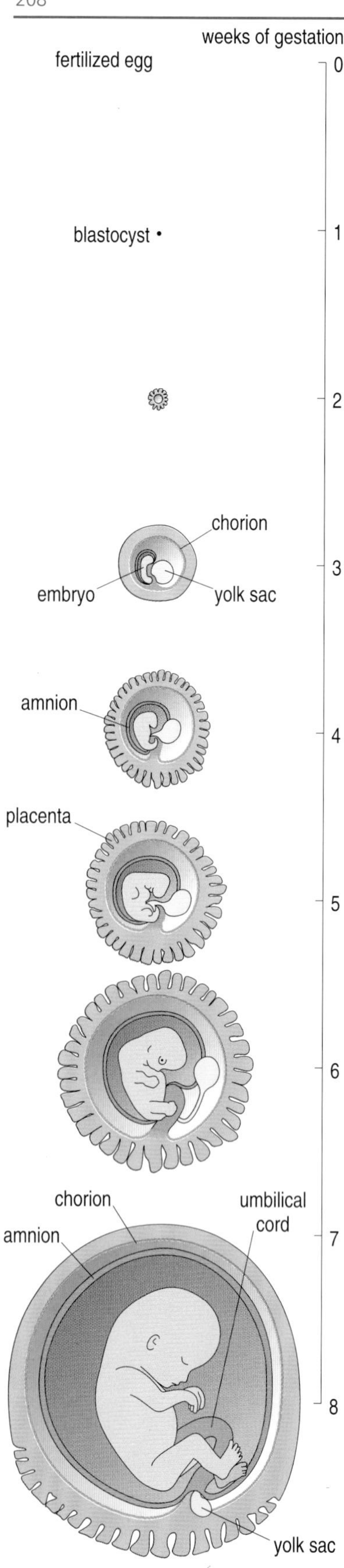

initial groove, cells stick firmly together to form the notochord, which will become the backbone. Paired somites, which develop from the mesoderm into the back muscles, appear on either side. As they form, the overlying cell sheet bulges up and folds together to produce the neural tube, from which the nervous system develops. Fusion of the folds starts in the region that will form the upper neck and proceeds towards the head and the tail. As this 'zipping' action proceeds, the somites become more prominent, appearing like paired beads on either side of the neural tube.

Genetic and environmental influences can disturb the formation of the body axis and closure of the neural tube towards the head or the tail, resulting in abnormalities of varying degrees of severity. These influences include genes that appear to act on basic cellular properties such as cell division rates. Amongst environmental effects, folic acid (a vitamin) substantially reduces the recurrence of neural tube defects. Many other factors can affect early fetal development, genes and environmental influences interacting in various ways so that there are rarely single, simple causes of birth defects.

5.4 Formation of the circulatory system

The tailed vertebrate of Figure 5.16 undergoes a dramatic series of transformations over the next four weeks of development, shown in Figure 5.17, resulting in the distinctive form of the human at eight weeks. The brain grows dramatically, eyes, nose, ears and mouth appear, the limbs develop and the tail gets smaller, and eventually disappears. Inside the embryo more dramatic changes take place, with the emergence of all the major organs – heart, lungs, liver, kidneys, sex organs, stomach and intestines (both of which make up the gut). As the embryo grows and develops, so does the enveloping placenta and the membranes that surround the embryo. So that you can appreciate the changes that occur over the whole eight-week period, Figure 5.17 shows the actual size from the blastocyst to the fetus. (The egg is actually invisible at this scale.) By week 8 the embryo has developed the distinctive characteristics of the human species. All the major organ systems have appeared, though not all are functional and many still have a lot of growth and complex differentiation to undergo. The same basic principles of morphogenesis are followed by all the organs, despite great differences of detail. A general outline of further development will be presented, with some focus on two systems – the circulation of the blood (in this section) and the development of the limbs (in Section 5.6) – to illustrate further the connections between embryonic development and health.

Figure 5.17 The changes in size and external form of the developing human and of the structures outside the embryo, from fertilization to eight weeks of gestation.

You have seen in many of the earlier figures a large cavity lying next to the germ disc where the embryo develops. This is the yolk sac, about which little has yet been said. As you might expect from its name, the yolk sac plays a nutritional role in early development. However, nutrition of the embryo cannot occur without a means of delivering nutrients and removing waste products, which is a primary function of the circulatory system. It is the yolk sac which is initially responsible for producing blood and contributing to the development of a system for circulating it through the embryo.

By the time that the embryo has reached the external form shown in Figure 5.15d (24–25 days), the circulatory system has developed to the extent shown in Figure 5.18. Here you can see the relationship of the yolk sac to the embryo and the continuity of the blood vessels between them. The intermingling of the yolk sac vessels with those connecting the embryo to the placental circulation is evident: they form a single, enclosed system, despite the extensive branching of the vessels in the outer layer of the placenta that results in intimate proximity to the mother's circulation (not shown). Let us now look a little more closely at the functions of the placenta before returning to the circulatory system and the formation of the heart.

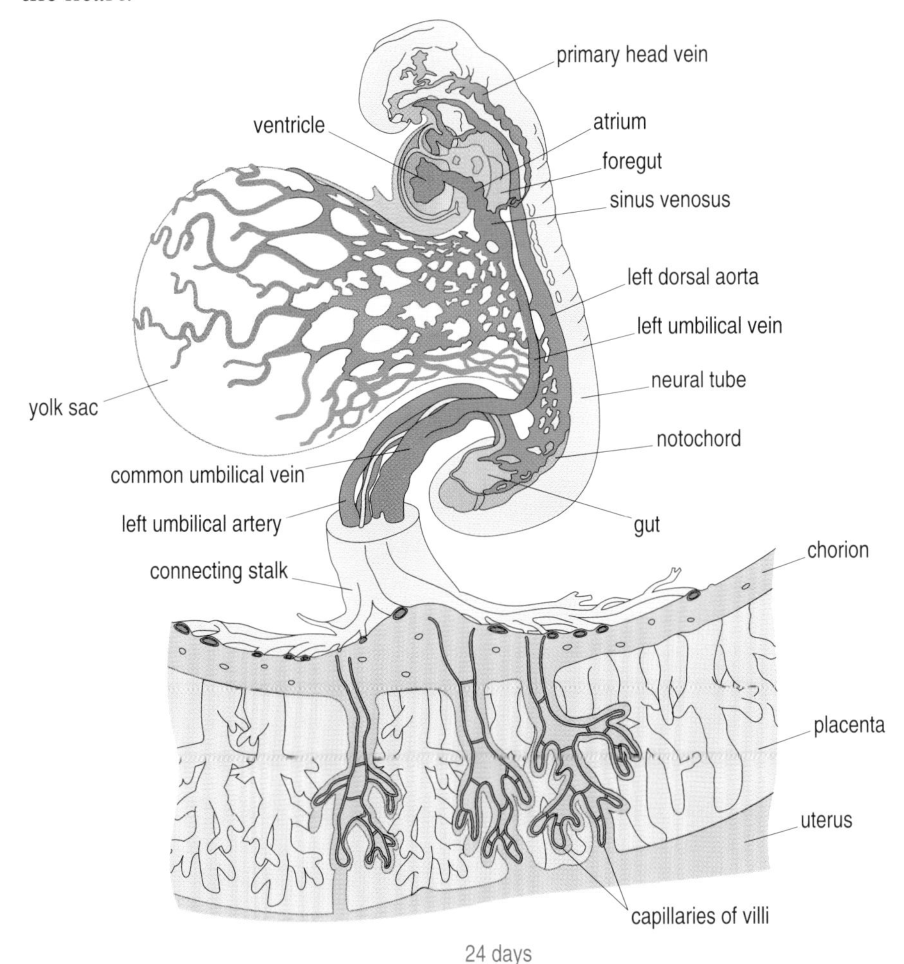

Figure 5.18 The circulatory system at 24 days of gestation, showing the relationship between the blood vessels of the placenta and the yolk sac to the developing embryo. Oxygenated blood (shown in red) from the placenta flows to the embryo via the common umbilical vein which then divides into left and right branches (only the left one is visible here). These deliver blood to the yolk sac and then fuse to form the sinus venosus. From this enlarged vessel, blood flows into the single primitive atrium of the early heart and then into the single ventricle, both of which have begun to beat. Blood passes out of the ventricle into the single truncus arteriosus (not visible here; see Figure 5.19) from which arise paired left and right aortic arches. From these, vessels branch out to irrigate the developing embryonic tissues and organs, and blood travels through the paired dorsal aortas, collecting blood from the yolk sac, before flowing into the umbilical arteries (deoxygenated blood is shown in blue). These carry blood back to the placenta, where it gets reoxygenated by the mother's circulation (not shown).

5.4.1 Interactions between fetal and maternal blood systems

As described in Section 5.2.1, the placenta plays a major role in supplying nutrients to the developing embryo and removing its waste products. If the embryo is not receiving enough of the essentials for its metabolism (nutrients and/or oxygen) because the mother is undernourished and/or anaemic, the placenta has the capacity to grow larger, increasing the area of intimate contact with the mother's circulation and so obtaining more of these basic materials. The developing embryo is thus provided for at the mother's expense.

❑ What is the hormone produced by the placenta that maintains the pregnancy?

■ Chorionic gonadotropin.

A further function of the placenta is to act as a screen for the embryo against harmful agents that might be present in the mother's blood.

❑ Can you suggest what these harmful agents might be?

■ Disease-causing organisms (pathogens) and harmful chemicals.

This is one reason why it is essential that maternal and fetal blood should not be in direct contact. The existence of a barrier of a few layers of cells means that many organisms, and even some large molecules, are just too big to pass from one side of the placenta to the other. The placenta does a marvellous job of protecting the embryo from these agents. However, it is not perfect, and sometimes harmful agents *can* get through and damage the embryo.

❑ From general knowledge, can you name one common infectious disease and one drug that can cross the placenta?

■ The German measles, or rubella virus is the most well known. Drugs that can cross the placenta include alcohol, nicotine and thalidomide.

The rubella virus can easily pass across the placenta. If this happens during the first three months of pregnancy it can seriously damage the embryo, leading to congenital abnormalities such as deafness or blindness, even death. (**Congenital** means a condition that the baby is born with; it includes, but is not limited to, inherited conditions.)

Although there are some harmful agents that *do* get across the placenta, it is fair to say that an important function of this organ is to protect the embryo from a possibly hostile environment. It is tempting to call this the 'external' environment, but this would not be accurate: much of the threat to the embryo comes from the mother herself!

❑ Why should this be the case?

■ As far as the mother is concerned, the embryo is 'non-self', and therefore a candidate for attack by the mother's immune system.

As described in Section 4.7 of Chapter 4, the developing embryo is a new individual with a distinct biological 'signature' in the form of recognition molecules that are different from those of the mother, since they are made by the embryo's genes which are a mixture of maternal and paternal genes. The question of why the embryo is not destroyed by the mother's immune system, but tolerated throughout the pregnancy, is a very interesting one. It is not the case that the mother does not 'notice' the embryo (in an immunological way!): interaction with the blastocyst is essential to *establish* the pregnancy in the first place. Nor is it the case that the embryo is not recognized as non-self: if pieces of embryonic tissue other than the trophoblast are transplanted to other sites in the mother, they will be rejected straight away, whereas transplanted trophoblast cells are *not* rejected. The trophoblast is 'privileged' in some way that protects it from the immune system. It seems likely that fetal recognition molecules are not present on the surfaces of the trophoblast cells, so that these cells, which are in intimate contact with the mother's circulation, are not recognized as being non-self. Antibodies made by the mother's immune system normally circulate in the blood, but since maternal and fetal blood are separated, there is little chance of the antibodies encountering embryonic cells. However, like a lot of substances which seem to be able to 'leak' through the placenta, certain kinds of antibodies *can* pass from the mother to the embryo via the placenta. These protect the embryo against many of the pathogens that might cross and cause damage, and is another function of the placenta.

However, there is a major exception to the general rule that maternal antibodies and embryonic cells do not meet, and this concerns a protein sometimes present on red blood cells, called the rhesus factor. People who have the rhesus molecule on their blood cells are referred to as 'rhesus-positive'; those lacking it are called 'rhesus-negative'. If the mother is rhesus-positive, she will recognize any rhesus molecules as 'self', and will not produce antibodies against them. But if she is rhesus-negative, any encounter with the rhesus molecules will provoke an immune response.

❑ What would you predict to happen if the mother was rhesus-negative, but the embryo was rhesus-positive?

■ If there was any contact between the mother's immune system and the rhesus molecules on an embryo's red blood cells, the mother would make anti-rhesus antibodies which might cross the placenta and attack the embryo.

Indeed, this is exactly what happens in some cases and, where the immune response of the mother has been particularly severe, most of the embryonic blood cells may be destroyed. This is likely to cause the death of the embryo. The rhesus problem is usually not evident with the first rhesus-positive baby of a rhesus-negative mother, because there is little, if any, escape of fetal blood cells until late in pregnancy. But if the mother has a subsequent child who is rhesus-positive, then the immune response will be rapid and large. This is because the immune system adapts to non-self molecules it has encountered before, and makes a stronger response the second time around. (The immune system is described in detail in Book 2, Chapter 5.)

We have now covered all the major functions of the placenta and these are listed below.

1 To supply the developing organism with nutrients.
2 To remove embryonic and fetal waste products.
3 To produce hormones to maintain the pregnancy.
4 To screen out most pathogens and harmful chemicals.
5 To provide the embryo and fetus with some antibodies that protect it from pathogens.
6 To protect the embryo and fetus from immunological attack by the mother by means of a layer of trophoblast cells which do not carry fetal identification molecules.

5.4.2 Factors involved in generating the circulatory system

The circulatory system shown in Figure 5.18 is in two colours, following the accepted convention: red for the oxygenated blood (from the placenta to the fetus) and blue for deoxygenated blood (from the fetus back to the placenta). The heart is not yet defined as a separate, distinct organ, being no more than a tube within which are slightly enlarged and thickened regions; these will later become the chambers of the mature heart. Two of these are visible in Figure 5.18, labelled **atrium** and **ventricle**. They have already begun to beat. The paired umbilical arteries (left and right – but only one visible in the figure) conduct blood from the embryo to the placenta, while the common umbilical vein carries the blood from the placental vessels to the embryo, splitting into right and left umbilical veins which deliver oxygenated blood to the yolk sac and to the tubes of the future heart.

After birth, blood flow from the placenta ceases and the baby's lungs take over the function of oxygenating the blood, as described in Chapter 6 (Section 6.8).

How are we to understand the development of a complex system such as the blood vessels and the heart? Is the design all in the genes, or is there another source of order that contributes to the formation of embryonic structures?

Notice the system of blood vessels that covers the surface of the yolk sac. It looks like a river system in a flat delta just before emptying into the sea. A river flows in channels which it makes for itself under the action of gravity as it flows downhill. The curving, meandering pattern is an expression of the properties of liquids: they naturally form wavy patterns like a snake in motion, and at the same time they find paths of least resistance for their flow.

The blood produced by the yolk sac first forms as little pools of blood, enclosed by cells, that gradually coalesce into continuous blood-filled tubes which curve and meander like a slowly flowing river. But the force that moves the blood in one direction isn't gravity but arises from the elasticity of the little tubes, the blood vessels.

❑ What gives the cells their elastic properties?

■ The presence of molecules such as collagen and elastin, as in any connective tissue.

As blood forms and fills these vessels, they expand but then resist, exerting pressure on the blood. This process of expansion and elastic resistance has a natural tendency to become rhythmic as blood moves along the system of tubes, a process called *peristalsis* which is also the way that fluids move along another important body tube, the gut (see Book 3, Chapter 3). The whole system begins to exert a kind of gentle pumping action on the blood. The larger vessels experience this periodic flow more strongly because of the greater volume of fluid. The cells making up the walls of these vessels differentiate in response to this stimulus, producing more elastin and collagen. The vessels therefore become better able to resist the pressure and exert a greater rebound force after a pulse of pressure expands the vessel. The primitive atrium forms where there is the greatest volume of blood flow, the blood flowing from the placenta via the umbilical veins being joined by blood from the head, which converges through the sinus venosus into the first heart chamber, the atrium (see Figure 5.18). At first, the heart chambers are just swellings in the conducting tube to accommodate the blood volume. However, if the volume of liquid flowing in a tube is sufficient, it will start to swirl, forming a vortex which is another natural tendency of liquids. So the large blood vessel begins to curve and grow larger under this swirling influence and the characteristic shape of the enlarging ventricle emerges. The ventricle then becomes the main source of pumping activity, accounting for 80% of the power in each beat of the mature heart, while the elastic properties of the blood vessels continue to play an important role in the circulatory process.

Genes contribute substantially to all of these processes by amplifying and consolidating the natural tendencies of blood flow and elastic properties of tubes, helping in the reliable production of a coherently organized whole. The increasing elasticity of blood vessels in response to increased blood pressure is due to the synthesis within the cells of the blood vessels of fibres such as elastin and collagen, which we encountered in dermal cells in Chapter 3 (Sections 3.8 and 3.9). This depends on gene activity, producing more elastic proteins in reponse to pressure signals. The formation of the heart involves a great deal of specific gene activity and cell differentiation to produce the detailed structure of its different parts. However, throughout this process there is an intimate interplay between physical forces and biological processes. These processes, involved in the development of the blood vessels of the body, could account for the link between placental to birth weight ratio and later hypertension (high blood pressure), as proposed by the Southampton Environmental Epidemiology Unit in their programming hypothesis (see Chapter 2).

❑ What is the programming hypothesis that links the embryonic and fetal experience of babies whose placental to birth weight ratio is large, to the development of hypertension in later life?

■ The proposal is that a large placenta relative to body size indicates conditions of low oxygen (hypoxia) or reduced nutrient supply during pregnancy, the placenta compensating by increased growth with the result that it accumulates these essentials more effectively. Under these conditions of deprivation the blood is preferentially shunted to the developing brain and the body arteries experience reduced blood pressure. These arteries then fail to develop normal elasticity, remaining relatively thin-walled and inelastic, which makes the individual prone to hypertension in later life.

Of course, individuals vary greatly in their capacity to recover from physiological stress during development, some never becoming hypertensive, while others do.

Summary of Section 5.4

The embryonic circulatory system forms independently of the mother's and remains separated from it, despite the intimate relationship which the two systems have in the placenta. Embryonic blood forms initially in the yolk sac and the circulatory system develops as an interconnected network of elastic vessels whose shape and behaviour reflect the properties of blood as a fluid. This is evident in the sinuous patterns of the smaller vessels and the rhythmic expansion and contraction that arises from blood flow and the elastic properties of the vessels, due to the presence of collagen and elastin fibres (as in connective tissue). The heart develops from enlarging regions of the system where there is convergence of the blood brought in the veins from the placenta and the head. The heart develops its characteristic shape by twisting and folding in response to the swirling vortices of the blood, together with the coordinated activities of genes.

Hypoxia or malnutrition of the mother results in compensatory overgrowth of the placenta and a diversion of blood to the brain at the expense of the body. As a consequence, the blood pressure in the vessels of the body is reduced, which is likely to result in a failure of these vessels to develop normal elasticity. The vessels then tend to be thin-walled and inelastic, which may account for the observation that babies with large placental to body weight ratios are at risk of developing hypertension in later life.

5.5 Formation of the major organs of the body

Between days 24 and 28 of development, the embryonic heart undergoes a series of folds and twists that turn it into an organ with the basic structure of the adult heart, as described above. By 26 days, the regions labelled atrium and ventricle in Figure 5.18 have folded into intimate contact with one another and take the shape shown in Figure 5.19. The drawing on the left gives a front view: blood flows via the sinus venosus into the primitive atrium and out of the truncus arteriosus. The truncus arteriosus gives rise later to the major arteries – the pulmonary artery (to the lungs) and the dorsal aorta (to the body) – when the ventricles separate internally into right and left. The side view on the right shows more clearly the veins below and the arteries above, inflow and outflow channels. The chambers of the adult heart arise by separation of the primitive atrium into two by the formation of a dividing membrane, and similarly for the ventricle, giving right and left chambers to each. This separation is already becoming externally visible for the atria (plural of atrium) by day 28, as shown in Figure 5.20, but internally there is still only one chamber. The embryo is vigorously pumping blood through its rapidly growing and developing tissues, through the yolk sac, and through the placenta. The heart is barely a millimetre in size at this stage. Nevertheless, it is very effectively irrigating the placenta, the yolk sac and the embryo. The brain, now growing rapidly, has a particularly rich blood supply. There is extensive blood flow back from the placenta via the umbilical vein, carrying nutrients and oxygen from the mother.

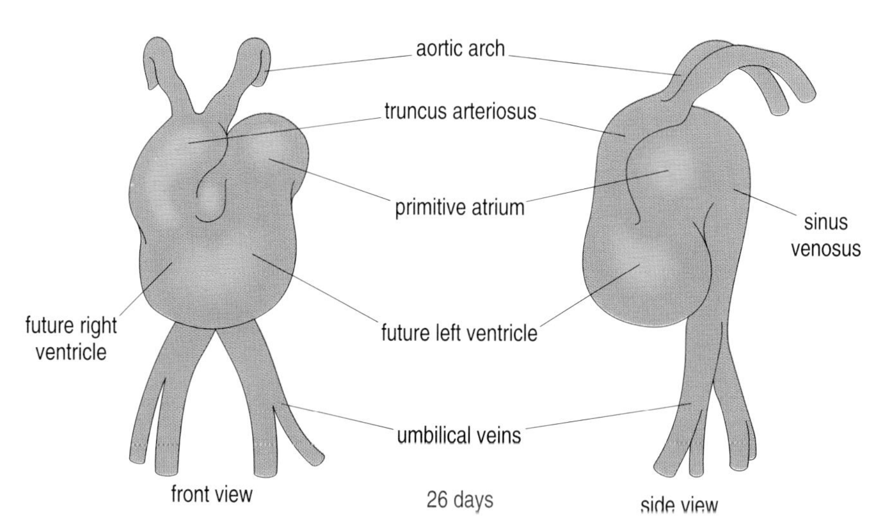

Figure 5.19 The embryonic heart at 26 days is a folded, enlarged tube without internal divisions. After fusion of left and right umbilical veins, blood flows via the sinus venosus into the primitive atrium, through the single ventricle and out via the truncus arteriosus to the aortic arches. Later, further pairs of aortic arches appear; the earliest-formed ones contribute to the dorsal aorta and the later ones to the pulmonary arteries. (The relationship of the embryonic heart and arterial system to that of the adult is very complex, so we have not gone into the details here.)

You can see in Figure 5.20 some of the other embryonic organs that are developing, particularly the prominent liver, the not-so-prominent stomach which has little to do at this stage, and the tiny embryonic lung, tucked under the vein from the head and between the stomach and the heart. By comparison with these organs, which will later on all exceed it in size, the heart is large and prominent, performing its essential function. Only the brain, extending from the forebrain with the distinct eye vesicle to the hindbrain, is larger by several-fold. This prominence of the brain at a mere four weeks of development is characteristic of all primate embryos (chimpanzees, gorillas, monkeys, gibbons, baboons and the like), but in humans the large size of the brain relative to the rest of the body persists throughout development and into adulthood.

The dramatic increase in relative head size is evident if we now look at the 48-day embryo which is shown in Figure 5.21, whose actual size and external form are shown at the upper right of the figure. Here the head is nearly as large as the rest of the body, with a very extensive network of blood vessels irrigating it. The collecting veins (blue) are on the outside while the arteries supplying blood from the heart are less visible underneath. The heart is taking on its adult form externally, with separate atria, but the ventricle is still single-chambered. The lungs, adjacent to the heart, have now developed their characteristic branching structure, and there is a small pulmonary artery which provides nutrients and oxygen for the developing lung tissue. However, the lungs, attached to the digestive tract and so connected to the developing mouth and nose, are still non-functional since the embryo, bathed in its watery environment, has no air to breathe.

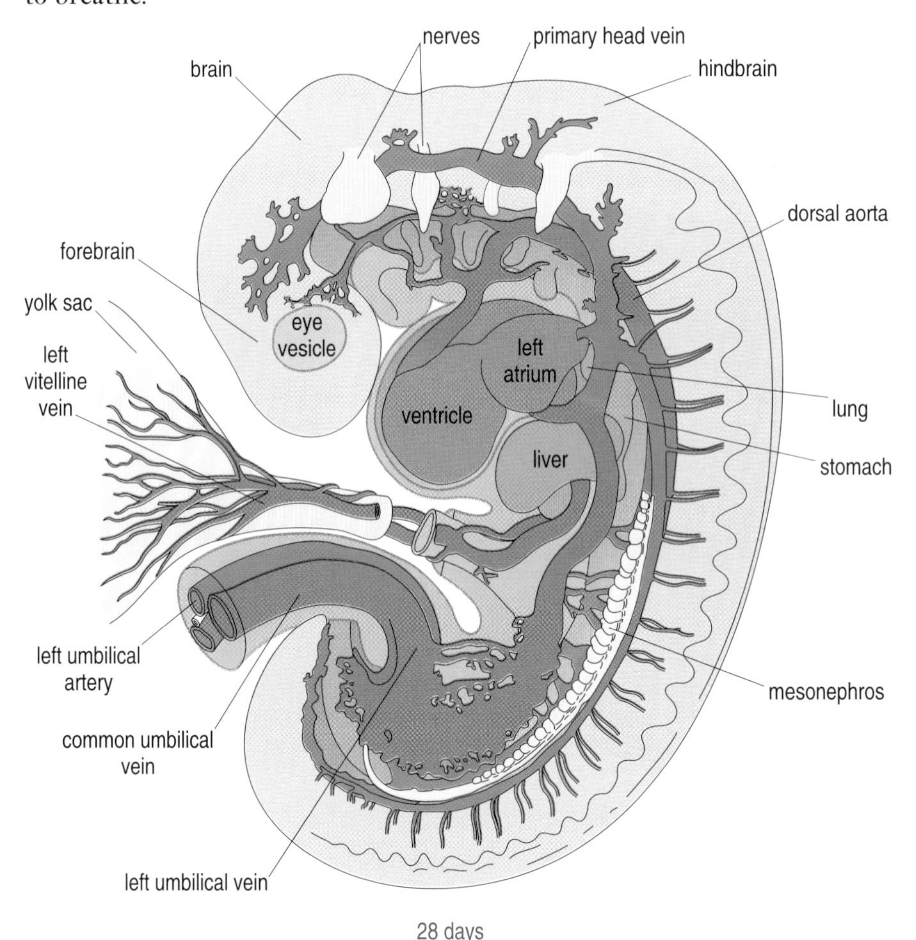

Figure 5.20 The internal organs of the embryo at 28 days. Right and left atria are beginning to separate externally, but internally they have a single chamber, as do the ventricles. There is now an extensive system of blood vessels to the brain and other developing organs such as the lungs, the stomach, and the liver. The yolk sac is beginning to get smaller and will contribute to the formation of the gut.

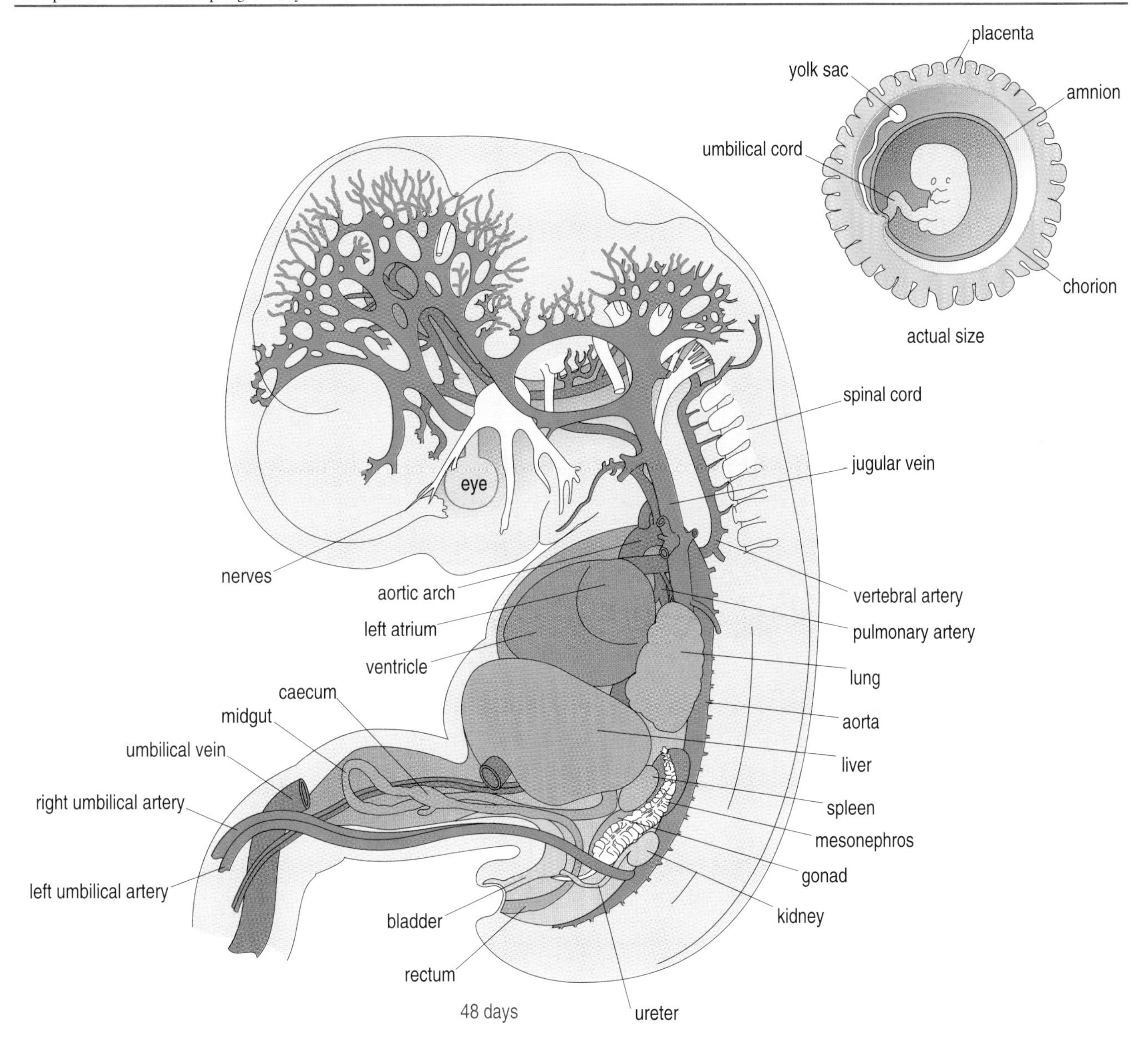

Figure 5.21 The embryo at 48 days, showing the large head with an extensive system of blood vessels and developing structures such as the eye, the heart which is assuming its adult form, rapidly growing liver and lungs, spleen, kidneys and gonads (reproductive organs), with the transient mesonephros now decreasing in size. The actual size of the embryo is at the upper right, which shows the umbilical cord, the placenta, and the remnant of the yolk sac which acts as a receptacle for embryonic wastes.

It is evident from Figure 5.21 that the liver is now very prominent. It is also highly active, producing blood cells and many of the proteins that circulate in the blood. The spleen is also present, and the kidney is developing adjacent to an embryonic organ called the *mesonephros* (plural: mesonephroi) which consists of a series of paired structures. If you look back at Figure 5.20 you will see that, at 28 days, the mesonephros extends over a considerable length of the spine. There are two mesonephroi, one on each side of the spinal cord. These are structures that we share with all other vertebrates, including fish, where they function as kidneys, clearing the body of wastes and passing them out of the anus. In humans this organ is transient, serving as a blood cleaner and passing the wastes to the remnants of the yolk sac. You can see this in the actual-size inset in the upper right of Figure 5.21, looking like a ball and chain which is connected to the embryo via the *umbilical cord* (the stalk that joins the embryo to the placenta)

The mesonephroi gradually disappear as the kidneys take over their function. However, they are also associated with the formation of another pair of crucial organs – the *gonads* (reproductive organs). You can just see these behind the mesonephros in Figure 5.21. The gonads, at this stage identical in males and females, arise adjacent to the mesonephroi. The way in which the different female and male gonads develop from initially identical organs and the factors involved in this process will be described in Book 4, where sexual differentiation is discussed in some detail.

Summary of Section 5.5

Most of the major organs of the body have emerged by the end of the fourth week of gestation (28 days). The heart has developed its characteristic curved form and is actively pumping blood through a single, enlarged tube, internal separation into four chambers happening later. The brain is already very prominent and developing rapidly, the liver is relatively large and active in blood production, while the lungs and stomach are just beginning to develop.

By 48 days (end of the seventh week) the external form of the heart is like that of the adult, with separated atria, but the ventricle is still single-chambered. The brain continues its rapid rate of growth while the liver, lungs, spleen and kidneys are developing their characteristic adult structures. The reproductive organs develop in association with transient structures, the mesonephroi, but at this stage of development there is no morphological difference between male and female embryos.

5.6 Formation of musculature and limbs

5.6.1 Pattern of normal development

Embryonic muscles start forming very early, as soon as the basic structure of the vertebrate form begins to emerge. We saw this happening during the third week of development (Figure 5.14) when the neural tube was being produced by the coming together of the neural folds along the midline of the body.

❑ What are the muscle-forming structures that first appear in conjunction with neural tube formation?

■ The somites, which arise in pairs on either side of the neural tube.

The somites form the back muscles that do so much work in maintaining our upright posture as adults and the bones of the skeleton form in close association with muscle tissue. In this section we shall be examining these processes primarily in relation to limb development.

If you look back at Figure 5.17, you will see that at five weeks of development the tailed vertebrate with its prominent head and curved back is showing lateral outgrowths, one near the tail and the other further forward. These outgrowths, the limb buds, are also present on the other side of the embryo, and will form the limbs, as shown in the sequence of stages to eight weeks of development. Paired structures, such as somites and limbs, give our bodies the overall form that is described as **bilateral symmetry**, the two sides having a structure that would result from reflection of either side in a mirror.

❑ What other structures in the human body occur in pairs?

■ Eyes, ears, nasal passages, breasts, testicles, kidneys, ovaries, lungs.

❑ Which internal organs are not paired?

■ Spinal cord, heart, liver, stomach, intestines, bladder.

If we look inside the embryo at the region where the arms will form, then on the 22nd, 26th, and about 29th days of development we see the structures shown in Figure 5.22. At first there are only the somites which have separated into two distinct parts, the lower ones called myotomes (muscle producers) while the upper parts are called dermatomes (described in Book 2, Chapter 3).

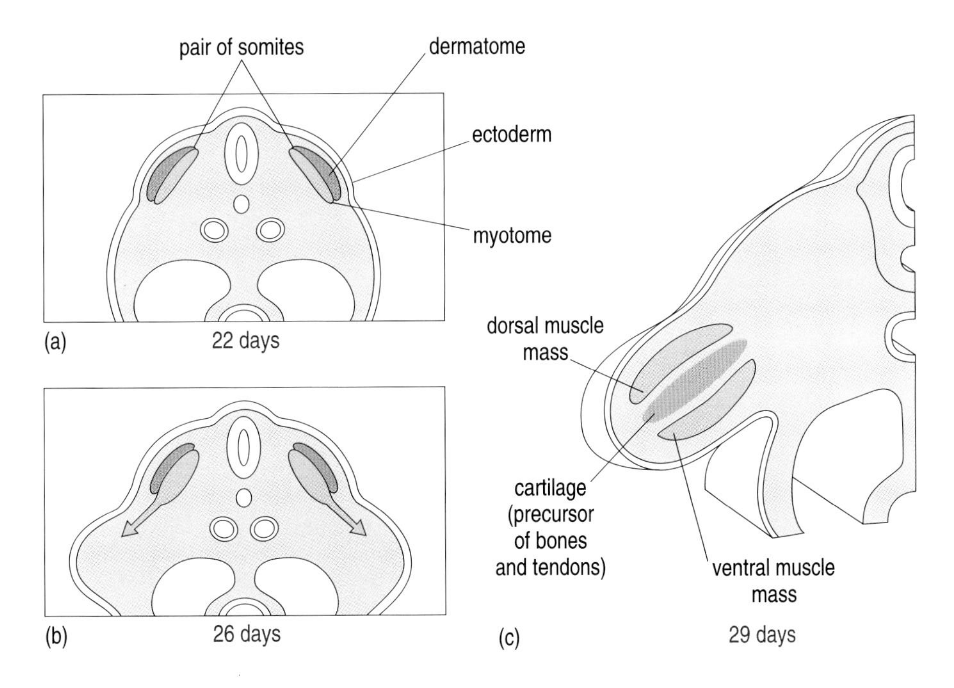

Figure 5.22 A schematic description of early forelimb development, showing (a) one of the four pairs of somites associated with the formation of the arm, whose muscles will develop from cells coming from the lower half of the somites (the myotomes). These cells migrate into the developing limb bud (b, arrows), whose growth depends also on the migration of other cells from the flank of the embryo. The myotome cells aggregate into the dorsal and ventral muscle masses in the developing limb (c), with a region between them where cells condense to form cartilage, precursor of the limb bones and tendons.

As the limb buds form, muscle cells migrate into them from the myotomes, as do other cells from the flank of the embryo. The muscle cells form two muscle masses with a region between them in which cells become very adhesive, forming a compact mass. The condensed central region becomes **cartilage**, a strong but flexible tissue which later is replaced by bone and tendon, with muscles above and below. Skeletal muscles (those associated with bones, which make up the skeleton) usually come in opposing pairs of this kind, which together give fine control over body movements, allowing us to raise and lower our limbs with precision. (This topic is dealt with in Book 2, Chapter 4.)

A look at the whole embryo at 37 days ($5\frac{1}{4}$ weeks), with focus on the muscles of the body axis and the limbs, gives the picture shown in Figure 5.23. The paired muscles of the back are prominent, extending to the tip of the tail. The limbs are growing fast, with the precursors of bones forming in the centre of the muscle masses. The first ones to form are those closest to the body – the humerus in the arm, the femur in the leg. Arm development is ahead of leg, and the end of the arm is beginning to flatten into the hand.

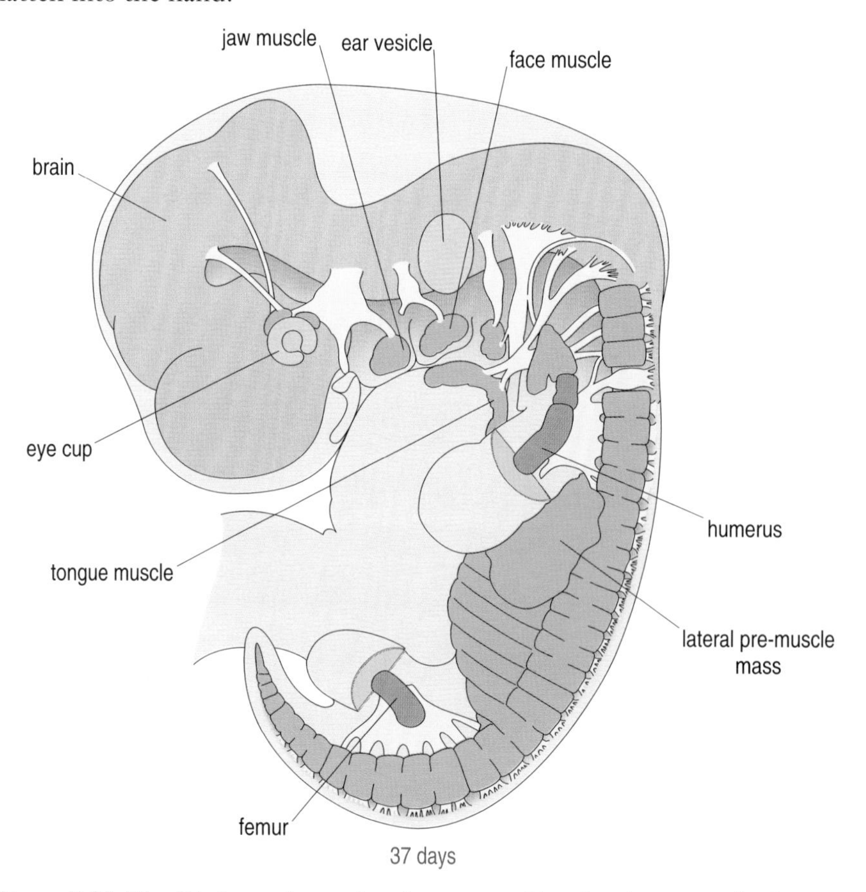

Figure 5.23 The 37-day embryo, showing arm and leg development, the latter slightly delayed relative to the former. The muscles and bones closest to the body form first (humerus in the arm, femur in the leg). Muscles associated with the developing eye, jaw and tongue are shown in the head, with the system of nerves arising from the spinal cord that innervate these and the muscles of the limbs.

By 41 days (nearly six weeks) the arms have many of their basic components and the legs are not far behind in their development (see Figure 5.24). Fingers and toes are beginning to emerge from the paddle-shaped structures (plates) at the ends of the limbs. This sculpting involves cell death in the tissue between the digits, a process called *apoptosis*. The capacity to die is another property of cells that contributes to the formation of detailed embryonic form and is under the influence of specific genes, resulting in specific cell death in particular regions of the embryo, as between the developing digits. In other species, such as frogs and ducks, the epidermal cells between the digits do not die, with the result that their appendages are webbed. This also occurs in some children where the genes fail to produce apoptosis. By the stage shown in Figure 5.24 the complex array of muscles which allows us to produce the extraordinarily fine manual control required for, say, writing, knitting, or playing a musical instrument is in place. Other muscles are forming in association with structures such as the eye, the jaw, and the face. Also evident is the prominent array of nerves along either side of the spinal cord. These include the *motor nerves*, conducting the electrical signals that control contractions of the skeletal muscles. (Book 2 covers the nervous system and motor control in some detail.)

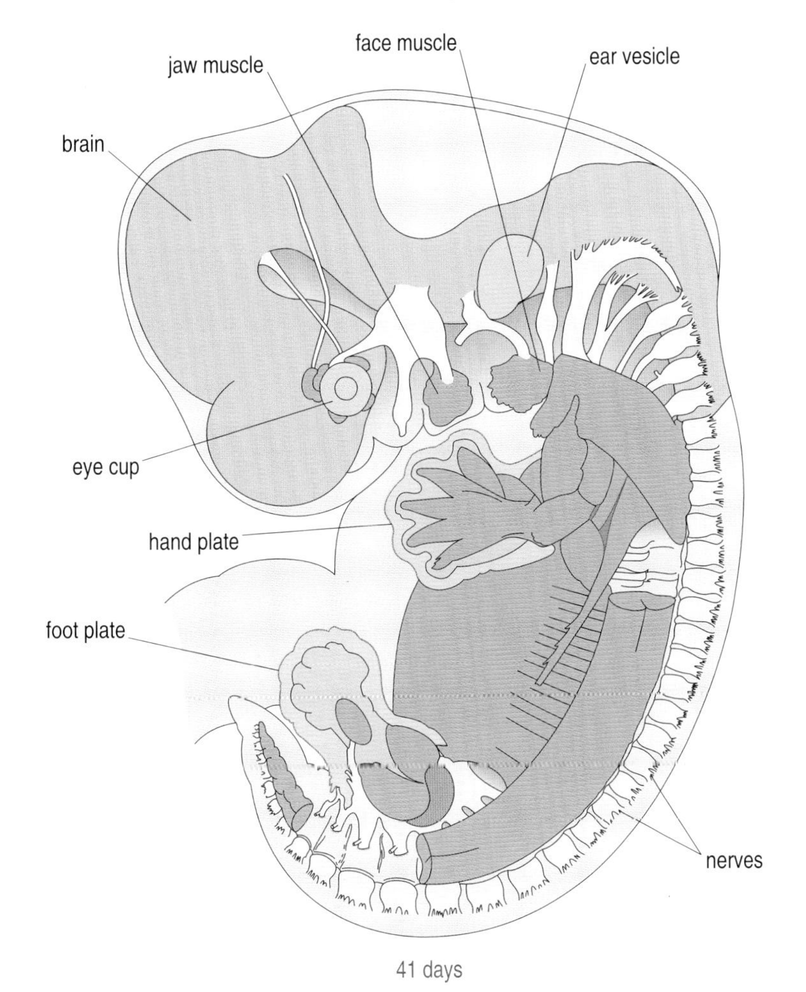

Figure 5.24 At 41 days the arms have many of their bones and muscles, though there is still extensive reshaping and differentiation before the structure of the mature limb is achieved. The fingers are emerging (from the hand plates) by the death of cells (apoptosis) between them, while the feet are still paddle-shaped structures (foot plates) with digits just beginning to emerge.

Focusing on the skeleton of the embryo a few days later (44 days), we see the array of structures shown in Figure 5.25. The tissue that defines the outline of the limbs and the backbone with protruding ribs between the limbs is made up of cells embedded in an elastic, fibrous jelly called mesenchyme. (This is similar to the loose connective tissue that makes up the dermis of the skin, which also develops from mesenchyme.) Inside the mesenchyme are the structures that will become the bones. At this stage, these structures are made of cartilage. This consists of densely packed cells embedded in a tough matrix which includes a lot of collagen. Cartilage is both strong and flexible, so it provides a structure with the mechanical properties required to give some rigidity to an otherwise floppy embryo. The cartilage will later be replaced by mature bone, which is much stronger (but also much more brittle, fracturing under excessive force). This process, called **ossification**, is gradual and progressive, starting in the eighth week. At birth many of the bones still have considerable cartilage in them, making them quite flexible, and this aids delivery.

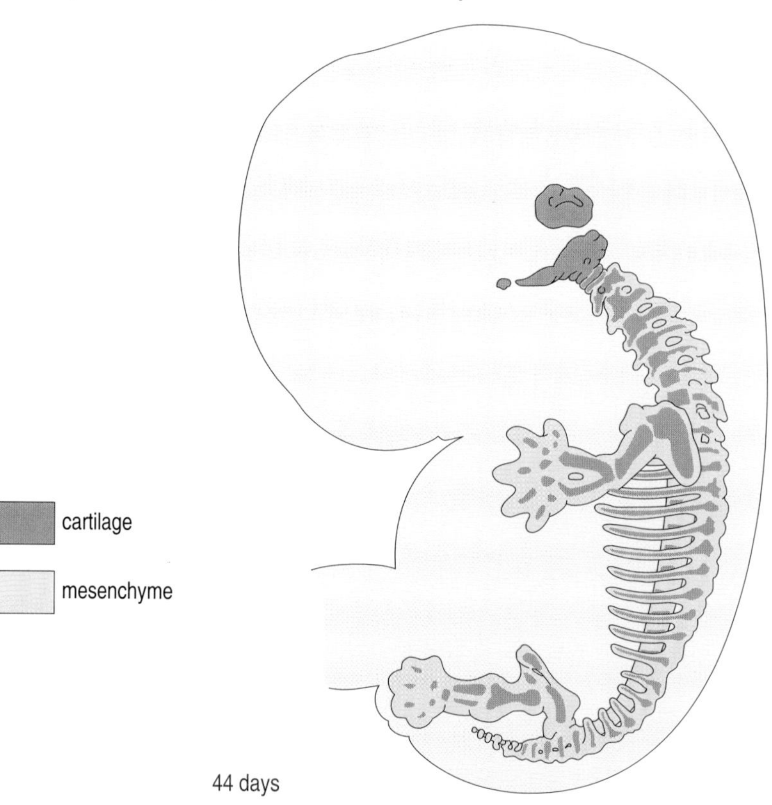

Figure 5.25 The embryonic skeleton at 44 days. The bones of the limbs form as condensations of cells that produce cartilage, within the surrounding mesenchyme. The same process is occurring in other parts of the body: the shoulder and hip, the backbone and associated ribcage, and regions of the head.

At nine weeks the body skeleton has all of its components, as shown in Figure 5.26. However, the skull bones are still forming and much of the brain is without any skeletal envelope. The replacement of cartilage by mature bone in the limbs progresses from the centre of the skeletal components to the ends. There is a complex array of bones in the hands and feet. You can find most of these in your own hands simply by feeling your way from finger tips to wrist and tracing the bones along their length, though the small bones of the wrist are very hard to detect separately.

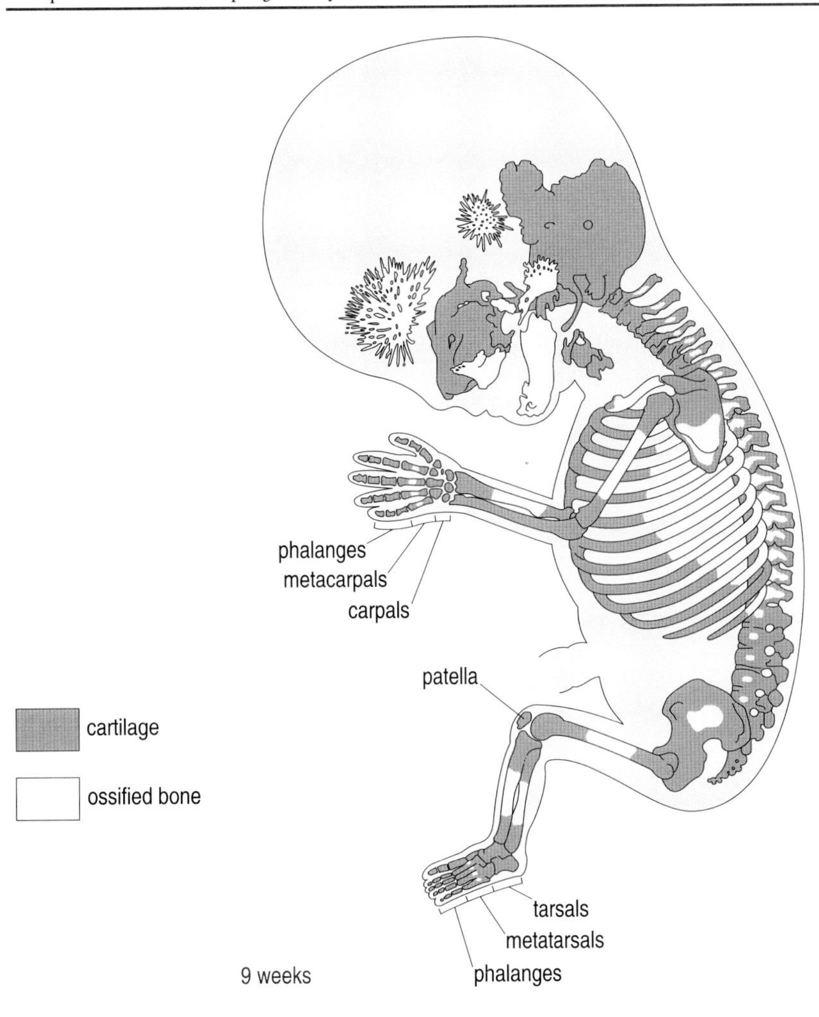

Figure 5.26 At nine weeks the body skeleton of the fetus is complete but the skull is still developing. Replacement of cartilage by bone is occurring in many parts of the skeleton.

❑ How many bones do you have in each finger and what are they called?

■ Three, called the phalanges (Figure 5.26).

❑ Locate the other longish bones, the metacarpals, in Figure 5.26.

■ They are in the palm.

❑ What distinguishes the thumb structurally from the fingers?

■ It has only two phalanges, whereas the fingers have three each.

5.6.2 Disturbances of limb development

A process as complex as limb formation can go wrong in a variety of ways, most of which result in minor defects that do not interfere with basic function. One of the most common is a failure of toes or fingers to separate completely, and you may well have seen this condition on some people's feet. When it happens to fingers it is something of an impediment to normal dexterity, so surgery is used to separate the digits. This is possible if the fingers are joined only by skin, resulting in webbing, as described earlier. However, if they share a bone, such as the first of the three phalanges that normally define a finger, then no separation can be carried out. Depending on the fingers involved, this need not be a serious handicap. But there are more severe malformations and, like neural tube defects, it has been discovered that some of these run in families and so have a significant genetic component. Even in these cases, it is clear that environmental factors contribute to the probability of a defect appearing and to its severity, so that disturbance to the normal process results from an interaction between external influences and the individual's genetic make-up.

As we change our environment more and more, and make increased use of drugs that are designed to affect specific body functions, the probability of exceeding the range of the body's normal range of tolerance is likely to increase in unpredictable ways. This is because of the complex pattern of interactions that are the very basis of living processes, so that affecting one function with a drug or an environmental stress may have unexpected consequences on others. One of the most dramatic and tragic instances concerned the use of the drug thalidomide. This was first marketed as a mild sedative in 1957 under the trade name Contergan. For a few years it was thought to be virtually free of side-effects and its use spread widely. There was at that time very little evidence for drug-induced malformations in humans. However, by 1962 two investigators (Walter Lenz and William McBride) had independently presented evidence that thalidomide caused an enormous increase in a previously rare syndrome of abnormalities in which the limbs are much reduced in size. Several thousand affected infants were born to women who had taken the drug, and a woman need only have taken *one* tablet to produce a child with all four limbs deformed. Other abnormalities included heart defects, absence of external ears, and malformed intestines. The drug was withdrawn from the market in November 1961.

Thalidomide was first manufactured in a town called Stolberg, in West Germany. On 25 December 1956, a girl with no ears was born in the town. Her father worked for the manufacturer and had been given free samples of the new drug for his pregnant wife. Approximately one year later the epidemic of congenital limb deformations commenced throughout West Germany. Hindsight shows that the incidence of these malformations exactly paralleled the West German sales figures of thalidomide, with a lag of between seven and eight months, as was found in 20 other countries where the drug was marketed. It is estimated that approximately 5 850 infants were affected, of whom 40% died, leaving some 3 900 survivors.

This tragedy revealed the limitations of using animals to test for the safety of drugs used on humans. Different species, and even different strains within species, metabolize thalidomide differently. Pregnant mice and rats – the animals usually used to test such compounds – do not have malformed pups when given thalidomide. Rabbits produce some malformed offspring, but the defects are different from those seen in human infants. Marmosets (a species of monkey) do seem to mimic the human malformations, and they have been used in an attempt to identify how the drug acts. The suggestion is that the primary target is the developing nervous system, particularly the nerves that innervate the limbs and maintain them. However, there is also an indication that thalidomide interferes directly with cartilage formation, so that the development of the skeletal components themselves is suppressed. As we might expect from the spectrum of abnormalities, the developmental processes affected are likely to be diverse. However, a study of the age of the embryo at which thalidomide produces different defects showed that there is a coherent and interpretable pattern of effects. This is shown in Figure 5.27. The drug was found to be a **teratogen** (a cause of abnormalities) only if the mother used it during the period between 20 and 36 days after fertilization. The effects on limb development start at 24 days.

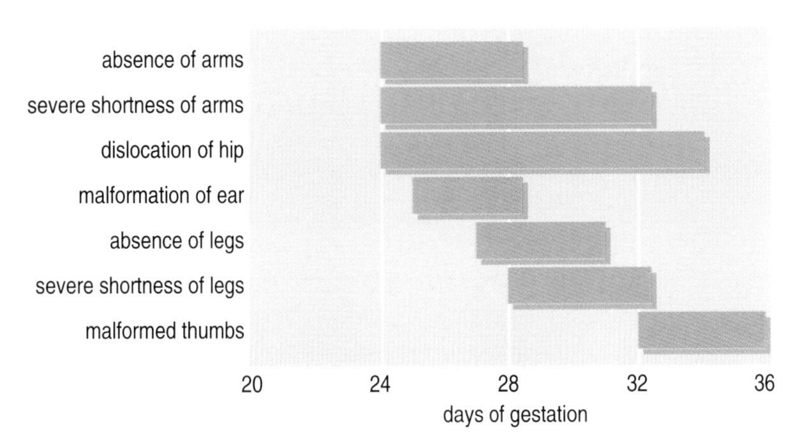

Figure 5.27 Timing of susceptibility to the teratogenic effects of thalidomide on development.

❑ What is happening with respect to arm development at 24 days? (Look back at Figure 5.22.)

■ This is a very early stage of arm development, in between those shown in Figure 5.22a and 5.22b, when the cartilage and the muscle masses are just beginning to form in the upper limb.

❑ What can you deduce from Figure 5.27 about the time difference between equivalent stages of arm and leg development?

■ There is a difference of about three days, since absence of arms first occurs when the drug was taken at day 24, and absence of legs if it was taken at day 27.

There was another important lesson highlighted by the thalidomide tragedy: the effects of substances on embryos are different from those on adults. This is primarily because the *construction* of an organ can be affected by chemicals that have no deleterious effect on the normal *functioning* of that or other organs. Several medicines for adults have turned out to be teratogenic for embryos. These include methotrexate (a drug used to stop tumour growth), anticonvulsants such as trimethadione, and anticoagulants such as warfarin (used to prevent blood clotting).

These results all point in the same direction: drugs need to be introduced and used with extreme caution, because there is no way of testing for their effects that can ever guarantee that they will cause no harm. This is because each species can have a distinctive response pattern to a drug, and what happens to adults is in general not the same as what happens in the developing embryo. It is only when the drug is actually used that its effects will be revealed, so first use will always entail risk. However, the potential dangers can be substantially reduced by testing on other species.

Summary of Section 5.6

The first muscles, the paired somites, are formed in association with the backbone during the third week of development, and this intimate relationship between muscle and bone formation occurs in all parts of the developing body. Nerves grow out from the spinal cord and make specific connections with muscles as they develop. The limbs arise at four weeks as little bulges on either side of the body, closely associated with particular somites. As the limbs grow, bones and muscles form within them, starting with those closest to the body and extending to the fingers and toes, the arms being about three days ahead of the legs in their development. The first bones are made of cartilage, which is stiff but flexible. Conversion to hard adult bone begins in the eighth week, though many bones are still largely cartilage at birth.

Drugs designed to affect adult functions in specific ways can have very different effects on developing embryos. A dramatic instance of this was the discovery that thalidomide, a mild sedative which was virtually free of side-effects in adults, caused severe abnormalities in babies when taken by pregnant women, its effects starting during the third week of development. Tests of the drug on mice and rats prior to use on humans revealed no defects in the offspring, while rabbit embryos are slightly sensitive but do not have limb deformations. Only marmosets seem to mimic human sensitivity in this respect, which is believed to arise from disturbances to both cartilage formation and establishment of nerve connections to muscles.

5.7 Twins

There are many species in which it is standard for several eggs to be released simultaneously at ovulation so that multiple fertilizations take place, followed by multiple implantations into the uterus. Each embryo then develops separately within its own protective membranes and has its own placenta, though some degree of fusion often occurs between adjacent outer layers (the chorions). Dogs, cats and pigs all have multiple egg releases at ovulation, with an average number that is characteristic of each species, so that at term dogs and cats will have from two to six young in a litter while pigs can have as many as 15.

In humans, it is usual for a single egg to be released at ovulation, but two is fairly common while higher numbers are increasingly rare. Two fertilized eggs that develop to term in the uterus, each with its own separate placenta are known as **dizygotic** (or **fraternal**) **twins**: each twin arises from a distinct zygote (fertilized egg). However, there is another type of twinning that can occur, which emphasizes a very fundamental property of developing embryos.

Look back at Figure 5.1 and recall the earliest stages of development after fertilization, from first cleavage to the formation of the inner cell mass. It has been suggested that after first cleavage the two identical cells produced can separate from one another while remaining within the zona pellucida, each one continuing to divide and producing a separate blastocyst. The two blastocysts then escape from the zona and implant at separate sites in the uterus. Subsequent development is just like that of two separately fertilized eggs which produce dizygotic or fraternal twins. This is shown in Figure 5.28. On the left is a sequence of stages in the development of dizygotic twins from two separate zygotes. On the right is a series of possible developmental pathways for a single zygote that produces two embryos – these are **monozygotic twins**. The pathway in which there is separation of the two cells arising from first cleavage of the zygote is shown following a similar path to that of the dizygotic twins, with separate blastocysts that implant at separate sites in the uterus. However, this has never actually been observed so that precisely how identical twins with separate placentas are formed remains unclear.

❑ What is the genetic difference between dizygotic and monozygotic twins?

■ Dizygotic twins are genetically distinct, since they arise from separate eggs and sperm. Hence they could be male and female, or both male, or both female. Monozygotic twins have identical sets of genes, since they arise from the zygote produced by the fertilization of one egg by a sperm. They are therefore of the same sex. (They are *clones* of one another as they have identical genes.)

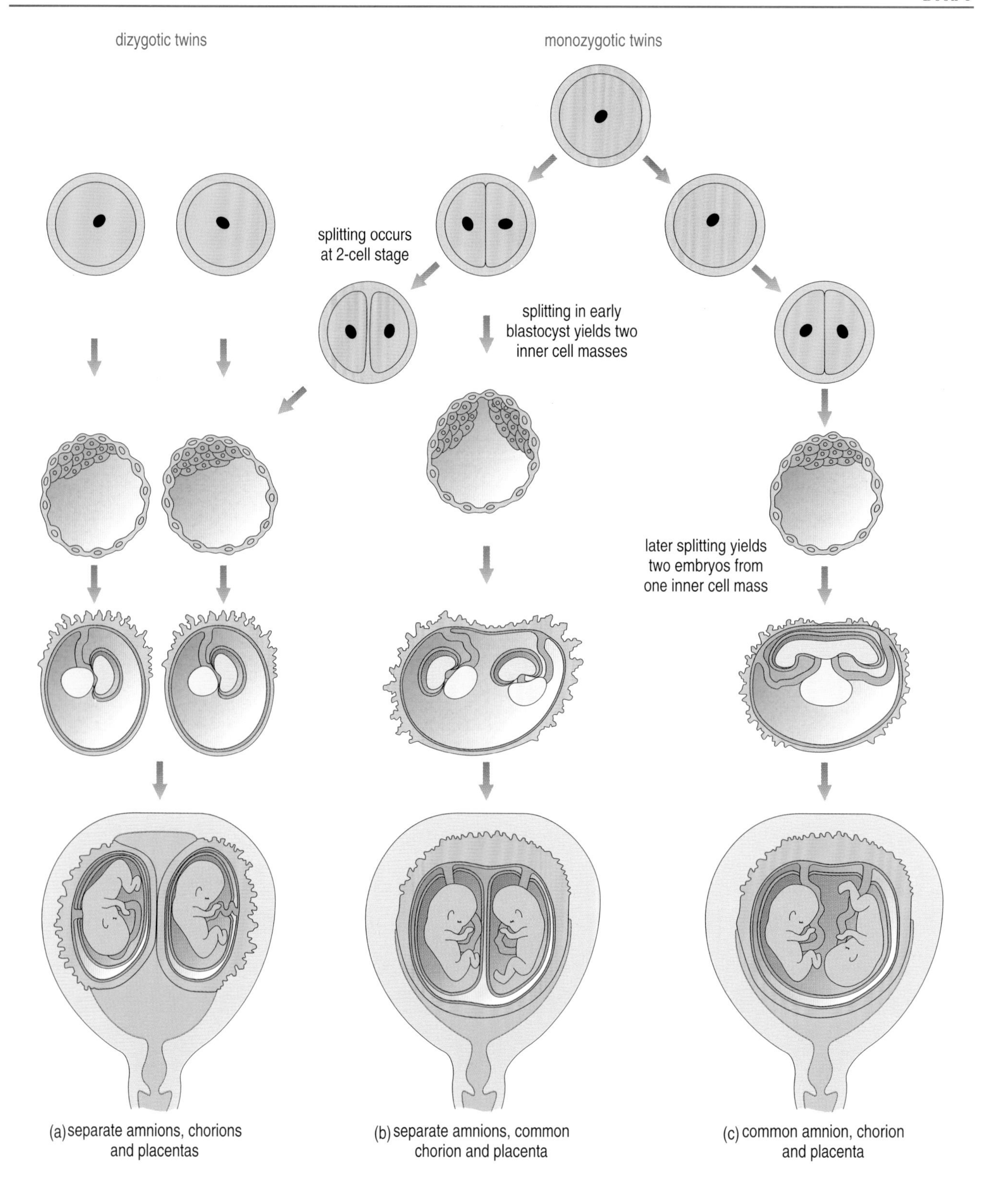

Figure 5.28 Pathways of development for twins. To the right are three possible ways of producing monozygotic (identical) twins, resulting in (a) complete separation of embryos, membranes, and placentas; (b) separate embryos and amnions but common chorion and placenta; and (c) common amnion, chorion, and placenta, with the possibility of partial fusion of embryos.

What happens if, instead of a separation at the two-cell stage, there is a separation of cells at a later stage of development? One possibility is shown in Figure 5.28b: the inner cell mass separates into two within the same blastocyst. Each of these inner cell masses then goes through all the subsequent stages of development, resulting in two separate embryos that each develop into an individual. In this case the two embryos develop within distinct inner chambers (amnions) but they share a common outer chamber (chorion) and placenta.

Separation of the inner cell mass can occur to different degrees and at different times, with the result that embryos are produced with varying degrees of separation between them. Figure 5.28c shows the situation that arises when the inner cell mass separates at some stage during the events early in the second week as the embryo is implanting and developing the amnion (Figures 5.2–5.4). The twins can develop within a common amnion as well as sharing a chorion and a placenta. Each embryo develops a distinct umbilical cord, but they share the same blood supply from the placenta. Since they are genetically identical, this causes no compatibility problems.

Twins arising from these later separations of the inner cell mass can result in even greater degrees of shared structure between the embryos, to the point where the bodies are themselves joined together to varying extent, resulting in 'Siamese' twins. Shared surface tissues like skin and connective tissue, and even partially shared organs, can be separated by surgery, but the more two become one the greater are the dangers and difficulties of separation. What these phenomena of monozygotic twinning tell us, however, is something very basic about embryonic development concerning the relationship between part and whole. The two cells which result from first cleavage would normally give rise to part of the resulting embryo, including the structures outside the embryo (amnion, chorion, placenta). However, each of these initial cells has the potential to make a whole, as occurs in the pathway in Figure 5.28a from the single zygote, through separation of cells after first cleavage, to the two fully separate embryos. Each part of the inner cell mass would likewise normally give rise only to a part of the embryo. However, if two parts of the inner cell mass separate, each one has the potential to produce a complete whole.

This capacity of parts to give rise to complex, integrated, complete wholes is one of the most basic properties of living organisms, on which our lives depend in two distinct but related senses. First, it is the basis of reproduction: an egg is produced as a part of a woman's body, and this part has the potential to become a whole new human being, when combined with the genetic contribution from a sperm, a part from a man. It is this potential of the egg that is transmitted to the cells produced by first cleavage, and the potential continues in the inner mass cells. Secondly, the capacity of parts to make wholes is the basis of healing, as we saw in the regenerative process that occurs in the healing of skin wounds. Making whole is the foundation of health. The embryo has this capacity in high degree, and it is manifested in the twinning process.

Summary of Section 5.7

Twins can arise in two different ways. Two eggs can be released at ovulation and be fertilized, the embryos then implanting and developing independently. Such dizygotic or fraternal twins can be of either sex.

Separation of a single embryo into two parts at any stage of early development can give rise to monozygotic twins, which are genetically identical. Depending upon when the separation occurs, the embryos will have different degrees of shared structure, from independent placentas to joined embryos. The formation of two complete embryos from one egg shows that a part of the embryo is capable of making the whole, an embryonic property that is reflected in all healing capacities.

Objectives for Chapter 5

After completing this chapter you should be able to:

5.1 Define and use, or recognize definitions and applications of, each of the terms printed in **bold** in the text.

5.2 Describe the basic types of cell property and movements involved in generating the inner cell mass, the amniotic and chorionic cavities and the yolk sac. (*Question 5.1*)

5.3 Explain how the backbone and the neural tube are generated and describe the basic types of abnormality that can arise from disturbances to this process. (*Question 5.2*)

5.4 Describe the functions of the placenta in nourishing and protecting the embryo and fetus. (*Question 5.3*)

5.5 Understand how the physical properties of the blood as a fluid, the elastic properties of the blood vessels, and the contractility of the heart all contribute to the formation and function of the circulatory system. (*Question 5.4*)

5.6 Explain the programming hypothesis connecting influences on the development of the circulatory system in the embryo to the later appearance of hypertension in the adult, describing the embryonic processes involved. (*Question 5.5*)

5.7 Describe the processes involved in the development of arms and legs, and the types of defect caused by thalidomide. (*Question 5.6*)

5.8 State the difference between dizygotic and monozygotic twins, and explain how they occur. (*Question 5.7*)

Questions for Chapter 5

Question 5.1 (*Objective 5.2*)

Summarize the basic cellular properties and activities which produce the structures of the early human embryo – the inner cell mass, the trophoblast, the amniotic cavity, the yolk sac and the chorionic cavity.

Question 5.2 (*Objective 5.3*)

How are the backbone, the somites, and the neural tube formed, and how can their formation be disturbed by genetic and environmental influences?

Question 5.3 (*Objective 5.4*)

How is the placenta formed and how does it protect the developing embryo?

Question 5.4 (*Objective 5.5*)

What are the factors, physical and biological, that are involved in producing the circulatory system and the rhythmic movement of the blood around the body?

Question 5.5 (*Objective 5.6*)

How could inadequate nutrition of an embryo result in hypertension in the later life of the individual?

Question 5.6 (*Objective 5.7*)

Using the example of thalidomide, explain the concept of sensitive stages during which the embryo is particularly prone to disturbance in the formation of specific parts of the body.

Question 5.7 (*Objective 5.8*)

Suppose you have just assisted in delivering twins and observe that they have separate placentas. Could you conclude that they are dizygotic (fraternal)?

References

Copp, A. J., Brook, F. A., Estibeiro, J. B., Shum, A. S. W. and Cockroft, D. L. (1990) The embryonic development of mammalian neural tube defects, *Progress in Neurobiology*, **33**, 363-401.

Laurence, K. M., James, N., Miller, M. H., Tennant, G. B. and Campbell, H. (1981) Double-blind randomized controlled trials of folate treatment before conception to prevent recurrence of neural-tube defects, *British Medical Journal*, **282**, 1509-11.

Lenz, W. (1962) Thalidomide and congenital abnormalities, *Lancet*, **1**, 45.

McBride, W. G. (1961) Thalidomide and congenital abnormalities, *Lancet*, **2**, 1358.

CHAPTER 6
FETAL DEVELOPMENT AND BIRTH

6.1 Introduction

The previous chapter described how complex processes, relying on simple developmental principles, led to the formation of the structures needed by a new individual, starting from the three germ layers: ectoderm, mesoderm and endoderm. In this chapter, we shall look at how the fetus grows and matures, enabling it to survive birth and enjoy an independent life. Although by the end of the first trimester (the first three months, or 12 weeks, of pregnancy) all the organ systems are in place in the fetus, they all need to undergo a more or less extended period of change so that they can be fully functional in the outside world. The growth and development of each organ system proceeds at its own pace through the last six months of gestation, so that by the time of birth the baby is fully functional – or, at least, as functional as a new-born baby can be. In the interests of clarity, we have elected to describe each of the systems in turn; this isolationist approach is not ideal because, as you know by now, body processes depend on *interactions*, and no system actually does – or even can – develop in isolation. We shall therefore try to end the chapter by giving a more integrated view of the overall process. In conjunction with further development, however, the fetus has to *grow*, and it is this aspect that we shall examine first.

6.2 Fetal growth

Growth is generally defined as an increase in mass (weight) over a period of time. The birth weight of the average baby is 3.2 kg (just over 7 lb), representing between 5 and 6% of the weight of the average mother. All this mass must be accumulated over a period of 38 weeks, the average length of a normal pregnancy. The increase in mass of the fetus is shown in Figure 6.1. The red line indicates the growth pattern of the *average* fetus; in Figure 6.1b, the pink-shaded areas on either side of the line are where the values for 95% of all fetuses fall.

How is the weight of the fetus actually determined? It is obviously not possible to take the fetus out of the uterus, weigh it, and put it back in.

❑ Can you think of a way of weighing the fetus?

■ One way would be to measure the *mother's* weight increase, and subtract her non-pregnant weight to get that of the fetus.

(a)

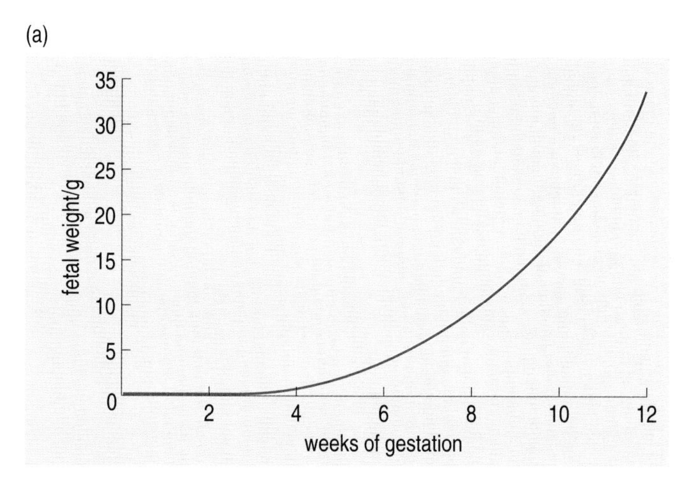

(b)

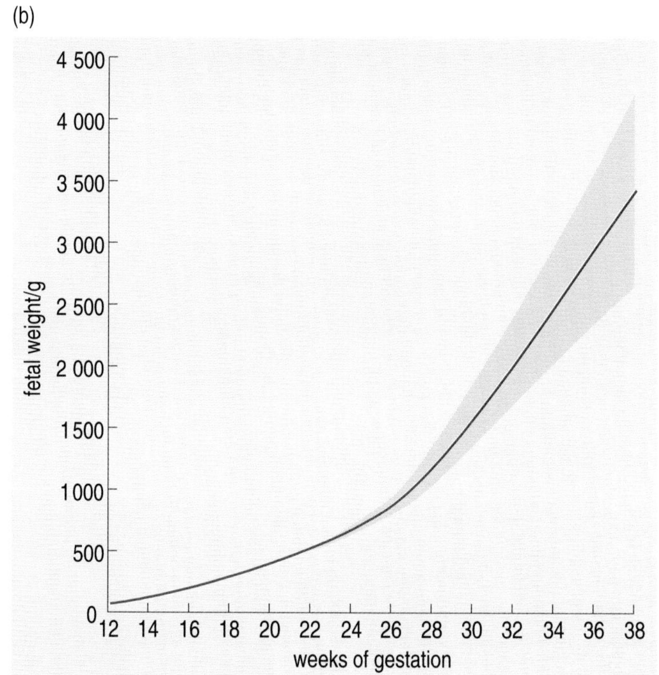

Figure 6.1 Graphs of fetal weight against gestational age: (a) the first 12 weeks; (b) the remaining 26 weeks. The shaded areas in (b) represent the range for 95% of fetuses.

However, this makes the assumption that any weight gain is due to fetal growth alone, and this is clearly not true.

❑ From what you have learned already, and from general knowledge, can you guess what else might be changing during pregnancy?

■ The placenta is growing, the amount of amniotic fluid is increasing, the mother's blood is increasing in volume, and hence weight, and she is accumulating fat.

So this approach does not look very promising. In fact, there is no simple way to assess the weight of the fetus without damaging it or its mother (although some information can be obtained by weighing aborted fetuses). So instead, weight is assessed *indirectly*. This is accomplished by the use of ultrasound scans, which allow various measurements to be made accurately. By applying a mathematical formula to these measurements, the weight can be determined reliably. Measurements made routinely include the head circumference, abdominal circumference, crown-to-rump length, and femur (thigh bone) length. Generally speaking, these will all increase in a predictable way during gestation (see Figure 6.2); if one or more of the measurements is not increasing normally, it may be an early indication that there is something amiss with the fetus.

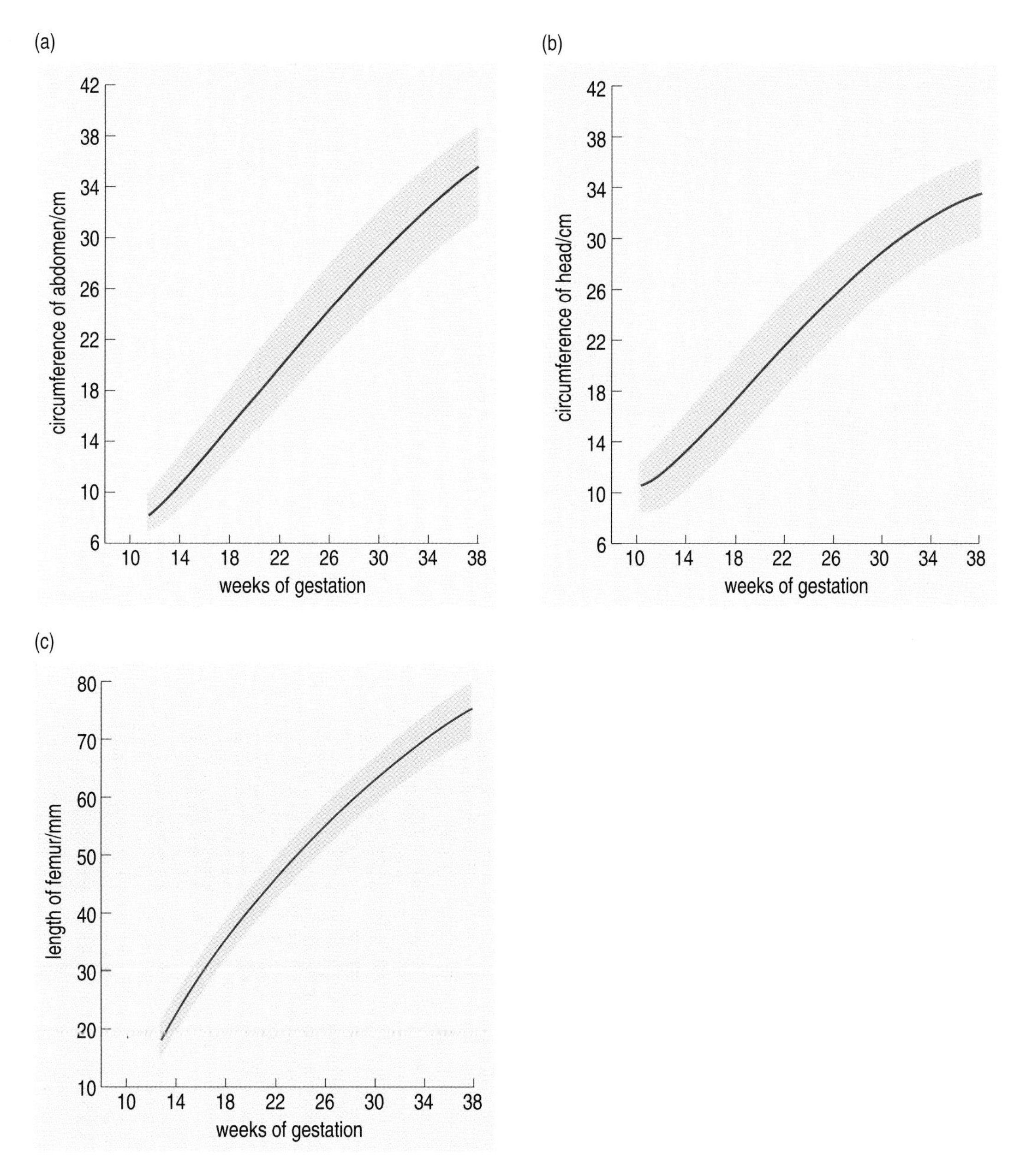

Figure 6.2 Increase in several variables measured by ultrasound through gestation. Shaded areas indicate the ranges for 95% of fetuses.

You might expect that all the measured parameters would increase in a **linear** way, i.e. by equal amounts in successive equal time intervals, but Figures 6.1 and 6.2 show immediately that this is not the case. Figure 6.2 shows that there is a slowing of the rate of increase with age for all three variables measured, even though weight continues to increase (Figure 6.1b). At later stages, individual variation becomes quite large, as shown by the divergence of the 95% lines. The smallest *normal* babies at birth can be the same size as the largest normal babies at about 32 weeks of gestation! This all goes to show that attempts to categorize babies by size alone are full of uncertainties. This is why not only are several measurements taken, but also some assessment of fetal maturity needs to be made. In Britain, nearly all women have an ultrasound scan at around 18–20 weeks after their last menstrual period, corresponding to 16–18 weeks of gestation. If abnormalities are suspected from this, the woman is asked to return a couple of weeks later for a further scan. By comparing the actual progress of the fetus with the predicted progress for a normal individual, growth problems can be diagnosed more reliably.

Fetuses grow primarily by increasing their cell numbers. All the energy and raw materials for fetal growth must come from the mother, via the placenta, so this places demands on a pregnant woman. As you saw in Figure 6.2, there is a significant variation in size between different babies. Part of this is due to genetic factors: some people come from families whose members are generally taller and larger than others.

❑ To what else is this variation in baby size attributable?

■ To the environment.

In the case of the fetus, 'the environment' refers mainly to the environment within the womb, which is obviously strongly dependent on the mother.

❑ Based on what you have learned so far, can you list any factors that might influence the conditions in the womb?

■ How well nourished the mother is, whether she smokes, whether she is taking any drugs, and whether she has an infection are all important factors.

In particular, the rate at which maternal blood passes through the placenta seems to be important for fetal growth. (This has been shown to be the case in various other mammals, but for ethical reasons it is not possible to do this experiment with women.) If the mother herself is undernourished, she may not be able to provide all the nutrients that the fetus needs. To maximize the chance of fetal survival, the placenta acts as a scavenger, actively removing what the fetus needs, even if there is very little in the maternal blood. The placenta has the ability to concentrate various substances, so that the fetal cells are bathed in relatively high concentrations of nutrients, regardless of the original concentration in the mother's blood. This is generally beneficial to the fetus, although in some cases it can be a disadvantage (see below). You have already seen how maternal nutritional status can affect the reproductive process, including the ability to ovulate and the development of organ systems in the embryo and fetus; even if all the organs have been successfully laid down, inadequate nutrition later in pregnancy can result in small or malformed babies. In the next section we go on to look at this in a rather more detailed way.

Summary of Section 6.2

1. Fetal growth is not a linear process; the rate slows towards the end of gestation.
2. Different fetuses grow at different rates, but 95% of all fetuses will be within a narrow range of the 'average' values.
3. Fetal growth occurs by an increase in cell numbers.
4. Fetal growth is influenced by the mother, by the placenta and by the fetus itself.
5. The placenta can act as a scavenger to provide the fetus with high concentrations of nutrients.

6.3 Maternal nutrition during pregnancy

Although it is relatively easy to study nutrition in laboratory animals by feeding them defined diets and looking at the outcomes, it is not as easy to obtain nutritional data relating to humans. However, the Second World War has provided a lot of evidence, thanks to meticulous record-keeping before, during and after the hostilities. During the mid-1940s, Dutch women experienced prolonged nutritional deprivation over many months. Yet the birth weights of their babies fell only by a few per cent, showing that the placentas had been able to scavenge enough nutrients for the fetuses. In these cases, fetal nutrition was at the expense of the mothers' body reserves; the mothers lost body fat and their bones and muscles wasted. During the siege of Leningrad between August 1941 and January 1943, both the quantity and the quality of diet was poor. There was no fuel for heating, and women often had to work outside in freezing conditions. Over this period, birth weights were reduced by more than 500 g (about 15%), crown-to-rump lengths were reduced, and the incidence of still-births and premature births increased, as, not surprisingly, did maternal mortality. It seems that these women had no more reserves to give their fetuses. Once conditions improved, after the war, these effects disappeared, and, indeed, post-war recovery has generally improved the European diet, and the size of new-born babies. More recently, a study of mothers and babies from a refugee area in

Karachi confirmed that malnutrition reduces both the length and the weight of the new-born. In particular, maternal diets deficient in protein seem to have a severe effect on fetal size, although maternal carbohydrate intake is important too.

❑ What is the main function of carbohydrates in the diet?

■ They provide a source of energy, so that the cells can carry out all the necessary metabolic reactions.

Dietary proteins are needed to provide building blocks for fetal cell growth. There is little evidence about the role of fat in the maternal diet with respect to fetal size, but experiments using rats, whose fat reserves normally increase by 30–40% during pregnancy, show that when the mother's diet is deficient in certain fats, the litter weight is smaller. Also, in the absence of adequate fat, the mother will not be able to make sufficient milk, which is very rich in fat (see Book 4), and so the offspring will not thrive unless supplementary, artificial milk is supplied. It is likely that a similar situation exists in human mothers.

6.3.1 Dietary quality

It is worth while at this stage taking time to think about the constituents of a diet. What is actually *meant* by a 'good' or a 'poor' diet?

❑ What would you say constitutes a good diet?

■ One that provides appropriate energy and raw materials for all the body's cells to carry out their functions efficiently.

You will learn more about diet as such in Book 3.

In the case of a pregnant woman, the diet should provide enough energy and nutrients for the woman herself *and* for her growing fetus.

The energy value of a diet can be translated into how much ATP it can provide (see Chapter 3) after it is digested and absorbed. In science, energy is measured in joules (J), but you may be more familiar with measurements in calories (cal), where 1 cal = 4.2 J. In this discussion we shall therefore adopt the use of calories. In fact the energy value of foods is usually measured as kilocalories (kcal), as the numbers are rather large (1 kcal = 1 000 cal). A kilocalorie is the same as the old, and rather confusing unit, the Calorie. Nowadays food labelling uses the unambiguous kcal and gives the equivalent in kJ as well. To put this into context, a slice of bread has an energy value of around 60 kcal.

❑ How many kilojoules, then, can be obtained from a slice of bread?

■ $60 \times 4.2 = 252$ kJ.

Most dietary energy, as you know, comes from carbohydrate, but a lot comes from fat. Protein *can* be used as a source of energy, but this is not very efficient, and protein only makes a significant contribution to energy supplies when carbohydrate is unavailable, either because of a particular diet, or during times of starvation, as in the Second World War examples mentioned above. In a body that is fed only protein, cells can convert the constituent amino acids to glucose. In a starving mother's body, tissues such as muscle – which is mainly protein – are broken down in an attempt to maintain the energy level needed by the brain, and because of the placenta's scavenging activities, the fetus gets the lion's share of the fuel. In contrast to the situation in the mother, whose tissues other than the brain can use various substances as a source of energy, the fuel actually used by fetal cells is almost exclusively glucose. Note that although protein does not usually supply much energy, it is always absolutely essential as a supplier of the amino acid building blocks for the protein synthesis involved in growth.

Dietary energy requirements differ at different times of life. On average, 5-year-old girls need around 1 550 kcal per day; moderately active 25-year-old men need 2 560 kcal, and men aged 65–74 need 2 330 kcal. A pregnant woman has a large demand for energy, requiring as much as 500 kcal extra per day over and above the 1 900 kcal that a non-pregnant, young adult woman needs. This represents a 26% increase. This extra energy is needed to support the growth of the fetus and the placenta, and the increase in various substances, such as blood and fat, that a woman produces while she is pregnant.

❑ The energy demand at any given age is not fixed at an absolute level. Why do you think this is?

■ Energy demand depends not only on basic body maintenance needs, but also on physical activity undertaken. For example, a builder needs more energy than a bank clerk of the same sex and similar size and age.

Although an extra 500 kcal should be ample for most women during pregnancy, there are several factors that could increase the demand beyond this. Women who are having a baby a short time after a previous one, those who are busy looking after an existing family or who are generally underweight or under stress may well need more energy. It is not difficult in our society for a woman to obtain the extra dietary energy she needs, particularly in view of the increase in appetite experienced by the majority of pregnant women. But dietary *quality* is another matter. Most people would agree that a diet consisting of bars of chocolate would give plenty of energy, but it would certainly not supply all the needs of a pregnant (or, indeed, non-pregnant) woman. Besides the basic raw materials which the body will use as building blocks to make its own molecules, the body requires a number of other components in a balanced diet.

❑ Can you suggest what these might be?

■ There are many answers to this, but all should include fibre, vitamins and minerals.

Dietary fibre, the undigestible parts of plants, is essential for good health as it provides bulk which stimulates the gut to contract and so allows the bowel to function properly, moving food along so that it can be fully digested and then absorbed. During pregnancy, there is a tendency towards constipation, because the growing fetus presses against the gut and suppresses its contractions, and hence the movement of material along it. Also, the high levels of some hormones, particularly progestogen, during pregnancy adversely affect the tone of *smooth muscle* (the kind of muscle that makes up the gut wall), thereby reducing the strength of gut contractions. Not only is constipation bad from a digestive point of view, it is also very uncomfortable. Therefore, during pregnancy it is all the more important to eat an adequate amount of dietary fibre to alleviate any tendency towards constipation.

Vitamins are a group of chemically diverse substances with various functions in the body (see Chapter 3). They are not broken down into smaller units within the body, but are used intact to facilitate cellular processes. Vitamins cannot be stored for long in the body: those that are water-soluble are removed by the kidneys within hours, although the others, which are fat-soluble, may be stored for a while in body fat depots. In general, however, a daily intake of vitamins is important. A balanced diet containing plenty of fruit and vegetables will contain enough vitamins for most healthy pregnancies. However, there is some evidence to suggest that if vitamin supplements are taken before and during pregnancy, there is a reduced incidence of defects in the formation of the neural tube (see Chapter 5 and the video programme *The human embryo: a developmental journey*).

❑ What is the most common form of neural tube defect?

■ Spina bifida.

❑ What is the vitamin believed to play a preventative role in the occurrence of spina bifida?

■ Folic acid.

Folic acid is one of the B vitamins. Women who are at high risk of having a baby with a neural tube defect are advised to take folic acid supplements even before they start trying for a baby. Women who suffer severe vomiting during pregnancy are also advised to take vitamin supplements to 'build them up', although what is actually being built up is the level of vitamins in the body!

The other essential class of dietary substance is minerals. These substances play defined roles within the body and so are essential for normal function. For example, calcium is needed to build bones and teeth, and for the normal functioning of the nervous system; iron is a component of haemoglobin, the pigment that allows red blood cells to carry oxygen around the body. Minerals come from the earth, and it is alleged that the urge to eat soil or coal which some pregnant women experience is actually a response to an insufficiency of minerals in the diet. Indeed, many of the dietary cravings that occur during pregnancy can be ascribed to this effect. Although the woman herself may know little or nothing about nutritional science, in some way her body 'knows' which foods are rich in particular nutrients she may be lacking. For example, a common craving during pregnancy is for dried apricots, which turn out to be a good source of iron. However, by no means all cravings can be explained in this way; another suggestion is that since many of the 'craved' foods are strong-tasting, they may be an attempt to rid the mouth of the bilious taste that many women experience during pregnancy. It is sometimes difficult to get enough calcium and iron from the diet, and women are often advised, as a precautionary measure, to take supplements of these minerals. The ideal diet for a pregnant woman is actually very similar to that for anyone. The only difference is that there should, in general, be *more* of all the constituents of a balanced diet, particularly with regard to folic acid, iron, calcium and dietary fibre.

We have looked at what a pregnant woman *should* ingest before and during her pregnancy. Let us now go on to look at what she should *not* ingest.

6.3.2 Harmful substances

As we have tried to emphasize in this course, our bodies do not exist and function in a neutral environment, but in a strongly interactive one. In daily life, we adopt patterns of behaviour that seem to us to be appropriate to our immediate circumstances, and these patterns can generally be changed in line with our surroundings. Thus, once a woman recognizes that she is pregnant, she may adopt behaviour patterns that seem to her to be more fitting to a pregnant woman; for example, she may try to rest more, or to drink more milk to increase her calcium intake. One pattern of behaviour that many people – pregnant and non-pregnant – try to change is their consumption of nicotine: smoking.

It is probably fair to say that most people in Britain today know that smoking is harmful. Certainly the evidence is incontrovertible: smoking causes heart disease and lung cancer, and contributes to many other unpleasant conditions such as bronchitis. It is not our intention here to go into an analysis of the effects of tar and nicotine; suffice it to say that because of its addictive nature, smoking is a hazardous activity, not to be embarked on lightly. Yet many people do so, and for a variety of reasons which seem valid to them.

- ❑ From general knowledge, do you know what effect a woman's smoking has on her fetus?

- ■ The baby is smaller at birth than it would otherwise have been.

The small size of babies born to women who smoke is due to two main causes: the nicotine in tobacco, and the carbon monoxide from the smoke. One of the effects of nicotine is to make blood vessels constrict, and so it can cause narrowing of the vessels supplying the placenta. This restricts the amount of nutrients reaching the placenta, and thus available to the growing fetus. This, in turn, restricts the growth and division rate of the fetal cells, resulting in fewer cells at the end of gestation. Since some developmental events involve growth, these babies may also be abnormally developed (see below). Carbon monoxide is a poison – it binds irreversibly to haemoglobin, and thus prevents it from carrying oxygen in the blood. Because the placenta can scavenge carbon monoxide very efficiently, whatever the level in the mother's blood, it is actually more concentrated in the fetus, where the effect of insufficient oxygen is extremely serious. So smoking reduces not only the supply of nutrients to the fetal cells, but also the supply of oxygen, and this is reflected in the weight of a smoker's baby: on average, 200 g lighter than that of a non-smoker.

As you saw above, babies vary a lot in size anyway, so you might imagine that it is no disadvantage to be born small. But, in fact, it turns out that it *is* a disadvantage: small babies are more likely to suffer disease, and to fail to progress normally. The smaller they are, the more likely they are to die, as shown in Figure 6.3. The reasons for this are far from clear, but a baby that is smaller than it *should have been* is generally rather immature, and its body defences are not fully developed, so it might be more likely to succumb to pathogens. Also, its immature systems are less able to cope with less-than-ideal environmental conditions which it might encounter.

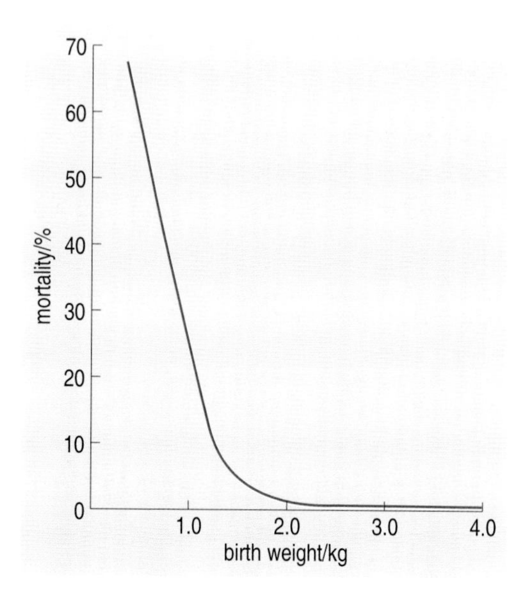

Figure 6.3 Mortality of babies according to birth weight. Mortality is measured as the percentage of new-born babies of a given weight who will die. This graph represents a simplification of the whole story: at any given weight, babies who are more mature will survive better than their less mature counterparts. Thus at almost any weight, a premature baby will fare less well than its full-term counterpart.

Babies born to women who smoke during pregnancy *are* smaller and more susceptible to disease than normal; their growth as fetuses has indeed been retarded. The damage to the placental blood vessels caused by nicotine means that smokers have twice the normal level of spontaneous abortions. The risk of having a child with congenital abnormalities is also increased among smokers, because the smaller, fewer fetal cells are not able to function properly during their sensitive developmental periods (see Chapter 5). Smoking during pregnancy is therefore not encouraged. Although it has been said that some women choose to smoke precisely so that their baby *will* be small and will give them an easier labour, in fact, there is no *evidence* that delivering a smaller baby involves a less traumatic labour than delivering a large one, despite common sense suggesting that this might be so. (Remember, though, that the woman's build must also be taken into account.) In practice, many women do give up smoking while pregnant, although few find this easy and some find it impossible. For women who are seriously addicted to nicotine, the stress and discomfort associated with trying to give up may have deleterious effects on the fetus for other reasons. Sadly, the effects on a fetus or baby can be seen even among non-smoking mothers, if they are frequently in a smoky environment. Also, babies who are raised in a smoky home are more likely than normal to suffer respiratory complaints and cot-deaths.

Another habit that people sometimes find hard to break is drinking alcohol. It is generally accepted that occasional alcoholic drinks do no *measurable* harm to the fetus, although even a small amount of alcohol does exert some effect. In fact, some women develop an aversion to alcohol during the course of their pregnancy, and do not consume any at all, which is best for the fetus. But if the mother-to-be is a heavy drinker, and persists with her drinking throughout the pregnancy, the fetus can be very severely affected, developing fetal alcohol syndrome. This is a collection of physical and mental deficiencies, caused by poor development of the fetus, which can never be compensated and remains with the individual throughout its life. The most sensitive period is between 6 and 12 weeks, and even a single 'binge' during this time can have serious consequences. Alcohol, and its metabolic product acetaldehyde, are both poisons, and kill cells by destroying their membranes. Alcohol is a small molecule, and passes unhindered across the placenta. That fetal cells are killed (as well as those of the mother) is indicated by the fact that babies of women who drink alcohol during their pregnancies are smaller than the norm.

Drugs are also best avoided during pregnancy. This applies to prescription and over-the-counter drugs, and also refers to illegal recreational drugs. Even the occasional aspirin should be avoided, and the medical profession is generally very wary of prescribing any kind of drug to be taken during pregnancy.

❑ Why should drugs be avoided?

■ Drugs, like nutrients and carbon monoxide, may be scavenged and concentrated by the placenta (see Chapter 5). The levels at which they then occur in the fetal tissues can be effectively overdoses.

No drug is entirely 'safe'; the prescribed dose of a drug is a compromise between stopping a disease and harming the patient. High concentrations of drugs in a healthy fetus will only do harm: paracetamol taken by the mother may cause liver failure in her fetus.

It is clear that a successful pregnancy depends to a large extent on the environment provided by the mother. A good, balanced diet is important, but so too is the avoidance of substances that may harm the fetus. Often, the pregnant woman's body 'knows' what is good and what is bad, and the woman develops cravings for, or aversions to various substances. But this is not a reliable measure for everyone, and the exercise of common sense, in tandem with medical advice when necessary, is an important part of being pregnant.

Summary of Section 6.3

1. The mother supplies all the nutrients used by the growing fetus. This puts a large demand on her body, so her level of nutrition must be good.
2. A good diet is one that supplies an adequate amount of energy, and includes carbohydrates, proteins and fats. It should also contain vitamins and minerals, and be high in fibre.
3. Minerals particularly important during pregnancy include calcium and iron.
4. Vitamins can be important in preventing deformities of the fetus.
5. The mother should avoid ingesting substances that will harm the fetus, such as alcohol, nicotine and most drugs.
6. Small babies are less likely to survive and do well than are large babies.

6.4 Control of fetal growth

Fetal growth is not simply a matter of all the fetal cells dividing as fast as they can to increase bulk. As you saw in Chapter 5, different populations of cells behave in very different ways with respect to division, with some dividing rapidly, others dividing slowly, and still others undergoing programmed death at an appropriate stage of development. Thus fetal growth is a highly controlled process, with net growth occurring in a coordinated way so that the fetus maintains correct and predictable proportions.

❑ What else is growing in a coordinated way?

■ The placenta.

The control of growth is actually a two-way process, with signals passing in the direction mother → placenta → fetus, and other signals passing in the opposite direction. The placenta clearly plays a central role in all this, and itself produces many of the growth control substances. At the level of individual cells, growth is controlled by a variety of **growth factors**. Some of these, such as *insulin-like growth factors*, IGFs, act in a general way to increase growth of all tissues fairly non-specifically; others, e.g. *nerve growth factor*, NGF, act only on particular target cells. Growth factors are generally named after the tissue in which they were first discovered, so the name of a particular growth factor does not necessarily tell you much about its function in the fetus. There is a vast number of growth factors, and the number is still increasing as more are discovered, so do not expect to be familiar with them all! Some are shown in Table 6.1; this is for information only – you do not need to memorize it. Please note that the functions shown are not necessarily the *only* ones, but they are the most well known.

Table 6.1 Some growth factors involved in fetal growth.

Growth factor	Function
IGF I (insulin-like growth factor I)	tissue repair, post-natal growth
IGF II (insulin-like growth factor II)	pre-natal growth
EGF (epidermal growth factor)	epithelial growth; tooth eruption
PAF (platelet-activating factor)	possible trigger of labour
PDGF (platelet-derived growth factor)	stimulates cell division
NGF (nerve growth factor)	nerve development
CCK (cholecystokinin)	gut development
TNF (tumour necrosis factor)	programmed cell death

In addition to growth factors, fetal growth is influenced by various other hormones.

❑ You already know of some hormones that are important in pregnancy. Can you name them?

■ Progestogen is important for implantation and the maintenance of pregnancy; chorionic gonadotropin is also important.

Thyroid hormones, produced by the thyroid gland and used throughout life, are also essential for growth of the fetal skeleton and brain. A lack of thyroid hormones results in cretinism, which is characterized by abnormal growth, mental deficiencies and poor coordination. Perhaps the most influential hormone in terms of fetal growth is *insulin*. Insulin, as you learnt in Chapter 3, is made by cells in the pancreas. It is involved in the regulation of blood glucose levels. Since in the fetus glucose is the energy source (Section 6.3.1), it is important to regulate it properly. There is a close correlation between fetal growth and insulin concentration in fetal blood.

The major hormones involved in fetal growth and their principal effects are shown in Table 6.2. Again, you do not have to learn this table. Note that several of the hormones are involved not just in growth, but also in the processes of maturation that most fetal body systems have to undergo. Some systems actually undergo little maturation as such: the musculo-skeletal system, for example, is laid down in a fairly complete form quite early on, and apart from the skull – which has to undergo a face-structuring developmental process – does little but grow, although the ossification which produces bone from cartilage must take place. However, many body systems require an extensive amount of modification to enable them to function efficiently when the baby becomes independent of the mother. Let us now look in more detail at these maturation steps. Remember that although we are dealing with them system by system, they are all coordinating with each other, and influencing each other's environment. Remember also that the time-scale we use is based on a 38-week gestation period (266 days), and we shall be referring to weeks post-fertilization, *not* weeks since the last menstrual period.

Table 6.2 Some of the hormones involved in fetal growth and development.

Hormone	Site of action
adrenocorticotropic hormone	adrenal glands
gonadotropins	gonads
prolactin	lungs and liver
placental lactogen	liver
insulin	general growth and laying down of fat stores
oestrogens	lungs, skeleton, gonads and brain
cortisol	lungs, liver, brain, retinas, gut and placenta
gastrin	gut
thyroid hormones	thyroid, brain, liver, skeleton and skin

Summary of Section 6.4

1 Fetal growth is regulated by many factors, some of which act in a generalized way, other of which are more specific in their action.

2 Insulin and insulin-like growth factors are particularly important in the control of fetal growth. This may be related to the role of insulin in the regulation of blood glucose.

3 Many of the regulators of fetal growth act also on the maturation of various body systems.

6.5 Maturation of the gut

The fetal gut does not have a role in nutrition.

❑ How does the fetus obtain its nutrients?

■ From the placenta, via the umbilical blood vessels.

Nevertheless, the gut has to mature so that the baby can feed immediately after birth. You will learn about the processes of digestion in Book 3; for now, all you need to know is that digestion involves breaking down ingested food into its small-molecule building blocks, which are then absorbed across the gut wall into the blood. Many of the processes of digestion and absorption involve enzyme-catalysed reactions, so you can guess that an important part of gut maturation must be the acquisition of the ability to synthesize appropriate enzymes. By the end of the first trimester, when you last encountered the gut, some of the enzyme systems are in place, but the whole set of digestive enzymes appropriate for a new-born baby is not present until 24–28 weeks. Because their diet is exclusively milk, babies' stomach enzymes are different from those found in the adult. Only the 'baby' enzymes are produced by the fetal gut.

Another important function that the gut must develop is the ability to carry out **peristalsis**. Peristalsis is the name given to the waves of contraction followed by relaxation which pass from the top to the bottom of the gut. It is what moves food through the gut: without it, the gut would simply get filled up until it was impossible to eat any more. Peristalsis is a muscular activity, but it is not one that we can consciously control: it is an **autonomic** function. Peristaltic movements first become evident in the fetal gut by about 14 weeks. From then until birth the gut is active, producing waves of contraction and relaxation, which improves the muscle tone so that the muscles around the gut are well developed by the time they have to move real food. But in the absence of food, what fills the fetal gut? The answer is amniotic fluid. By 16 weeks, the fetus is able to swallow, and it does so with gusto, swallowing one-third of the total volume of amniotic fluid every hour. The fetal kidneys produce large amounts of urine which is excreted into the amniotic fluid, so this is also swallowed by the fetus. Though this may seem distasteful, it does no harm. As the intestine matures, cells along its length begin to produce mucus, which, after birth, will help to lubricate the food, and particularly the stools, during the passage through the gut. As the digestive enzymes become active, the mucus will be digested, and will leave a sticky, black residue in the bowel. This is the *meconium*, which is passed by the baby shortly after birth. Occasionally, meconium may be passed before birth, into the amniotic fluid. This is evidence of fetal distress, and is occasioned by the stress hormone, adrenalin. (Adrenalin causes contractions in adult bowels, too.)

Babies feed by sucking, which is a voluntary activity in adults, but a reflex in babies. As such, sucking cannot be done until all the facial muscles are in place and working properly. The fetus is often seen apparently sucking its thumb in ultrasound scans taken mid-way through gestation. It is doubtful if proper sucking is going on at early stages, although it has been suggested that some fetuses actively seek their thumbs for comfort sucking. In the new-born, sucking, swallowing and peristalsis must be coordinated to avoid choking. This coordination is not achieved until 34 weeks of gestation.

❑ Try sucking your thumb. Besides coordination, what else do you think is needed for successful feeding?

■ Sucking involves a squeezing movement between tongue and palate, so these must also be present.

The tongue is muscular, and the palate is developed as part of the skeletal system, so these must be growing correctly too for the baby to be a competent feeder.

A major organ associated with the gut is the liver. In the adult, the liver's function is largely one of carbohydrate, fat and protein management. But for the first two-thirds of gestation, the liver has a different function: its role is to produce red blood cells. As you learnt in Chapter 5, in the early part of development all fetal red blood cells are made in the yolk sac, but production moves to the fetal liver once this organ is established. After the 24th week, blood cell production in the liver drops sharply and enzymes characteristic of the new-born liver begin to appear.

❑ Does this mean that no more red blood cells are produced in the fetus?

■ No, but they are made elsewhere.

In fact, from late gestation throughout life, red blood cells are made in the bone marrow, which fills the cavity down the middle of long bones such as the femur. The cavity cannot appear until the bone has reached a minimum length, so this is another case of the development of one system having to be coordinated with the growth of another.

One of the most important enzymic functions of the liver is the synthesis of glycogen. Glycogen is a large molecule made from glucose building blocks (see Chapter 3). Glycogen synthesis is the body's way of storing glucose for a short period of time, for example after a meal until the glucose is needed. In fact, one of the reactions controlled by the hormone insulin is the one which converts glucose to glycogen in the first place. Glycogen is an important energy storage molecule, but to be useful it needs to be converted back to glucose again. This also takes place in the liver. In the adult most glycogen is stored in the liver, and over the final nine weeks of gestation glycogen starts to appear in the fetal liver too. The glycogen will play a vital role at the time of birth: it will provide a source of glucose to tide the baby over between the time when the placental supply is interrupted and the first meal. A reserve of fat is also laid down. Part of this is derived from maternal fatty acids, but the fetal liver can synthesize a limited amount by itself. This serves as a fail-safe device in case the first meal does not occur until after all the glycogen reserves have been depleted.

Although the liver's enzyme systems continue to develop through the last weeks of gestation, they are *not* complete by birth, but require a few days of post-natal life to be fully functional. As long as the new-born baby is well fed, this level of liver immaturity is not generally a problem, provided that the glycogen to glucose pathway is working. But overall, the gut must be capable of taking in and digesting milk by the time of birth, otherwise the baby will not be able to survive.

Summary of Section 6.5

1 The gut must be able to move food along, and digest it. Movement is achieved by peristalsis, and food is broken down by the action of enzymes.

2 The gut can perform peristaltic movements by 14 weeks of gestation, and peristalsis continues from then onwards. Mucus and other fluids are moved through the gut.

3 Gut enzymes appear by 24–28 weeks of gestation, and they are the enzymes used by a baby, not by an adult.

4 The liver is an important gut-associated organ. Its first role is to make blood cells. When the bone marrow has developed, blood cell production moves there, and the liver assumes its role in the new-born baby of managing fat, protein and especially carbohydrate metabolism. This happens in the last nine weeks of gestation.

5 Sucking, swallowing and peristalsis must be coordinated to avoid choking. This is achieved by 34 weeks of gestation.

6.6 Maturation of the lungs

The function of the lungs is to provide a pool of air inside the body from which the red blood cells can pick up oxygen, and into which the gaseous waste products of cell respiration – carbon dioxide and water vapour – can be discarded. At each breath, the pool of air is changed. Exchange of molecules of gases between the air pool in the lungs and the bloodstream takes place rapidly by simple diffusion (the movement of molecules down a concentration gradient – see Chapter 3, Box 3.2), as the distance to be crossed is only tiny. Nevertheless, diffusion can take place only if the surfaces involved are kept wet, as the gases must diffuse in solution, i.e. dissolved in a liquid. The lung itself consists of millions of air-filled sacs called **alveoli** (singular: alveolus) whose surfaces are covered by a thin layer of liquid (Figure 6.4). This gives a huge surface over which gas exchange can take place: the total area of all the alveoli in adult lungs is about the same as a tennis court! You will learn more about lung function in Book 3.

The complex, branching structure of lungs takes some time to develop. Up to about week 16 the bronchioles develop, then the vascular system (network of blood vessels) begins to grow around them, and alveoli start to form. However, these alveoli are not fully functional, and are simply small buds of tissue. Nevertheless, by week 28 they have started to form what will become air spaces, and premature babies born from this time on have an increasingly good chance of gaining full lung function, whereas prior to this time their chances of survival are very small. From 36 weeks onward, true, mature alveoli appear, and by birth there are about 50 million of them. As the child grows, so too do its lungs, so the production of alveoli continues throughout childhood, as the individual acquires the ability to exercise strenuously. During this period it is important to protect the immature lungs from environmental stresses such as cigarette smoke and other pollutants, which could interfere with normal lung function and development.

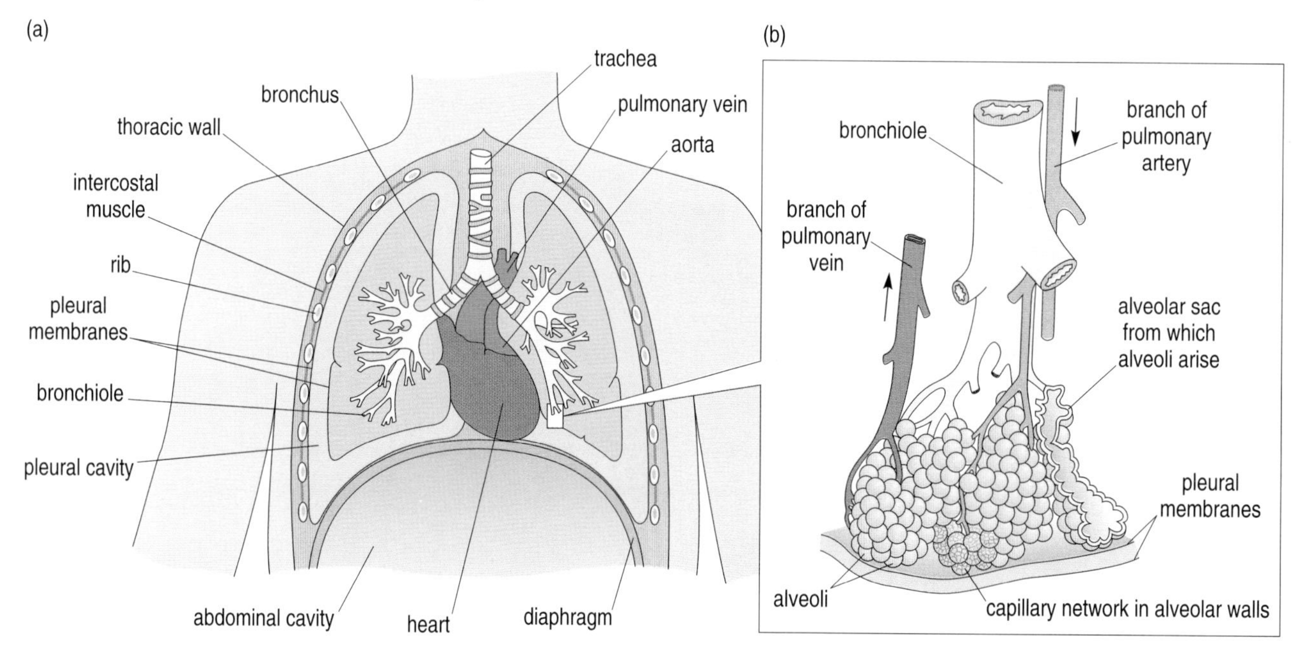

Figure 6.4 Lung structure. (a) From its opening at the nose and mouth, the respiratory tract consists of a single large tube, the trachea, which branches into two large bronchi. These branch progressively into ever-smaller bronchioles, each of which terminates at an alveolus, as shown in (b); it is here that gas exchange takes place. Note that the lungs are attached to the ribcage and to the large sheet of muscle, the diaphragm, lying beneath them. These attachments are essential for the breathing movements that pull air into the lungs and then expel it.

Throughout gestation the developing lungs are filled with fluid, and during the second half the fetus practises breathing movements. These are essential for proper lung development: as with peristalsis, the practising improves muscle tone.

❑ Look at Figure 6.4. What structures, apart from lungs, are necessary for breathing movements?

■ A ribcage and muscles (intercostals and diaphragm).

At birth, the lungs must empty of fluid and fill with air, and then continue breathing movements for the rest of the individual's life. The fluid that fills the lungs during gestation is constantly produced by the lung cells, and the practice breathing movements force it outwards from the lung to the amniotic fluid. Fluid production slows around the time of birth, and the stress hormone, adrenalin, which is produced in the fetus during birth, causes much of the fluid to be reabsorbed into the lung cells. Thus, there is normally only a small amount of fluid remaining in the lungs by the time of birth, and much of this is squeezed out during the birth process itself.

At the moment when the lungs are empty of fluid, but before the first breath has been taken, they are collapsed; the inner surfaces of the alveoli touch each other, and are held together by surface tension because the surfaces are covered with a thin layer of liquid.

❑ Why must there be a thin layer of liquid?

■ So that gas exchange by diffusion will be able to occur.

As the first breath is taken, the touching walls of the alveoli must be forced apart by the incoming air, so that there is enough alveolar membrane exposed to the air for gas exchange to take place. The force required to overcome the surface tension, separate the alveolar walls and inflate the lungs is very large. An analogy is to think of the effort required to separate two sheets of clingfilm that are placed one on top of the other, compared with that required to separate two sheets of paper similarly placed. The force must be generated by the action of the muscles involved in breathing – the diaphragm and the intercostals. This indicates the importance of the 'practice' breathing movements occurring prior to birth, which improve the strength of the muscles.

What lowers the surface tension within the alveoli, and thus makes them easier to inflate, is a substance called **surfactant** (*surf*ace-*act*ive *agent*, a slippery mixture of proteins and fats). Surfactant is made in the alveolar cells themselves, and spreads in a layer one molecule thick over the inside surface of the alveolar sacs. It starts to appear from about 18 weeks, then increases sharply in amount from about 24 weeks. This means that the development of mature alveoli is accompanied by a high level of surfactant. If the surfactant is deficient, the new-born baby will have trouble breathing, and may suffer from **respiratory distress syndrome**. Here, the lung cannot expand properly, so the air pool is too small to meet the infant's oxygen needs. It is possible to administer synthetic surfactant by nebulizer (a device that delivers a spray of tiny droplets into the lungs) to babies with this condition, and this may tide them over until their own level of production becomes adequate.

Thus for lungs to be working at birth, they need to have developed structurally, with an appropriate number of mature alveoli, and functionally, with the alveolar surfaces coated with surfactant, allowing inflation at birth and gas exchange thereafter.

Summary of Section 6.6

1 Gas exchange takes place in solution across the thin, moist walls of the alveoli.

2 Alveoli start to appear at 36 weeks of gestation, and continue to develop throughout childhood.

3 At the time of birth the lungs are collapsed, and the first breath must inflate them. This is helped by the production of surfactant.

4 Lack of surfactant can result in respiratory distress syndrome.

6.7 Maturation of the nervous system

To function correctly, many body systems need electrical signals, and most of these signals are generated by the nervous system. The nervous system consists of the brain, the spinal cord, and nerves running to and from most sites of the body. There are other important structures and components too, but we shall not concern ourselves with these now.

As you saw in Chapter 5, the brain and spinal cord are formed very early on in the embryo. But the nerves themselves pose something of a problem: they cannot develop fully until the structures they connect with have been formed. Thus, development of the nervous system is a highly dynamic process which goes on throughout gestation, and indeed for some time afterwards. Nerves initiated during the first few weeks of gestation retain the capacity to grow in length, so that as the fetus grows and develops, the nerve cells migrate to their final positions, and grow to keep pace with the whole organism.

❑ Can you name a factor that influences nerve growth?

■ Nerve growth factor (see Table 6.1). NGF stimulates nerve growth.

❑ How do you think the nerves can migrate to the correct site?

■ They may detect and follow paths defined by chemical and adhesive differences, sticking to some types of adhesive cells more than to others.

Most nerves are formed between weeks 12 and 16 of gestation, and after this time growth and final positioning take place. However, the brain is very far from complete at this stage.

Our brains are extremely complex, and it is this complexity that sets us apart from other mammalian species. The brain consists of many millions of cells, both nerve cells, or **neurons**, and supporting cells, or **glia**. Connections between neurons are called **synapses**, and it has been said that there are more synapses in the brain than there are stars in the Universe. You will learn much more about the brain and its workings in Book 2; for now we shall concentrate on its early development.

Although the brain first appears as an enlargement at one end of the neural tube (see Chapter 5 and the associated video programme), it very quickly grows into a recognizable, relatively large structure. From about 10 weeks after fertilization the brain rapidly increases in mass, and this is reflected in an increase in cell numbers. Neurons lose the capacity to divide shortly after birth, so a whole lifetime's worth of brain neurons must be produced by then. As shown in Figure 6.5, the brain continues to grow for some time after birth. Some of this growth is accounted for by an increase in cell number; however, much is due to the increase in size of individual cells. A new-born baby is actually neurologically quite immature, even though in terms of physical proportions it has a very big head. Babies' heads have to be big to accommodate their large brains, but if they are too big, normal birth will be impossible. If the baby's head is too large, it will get stuck, resulting in the death of both baby and mother. It is therefore probably necessary for the baby to be born at an immature stage, so that the head – the biggest part of the baby – will fit through the pelvis.

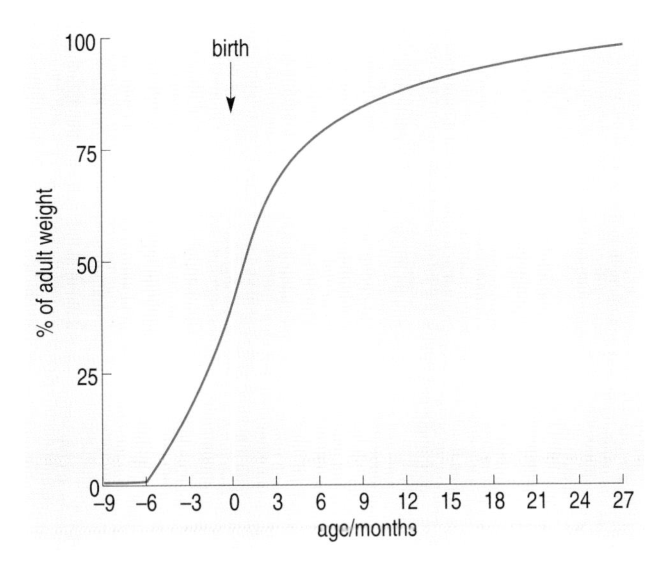

Figure 6.5 Increase in weight of the brain during pre- and post-natal development.

Although we talk about 'the brain', there are many different parts of this organ, all with different functions, growth characteristics, and developmental pathways. We cannot hope to describe all of them, so what follows will pick out only the major features of the growth and development of some parts of the brain.

There are many factors affecting brain growth.

❑ Can you suggest some?

■ Nutritional and hormonal factors are certainly important.

You might also have suggested oxygen supply, but, interestingly, the developing brain is highly *tolerant* of oxygen deprivation during gestation. This is certainly not the situation during and after birth, when even a small deficit of oxygen can have catastrophic results, such as fits and subsequent severe lack of coordination. Clearly, one of the steps involved in the brain's maturation is the development of sensitivity to oxygen levels.

❑ What would you predict to be the effect of undernutrition on brain development?

■ There would be fewer brain cells, and they would develop their functions later than usual, and perhaps abnormally.

Indeed, it is found that the babies of undernourished women develop movement and reflexes later than usual, and also show diminished learning ability. These effects are also observed in the babies of women who have taken harmful quantities of alcohol or some other drugs during their pregnancies.

Several hormones are known to affect brain development, in particular the thyroid hormones. These seem to have a major role in coordinating the development of all the parts of the nervous system, so that they develop together at an appropriate rate. If the thyroid gland itself does not function correctly, the effect on the brain will be severe. Either an underactive *or* an overactive thyroid gland can lead to lowered cell numbers in the brain, with damaging consequences.

Finally, genetic factors also influence brain development. This is shown by the level of development seen in cases of genetic abnormality, such as Down's syndrome, which is caused by an extra copy of chromosome 21. People with Down's syndrome have fewer cells than normal in certain regions of the brain, and they also seem to have an abnormal distribution of myelin, a fatty substance which coats neural (nerve) tissues. The above information suggests that the brain is a good example of the influence of both genetic and environmental factors on development.

You have seen that at birth the brain is immature in terms of its size, but what about its function? Among the functions of the brain are the senses, and we shall look at two of these: hearing and vision.

6.7.1 Hearing

Hearing depends upon the structure of a sensory organ, the ear (shown in Figure 6.6), which receives the sound stimulus, as well as on the neural connections which transmit information to the brain and permit the interpretation of that information. By the end of the first trimester, the middle ear is formed. (The middle ear comprises the three small bones involved in sound transmission: malleus, incus and stapes.) By 20 weeks, the inner ear, with its associated neural 'wiring', is in place. From about 24 weeks the fetus can respond to external noises – women who have been pregnant will be familiar with the fetal 'jumps' which follow loud noises. The sensitivity of the fetus will increase through gestation, as will the *range* of frequencies (high or low pitches) to which it can respond. The fetus is constantly exposed to a wide variety of sounds, some from its mother (heartbeat, intestinal gurgles) and some from outside her body. It is claimed by many people who study new-born babies that by the time of birth a fetus can recognize and respond to its mother's voice, and finds the sound and rhythm of her heartbeat soothing. This is probably one reason why babies enjoy being cuddled and carried around: they hear sounds that were familiar in the womb, and this is comforting. Fetuses in the final weeks of gestation can also associate sounds with particular activities: mothers who made a habit of resting and watching television at particular times while they were pregnant often find that after birth their babies will go to sleep when they hear familiar TV theme tunes! Clearly, then, hearing is quite well developed before birth, but it is not yet fully functional. Although the ear itself is mature by birth, the pathways needed to process the auditory signals continue to develop for about a month after birth. During the time of development, the immature system is susceptible to damage from loud noise (as well as by pathogens such as *Rubella* – see Chapter 5).

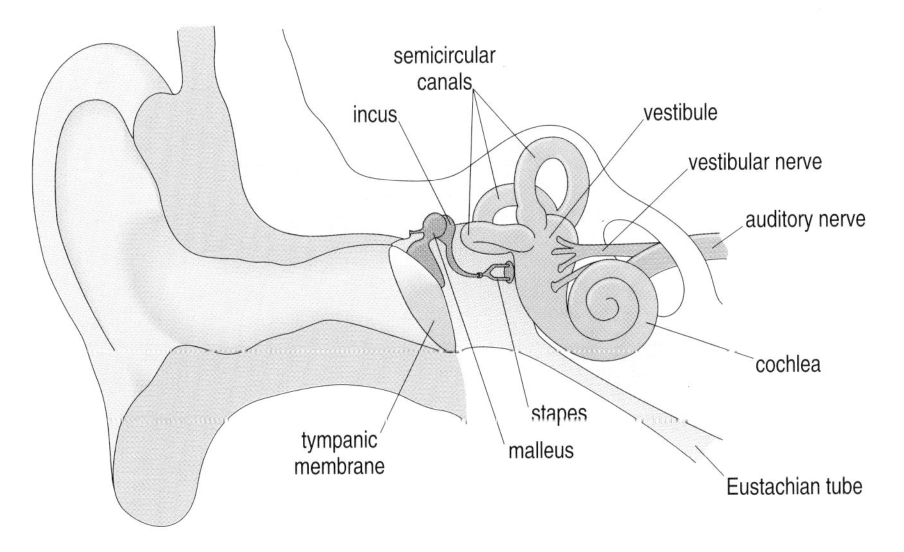

Figure 6.6 Structure of the ear. Sound waves received by the outer ear (the external visible part of the ear) are passed to the tympanic membrane (eardrum), causing it to vibrate. These vibrations are transmitted via the three small bones of the middle ear (malleus, incus and stapes) to the cochlea. In the cochlea the sound waves are transduced into electrical signals which are transmitted to the brain via the auditory nerve. The cochlea and semicircular canals (which are used for balance and posture) make up the inner ear.

6.7.2 Vision

As with hearing, the sight process depends on the functioning of both the sensory organ – in this case, the eye – and its associated 'wiring' within the brain. You will recall from Chapter 5 that the eye starts to develop very early in gestation. A large part of it is complete by about 22 weeks, but the light-sensitive elements (called rods and cones) do not begin to develop until about a week before birth, and continue their development for many months afterwards. It appears that visual experience is needed for the interpretative nerve pathways to be established, and since there is little if any light in the uterus, this can only happen after birth. Nevertheless, human babies are born with their eyes open, and are able to detect objects if they are close. In terms of visual ability at birth, this puts them among the more 'mature' mammals, many of which do not open their eyes for some time after birth (although their other senses, such as their sense of smell, may be very much better developed). It has been suggested that having their eyes open allows babies to find a breast, and therefore food, but this seems unlikely, as all baby mammals are able to locate their first meal.

6.7.3 Movement

The brain-controlled function that mothers are first aware of in their fetuses is movement. Fetal movements can be felt by about the 15th week of gestation, although of course the precise time will vary from fetus to fetus. In fact the fetus may well have been moving for several weeks before this, but was probably too small to be felt. Contraction is a fundamental property of muscle – even a group of muscle cells in a laboratory dish will contract spontaneously – but fetal movements of the sort that can be felt tend to be 'proper' movements, caused by electrical stimulation of the muscle via the nerves. The fetus will continue kicking, elbowing and head-butting its mother throughout gestation, and these movements are important for establishing muscle tone.

❑ What other cases do you know of where practice before birth is important for post-natal function?

■ Muscles used for peristalsis and for breathing must also be exercised.

6.7.4 The hypothalamus and pituitary gland

Up to this point in the course, we have mentioned hormones as controllers of body processes, without much explanation of where hormones are made or how their production is itself controlled. This is because this topic occupies quite a large part of Book 2. But any discussion of brain development would be incomplete if it did not mention this matter, however briefly. The pituitary gland and the hypothalamus are specialized areas of the brain (see Figure 6.7) which are responsible for the production of several hormones, and for controlling the production of many more.

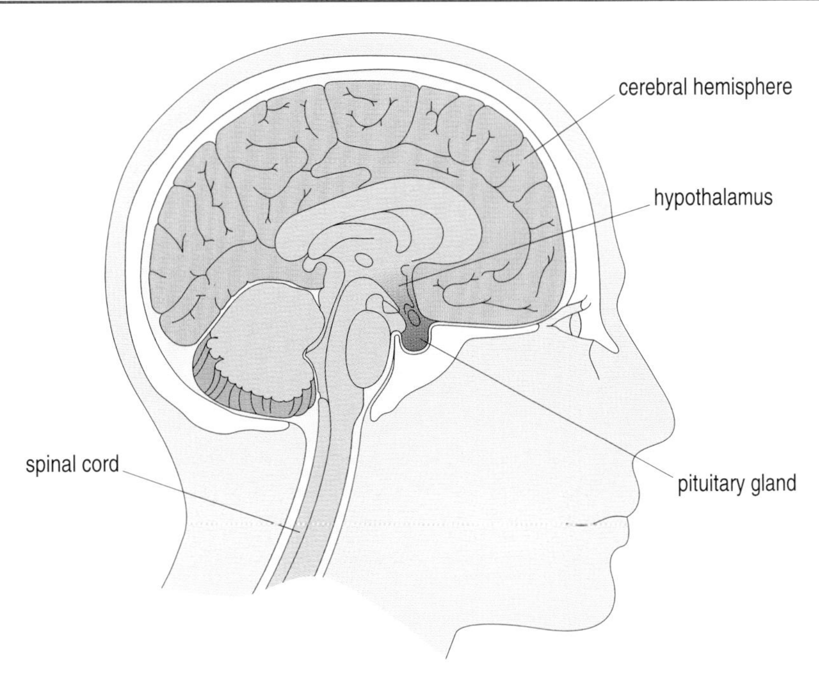

Figure 6.7 Diagram to show the position of the pituitary gland and the hypothalamus.

The pituitary gland and hypothalamus are almost completely formed by the first trimester of gestation, with the finishing touches in place by about 14 weeks. Although they produce several hormones involved in fetal growth and maturation, for now we shall limit our discussion to the production of **growth hormone**. Growth hormone influences the growth and repair of a large number of tissues, both in the fetus and, more importantly, at post-natal stages. Growth hormone can first be measured in the fetal blood from between 10 and 14 weeks. Between 14 and 25 weeks the level rises 10-fold. This coincides with the development of a major system of blood vessels supplying the part of the brain where the pituitary gland and hypothalamus are located. The blood can therefore take the newly made growth hormone round the fetal body, where it will affect its target tissues. After 25 weeks, the level of growth hormone declines.

❑ What does this imply about the action of growth hormone?

■ It exerts its major effect on the growth that is occurring between 14 and 25 weeks. Look back at Figure 6.1b to see how growth proceeds over this period.

The pituitary gland and hypothalamus produce other hormones involved in fetal development, in particular certain sex hormones. You have already met these in Chapter 4 and you will learn more about them in Book 4.

So, to summarize, the development of the nervous system shows a marked contrast to that of the gut or lungs in that it is very far from being complete by birth. Certain brain functions are needed at particular times during gestation, such as the pituitary gland/hypothalamic hormones, but for many 'higher' functions the end-points are reached only as the result of learning from experiences that occur *after* birth. Thus, from the outset, human sensory development relies on interactions with the external environment.

Summary of Section 6.7

1 The brain is apparent by 10 weeks of gestation, but continues to grow throughout gestation and post-natally.

2 Nerves are in place by 12–16 weeks of gestation, and grow to keep pace with the developing body. Nerve cells do not divide after birth.

3 The middle ear has developed by 12 weeks of gestation, and by 24 weeks the fetus responds to sounds.

4 The eye is largely complete by 22 weeks of gestation, but colour vision starts to develop at 37 weeks and continues post-natally.

5 Interpretation of sounds and visual information requires post-natal experience.

6 Fetal movements can be felt by 15 weeks of gestation.

7 Fetal movements continue vigorously throughout pregnancy. This improves the development of the muscles.

8 Although the pituitary gland and the hypothalamus are complete by 14 weeks, production of growth hormone does not peak until later, when a blood system has developed to deliver it to its target organs.

6.8 Maturation of the cardiovascular system

The **cardiovascular system** is the network of blood vessels throughout the body, and the heart which pumps the blood around it.

❑ From general knowledge, and from what you have learned so far, what are the functions of the blood?

■ It brings supplies of nutrients and oxygen to all body cells, and removes waste products from them.

In the fetus, where there is a massive amount of cellular activity going on, it is obviously essential to have an efficient blood supply. But the fetus is growing at such a rate that the cardiovascular system must develop quickly and extensively to keep up with it. The heart and major blood vessels are in place, as you know, by about 8 weeks, but they are very small, and the heart is relatively weak. Fetal blood pressure is about half that of the adult, so the heart has to work less hard to pump the blood round the fetal body. The placenta, through which all the blood must pass, does not offer much resistance to blood flow, and so does not cause any particular load on the heart. A simplified diagram of the fetal circulation is shown in Figure 6.8. Note that although a lung is shown, it does not, of course, have the gas exchange function that it will have after birth. Nevertheless, it has a blood supply to service the developing alveoli.

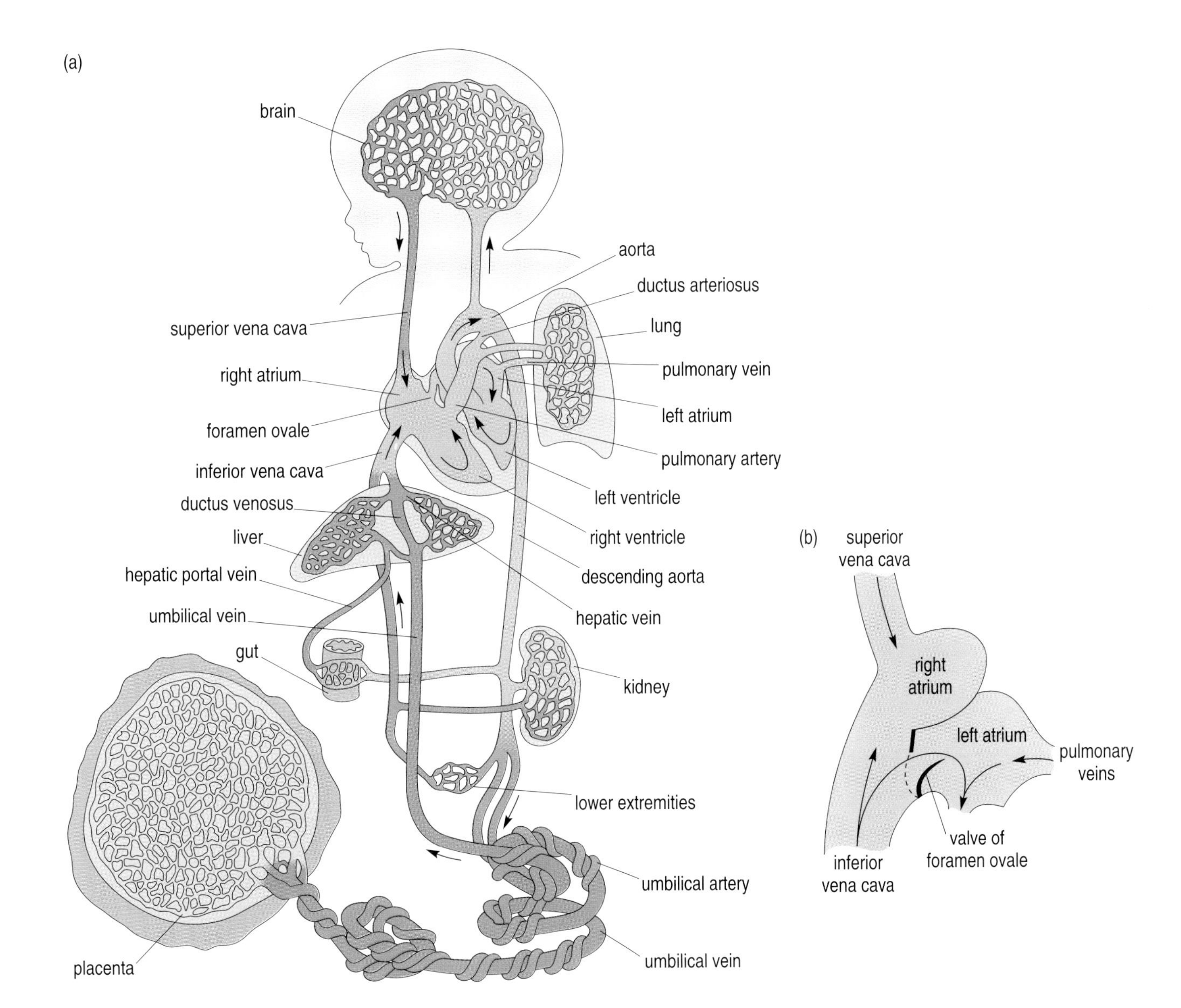

Figure 6.8 (a) The fetal circulation. Note the three vascular shunts (vessels that bypass particular areas where little blood supply is needed by the fetus): the foramen ovale between the right and left atria, the ductus arteriosus leading from the pulmonary artery to the aorta and the ductus venosus through the liver. (b) Enlarged view of the foramen ovale, which connects the two atria of the heart, allowing blood in the inferior vena cava, recently arrived from the placenta and therefore well oxygenated, to pass directly to the left atrium and hence straight out to the body again, so bypassing the lungs.

❑ Recall from Chapter 5 the difference between a fetal and an adult heart.

■ The adult heart has four chambers: two atria and two ventricles. The fetal heart has two atria, but the ventricles are not separate and so allow blood from the two sides of the heart to mix.

The heart matures by growing a membranous wall, or septum, which eventually separates the left and right ventricles. This means that after birth, the deoxygenated blood in the right side of the heart will not mix with the oxygenated blood in the left side of the heart, and thus the tissues supplied by the arteries will get the maximum possible amount of oxygen. Occasionally the septum fails to grow properly, so that after birth there is mixing of the oxygenated and deoxygenated bloods. Babies with this so-called 'hole in the heart' do not do well, because their cell metabolism is limited by lack of oxygen, but fortunately the septum can be repaired surgically with a good success rate.

In the fetal circulation there are three **vascular shunts**. These are temporary junctions between blood vessels which will later be completely separate. Vascular shunts allow blood to bypass certain organs (see Figure 6.8). This is important during fetal development. Two shunts, the *foramen ovale* between the right and left atria and the *ductus arteriosus* leading from the pulmonary artery to the aorta, shunt most of the blood away from the lungs, which need only enough blood for maintenance (see above). The foramen ovale and the ductus arteriosus are kept open by the high pressure of blood flowing through them. The third shunt, the *ductus venosus,* can direct blood away from the liver so that when there is not much oxygen or glucose around, what little there is can be used preferentially by the heart and the brain. By the time the baby is living outside its mother, these shunts need to be closed off so that the bypassed organs can receive the amount of blood they require for independent living, but this cannot actually happen until the fetus is separated from the placenta. Thus many vital changes take place *during* birth. Occasionally the closure of the fetal shunts does not occur, and the baby may be born with a heart murmur (so called because it sounds like a murmur instead of the crisp beat of a normal heart when heard through a stethoscope). This sometimes needs surgery to rectify it.

As the lung expands during the baby's first breath, its vascular resistance to blood flow drops sharply: the enormously expanded area allows the blood vessels to open widely, instead of being squashed closed as they are in an unexpanded lung. This encourages blood to flow into the lung, and the resulting pressure changes in the right atrium of the heart make the foramen ovale and the ductus arteriosus close. This is helped by the fact that the umbilical cord constricts, reducing the blood flow back to the right atrium. The closing of the foramen ovale stops blood going from the right to the left atrium, and makes sure that all blood entering the right side of the heart goes to the lungs. The precise control of all these processes is not completely understood. It is not known whether they are under nervous control, or whether the whole series of changes happens simply because of blood pressure changes. The blood pressure changes are caused not just by the large-scale differences in the circulation *pattern* (such as that associated with lung expansion), but also because of changes in diameter of individual blood vessels (such as the umbilical vessels). As you will learn in Book 3, blood vessel diameter makes a significant contribution to blood pressure. At birth there are large changes in the levels of various prostaglandins (see Chapter 4, Section 4.3.3). Prostaglandins are known to affect blood vessel diameter, so it seems as though prostaglandins may play a role in the rapid maturation of the cardiovascular system.

Thus the cardiovascular system shows a different maturation pattern still: the lung and most of the gut mature before birth, the nervous system after birth, and now the cardiovascular system shows that maturation can actually occur very rapidly *during* birth. Next we look at a system which never matures completely: the body's defence system.

Summary of Section 6.8

1. The fetal cardiovascular system develops early to service all the rapidly growing parts of the developing fetus.
2. There are major differences between the systems of the fetus and the new-born, which are related to function. Vascular shunts allow the blood flow to bypass the lungs and the liver, which are not very active before birth.
3. At the time of birth, blood pressure changes cause the shunts to close, and the new-born circulatory pattern is established. This is a rapid process, and is essential for the baby to thrive.
4. Prostaglandins may be involved in closing the shunts.

6.9 Maturation of the immune system

The immune system is extremely complex, and you will learn about it in Book 2. The cells that mediate the immune response are produced in the fetal liver, bone marrow and **thymus** (a small organ in the upper chest, derived from an outgrowing of gut endoderm), so these need to be up and running first. All the cells or their precursors are in place by the end of the first trimester. The function of immune system cells is to be able to recognize self and non-self molecules and, in the case of non-self, deal with them in an appropriate manner, 'neutralizing' them before they can do much damage. As we hinted in Chapters 4 and 5, the immune system is very efficient at doing this, and the implanting embryo needs to be protected from the mother's immune system so that it is not rejected. This is achieved by the absence from the trophoblast cells of the surface molecules that would normally signal fetal identity. But from the point of view of the *fetus'* immune system, the hard part is to 'learn' what is self, particularly when new versions of self, as identified by new molecules on the cell surface, appear with each differentiative event. Recognition is experience-based, and since this continues to occur throughout life, the immune system is never completely 'mature' in the sense of being finished. However, what we are interested in here is the development of the immune system's capacity for recognition and action.

The first cells of the immune system are, like blood cells, made by the fetal liver, and later by the bone marrow. Production of *some* kinds of immune system cells starts at about six weeks of gestation, but these cells are thought not to be fully functional because other components necessary for a full immune response are not yet present. Although production begins at around six weeks, the levels are very low, and increase only gradually. Even by birth, many of the components are present at only about half or less of the adult levels, and by the time the baby is a year old the various components of the system may still be only at about 75% of adult levels. During this time of immaturity, it is likely that most molecules encountered will be recognized as self. The precursors of the cells that actually recognize self and non-self are made in the fetal liver and bone marrow, but some migrate to the thymus. Once in the thymus, the precursor cells can differentiate into mature T (for thymus) cells under the influence of factors produced by the thymus itself. The cells that produce antibodies, the B (for bone marrow) cells, begin their maturation in the bone marrow; then, at about 17 weeks, they start to migrate to developing lymph nodes (small oval structures situated throughout the body along the lymph vessels). They do not complete their differentiation until they are challenged by a non-self molecule – you will learn about this in Book 2. Antibodies are not made until just before birth. The new-born baby is still very dependent on its mother for immunological protection, and receives, from her milk, antibodies to substances that *she* has been exposed to.

Summary of Section 6.9

1 The immune system is never fully mature.

2 By the time of birth the components of the immune system are present at only about half the levels found in adults.

3 Precursors of immune system cells are first made in the fetal liver. Later, two cell populations emerge: one, the B cells, are produced in the bone marrow and the lymph nodes; the other, the T cells, in the thymus.

4 The fetus depends for protection on its mother's immune system. The new-born also receives immune protection from its mother, via her milk.

6.10 Review

We have now reached the end of our discussion of how the fetus matures so that it will be able to survive after birth as a free-living individual. Before looking at the birth process, it may be worth while reiterating the main message of these sections. The first thing to say (again) is that none of the processes occur in isolation.

Pick three examples from Sections 6.4–6.9 of instances where developing systems must interact in order to mature. Draw a flow chart, or other simple diagram, of all three examples, showing any connections between them. Does this diagram suggest any further interactions that were not brought out in the text?

We hope that your diagram looked something like Figure 6.9, though you may well have chosen different examples. This kind of diagram can be useful to help you see at a glance the complex interactions which are an integral part of human development.

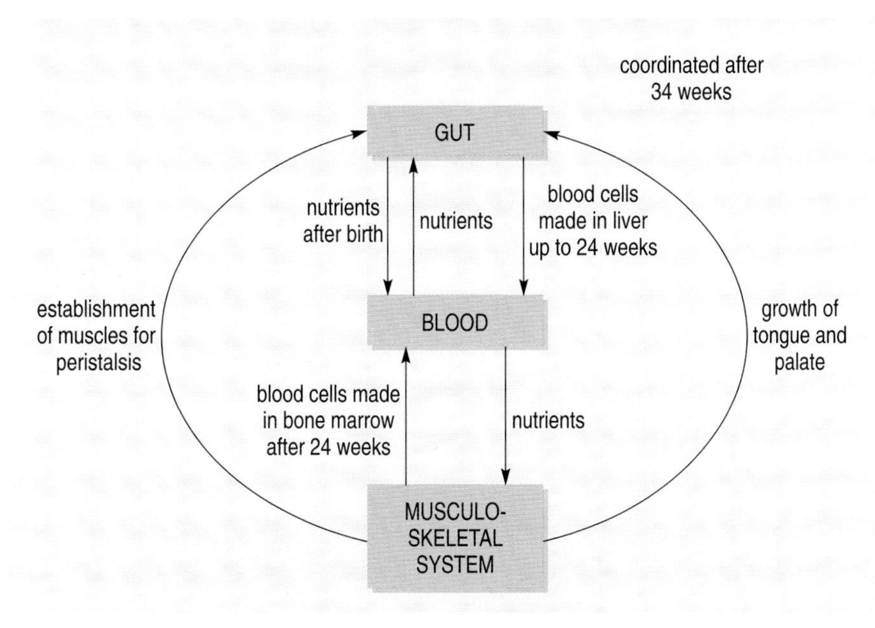

Figure 6.9 Flow diagram showing the interactions between the developing gut, blood and musculo-skeletal systems.

It is important to realize that the interdependence of the developing systems imposes time constraints on them. Although one system may be *able* to function, it may not actually do so because a different system is not yet sufficiently well developed to interact with it.

❑ Give an example of this.

■ The hypothalamus and pituitary gland are able to make growth hormone from week 10, but this hormone is not produced in significant amounts until week 14, when the circulation exists to transport it to its targets.

A week-by-week guide to fetal development is shown in Table 6.3. Although you do not need to remember all the details of this, you should at least remember the order in which different systems develop. This will enable you to estimate which, if any, systems might be damaged by harmful treatments at different times of gestation. Remember also that fetuses are *individuals*; although they all develop at approximately the same rate, they will show some individual variation in this, as in other, features. The week numbers quoted are for the 'average' fetus.

You have already seen that fetuses differ quite a lot in their growth rates and their size at birth. As well as varying also in the precise timing of their developmental steps, they exhibit a further difference, the uncertainty of which can make life difficult for expectant mothers: they differ in the time at which they are born. It is this aspect of reproduction which we look at next.

6.11 The onset of childbirth

Towards the end of pregnancy, many changes occur which signal that the development of the fetus is approaching completion and that birth is imminent. The woman herself will generally be very large. The extra load, consisting of baby, placenta, amniotic fluid and the mother's own increased blood volume and fat deposits, will make her very tired, and the large bulk pressing on her diaphragm will mean that she finds it difficult to breathe deeply.

❑ What are the advantages of these increases in blood volume and quantity of stored fat?

■ Blood volume increases to ensure a plentiful supply of nutrients to the placenta; some of the extra fat laid down is used for producing milk after birth.

You will learn more about pregnancy from the mother's point of view in Book 4, so for the time being we ask you to accept that the statements we make here about maternal physiology *are* actually based on evidence, and not just hearsay.

Just as the fetus practises moving before birth, so too does the uterus. During the last few weeks of pregnancy, many women notice that their uterus undergoes contractions similar to those that will eventually expel the baby. These practice contractions, called Braxton–Hicks after the people who brought them to the attention of medical science, are painless, and are evident as a tightening of the abdomen lasting several seconds. Their frequency and duration increase towards the end of pregnancy. The muscle in the uterus wall is smooth muscle, like the muscle surrounding the gut, and is not under voluntary control. Thus, no amount of wanting, or not wanting, labour to begin will actually cause, or prevent,

Table 6.3 A weekly guide to development.

Week of gestation	Gut	Lungs	Nervous system	Cardiovascular system	Immune system
12	digestive enzymes developing	bronchioles developing	growth hormone detected; middle ear complete	heart and major blood vessels formed	some immune cells present, but not functional
13					
14	peristalsis		hypothalamus and pituitary gland complete		
15			fetal movement felt		
16	swallowing	alveolar buds appear	most nerves formed by now		
17					cells migrate from bone marrow to lymph nodes and thymus
18		low level of surfactant			
19					
20			inner ear and auditory nerves in place		
21					
22			optic pathways in place		
23					
24	liver stops making blood; begins to develop metabolic functions	surfactant produced in large amounts	fetus responds to sounds		
25			growth hormone reaches maximum level, then declines		
26					
27					
28	all digestive enzymes present	air spaces appear			
29					
30					
31					
32					
33					
34	sucking, swallowing, peristalsis coordinated				
35					
36		alveoli mature			
37			rods and cones appear		
38				shunts close at birth	components present at 50% of adult levels; antibody production starts
post-natal	liver matures		brain growth continues; senses mature		development continues throughout life

any contractions, although the experience of contractions can be *influenced* by the woman's state of mind. Besides the physical changes a woman may experience, there are emotional ones too. Regardless of whether the pregnancy was initially welcomed or not, most women look forward to the prospect of the end of pregnancy. This is partly for the relief of physical symptoms which it offers, and partly because they want to see their baby after so many weeks of just feeling it. This expectancy is tempered by the knowledge that the birth process itself has to come first. Women approach childbirth with a whole range of feelings, but also with the knowledge that, no matter how they feel, the process is inexorable. So as the end of pregnancy approaches, many will experience feelings of trepidation and powerlessness. Other feelings, too, are common – in particular, 'nest-building' urges, which help the woman to prepare her living space for the baby. These feelings may range from decorating the baby's room during the pregnancy to frantic floor scrubbing while waiting for the midwife to arrive!

In spite of a great deal of research, it is not known precisely what triggers the process of labour. In the fifth century BC, the Greek philosopher Socrates claimed that the fetus dictates when it will be born, and this is still regarded as true. But *how* does the fetus initiate the series of powerful uterine contractions, distinctly stronger than Braxton–Hicks contractions, that will expel it? Many suggestions have been made, yet none can explain all the observations. To be a trigger for labour, a substance must either start to be made just before the onset of labour, or reach a threshold value at that time; alternatively, it could be inhibitory to labour, present throughout pregnancy but disappearing just before labour begins.

We shall look at some popular candidates for the trigger, and see how far they measure up to these criteria.

It has long been believed that the placenta, so central to the whole developmental process, must be involved in triggering labour. Although the placenta grows throughout pregnancy to keep pace with fetal demands, there comes a time when it will no longer grow, and its capacity to synthesize materials declines. So too does its ability to service the fetus. In such cases of *placental insufficiency*, it is clearly important for the baby to become independent of the placenta as soon as possible, or it will become starved of nutrients and distressed. Towards the end of pregnancy, it is not uncommon for the placenta to start showing signs of insufficiency, and it has been suggested that the declining placenta can signal the uterus to begin contractions. However, in the majority of cases of full-term labour the placenta is still fully functional, and could generally have continued to function for at least another couple of weeks if labour had not intervened. So it does not seem likely that any factor connected with placental insufficiency is responsible for triggering labour. Indeed, if it were the trigger, then fetal distress caused by placental insufficiency would never arise, because the baby would be delivered before any damage was done.

Many hormones have been suggested as candidates. For example, it is known that uterine muscle will tend to contract after it has been stretched, due to the production of a hormone by the pituitary gland, called **oxytocin**, which stimulates contraction. Special structures called **stretch receptors** are located in many tissues, including the uterus; their function is to detect, and notify the nervous system, when any tissue is being unduly stretched. When this happens, the pituitary gland is triggered to produce oxytocin. Perhaps the stretching of the uterus over a large fetus stimulates the production of enough oxytocin to start the contractions of labour. This would suggest that all births will begin when there is a specific ratio between the size of the fetus and the size of the mother, but in spite of many measurements, no such relationship has been found. Nor is there any evidence that the trigger is a nervous signal initiated by stretch receptors in the uterus. Another suggestion is that labour is initiated by a particular combination of fetal, placental and maternal hormones. It seems to be true that in some other mammals there is a sudden increase in the blood level of a specific prostaglandin, $PGF_{2\alpha}$, just before birth, but this is not the case in women, where the hormones studied so far show no discernible pattern that might suggest they act as a trigger. The level of $PGF_{2\alpha}$, a powerful stimulator of uterine contractions, *does* increase, but not until *after* the start of labour. This is another problem with oxytocin's candidacy, too: its increase also comes too late to be the labour trigger, although both it and $PGF_{2\alpha}$ play a part in *maintaining* labour once it has been started by something else. However, both oxytocin and prostaglandins have been used to induce labour artificially, so their effects *can* be enough to start off the contractions if they are present in sufficiently large amounts.

Some persuasive evidence suggests that the fetal pituitary gland plays a crucial role in initiating labour.

❑ What does the pituitary gland do?

■ It produces hormones.

One product of the pituitary gland is adrenocorticotropic hormone (listed in Table 6.2), which controls the synthesis of hormones from the adrenal cortex, a gland lying above the kidney. It is certainly true that the fetal adrenal cortex undergoes a big increase in size very late in pregnancy, and produces a large amount of a hormone called cortisol, together with oestrogen precursors (the substances from which the body makes oestrogen). These hormones affect the muscles of the uterus, encouraging contraction. Once again, this mechanism seems to be involved in maintaining labour, but whether it is the actual trigger for labour remains a mystery. A further candidate is a growth factor, platelet-activating factor (PAF) (Table 6.2), which is secreted by the fetus and accumulates in the amniotic fluid late in pregnancy. PAF can stimulate both uterine contractions and $PGF_{2\alpha}$ synthesis, and its presence in the amniotic fluid only as labour approaches suggests that it could act as a trigger once a threshold level had been exceeded. Again, no firm evidence exists to support this theory.

However, in 1995 a new candidate for the trigger of labour was suggested. The molecule is a protein, one of a large family of so-called 'second messengers', which propagate signals within cells. You will learn more about this sort of molecule elsewhere in the course. This particular protein, called $G\alpha_s$, is produced throughout pregnancy in high quantities; its function seems to be to relax the wall of the uterus, allowing it to stretch over the growing fetus. At the time of birth, $G\alpha_s$ is completely inactivated. This occurs not only in the cells of uteruses that have undergone a normal delivery, but also in those that have delivered prematurely. It is therefore associated with the actual end of pregnancy. Neither the exact mode of action of this molecule, nor what causes its inactivation have been established, but there is plenty of circumstantial evidence to suggest that it *is* likely to be involved in triggering labour.

Research on this topic is, of course, continuing. Whatever the trigger for birth turns out to be, the fact remains that at around 38 weeks after fertilization, the fetus has developed enough for it to be ready to survive outside its mother's body. The only certainty in the area is that, whatever the mechanism, the onset of labour usually occurs at the 'right' time for the fetus: as Socrates said, the baby comes when it is ready.

Summary of Section 6.11

1. Labour is believed to be triggered by the fetus itself.
2. Many candidates, including the placenta, the adrenal cortex, a hormonal or nervous signal, a prostaglandin, platelet-activating factor, and a second messenger, have been put forward as the trigger of labour, but there is no firm evidence to confirm any of them so far.

6.12 The three stages of labour

As you are aware, the healthy fetus moves around a great deal while it is developing. These movements are not just kicks and arm waves, but also include various rotations and somersaults. Towards the end of gestation these large-scale movements are severely restricted, for reasons of space; nevertheless, by the last few days before birth the baby must have rotated so that it is head down, and it must drop as far as it can into the pelvis, with its head (usually) pressing against the cervix. In this position the baby is said to be *engaged*, and this has to happen before birth can occur. Figure 6.10 illustrates this. Babies are usually born head first, and this ensures that the largest part – the head – is the first to exit, and the rest of the baby will follow through quite easily. However, in a significant number of cases the baby is born bottom first – a *breech* presentation – and this means that further effort is required on the mother's part to push the head out last. In practice, when a breech presentation is detected, efforts are made by medical staff to turn the baby inside the mother to make the birth easier, but this does not always work.

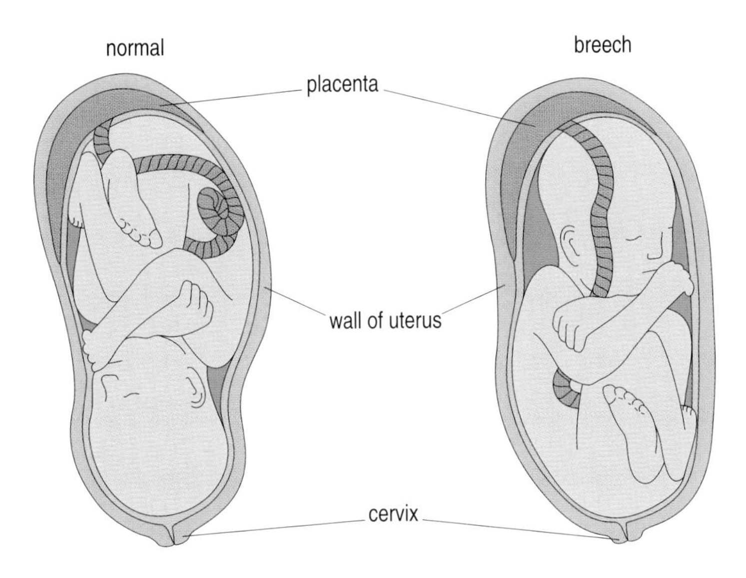

Figure 6.10 Different presentations: normal and breech.

The process which enables the baby to leave the comfortable confines of the uterus and arrive in the outside world is not without its problems. As you have seen (Section 6.7), the human baby's head is very large compared with the opening of the pelvis through which it must pass (see Figure 6.11). Thus, a certain amount of effort is required to force it out.

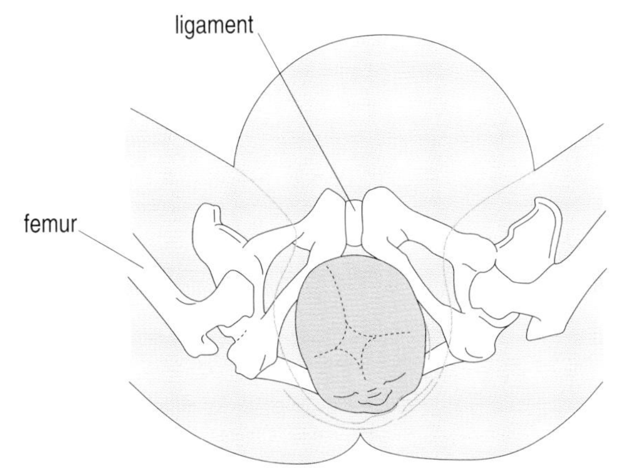

Figure 6.11 Diagram of a section through the pelvis, as though a woman, lying on her back with her legs raised, was being viewed from her feet. The baby's head is shown being delivered. Note that the two halves of the pelvis are joined ventrally (at the front) by a ligament, a piece of elastic material made from connective tissue. This can stretch a little, so that there is some 'give' in the pelvis. Ligaments will stretch under the influence of progestogen.

Before passing through the pelvis, the baby must first leave the uterus. To allow this, the hole in the cervix must open. During gestation, this hole has been blocked by a plug of mucus, and one of the first signs that labour is under way is often the appearance of the plug as it drops out of a dilating cervix. The early muscular contractions of labour are designed to pull the cervix open around the presenting part of the baby. This is why engagement is necessary for birth to take place. Only when the cervical hole has attained a diameter of around 10 cm, and becomes continuous with the walls of both the uterus and the vagina, will the contractions change so as to expel the baby. The contractions involved in opening the cervix are regular, and increase in strength, length and frequency as labour proceeds. This part of labour is called the first stage; it can last for more than 12 hours and tends to be longer for first births. Its end comes when the cervix is fully open (dilated) and the mother can begin to push the baby out during the second stage of labour, which generally lasts only one or two hours, and sometimes very much less. In fact, the mother's deliberate pushes only *help* the birth: the contractions of the uterus alone are strong enough to deliver the baby. But the urge to push that the mother experiences is irresistible, and, as long as she is conscious, she will be trying to push the baby out. The stages of labour are shown diagrammatically in Figure 6.12.

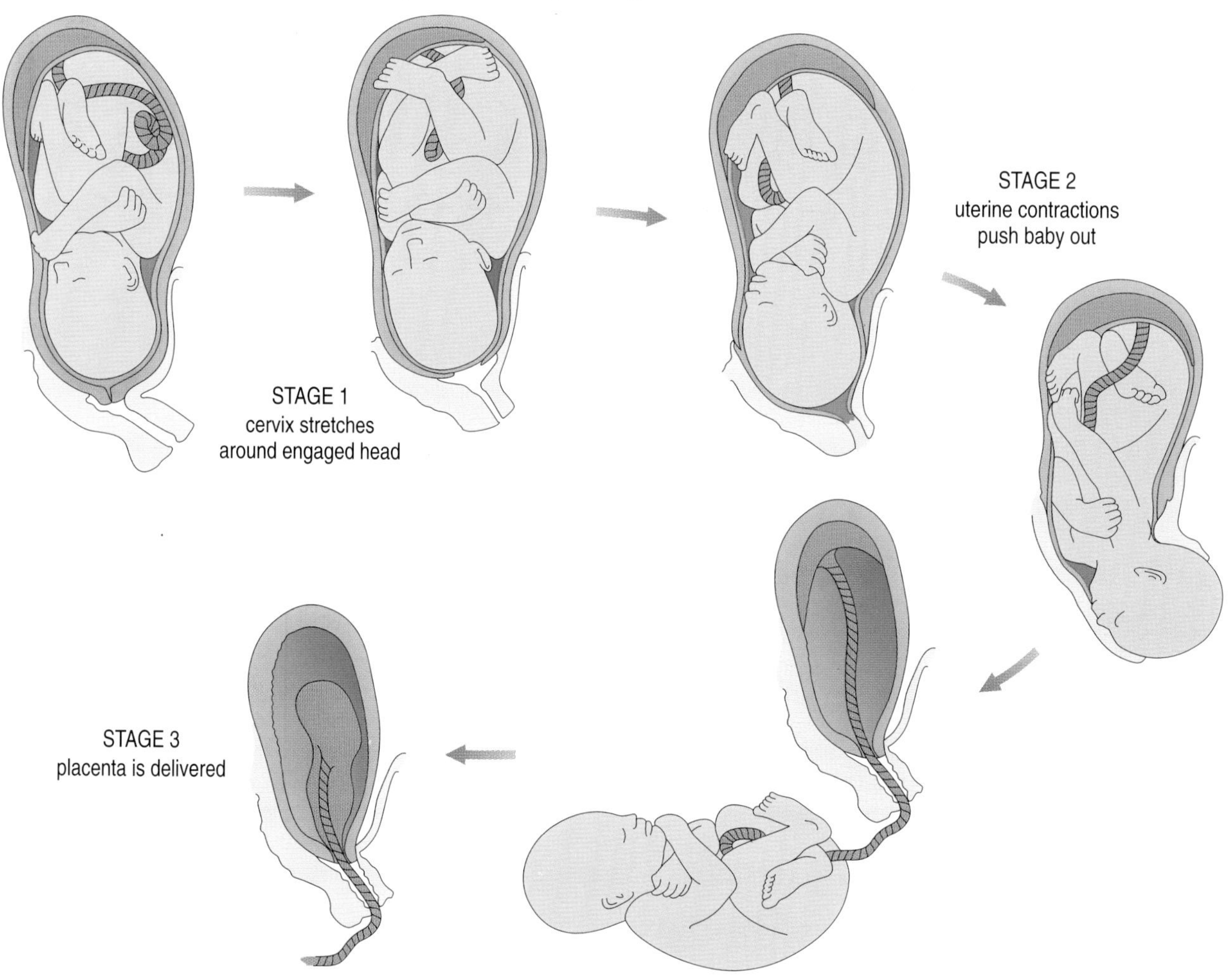

Figure 6.12 The stages of labour.

❑ Why do you think a 10 cm opening must develop in the cervix?

■ This is the approximate diameter of the baby's head.

Because the head is the biggest part, any hole that will let this part of the baby through will also allow the rest to pass through easily. However, the baby's head does not have a circular cross-section, and the birth canal (the vagina) stretches more easily in some directions than others. Moreover, a look back to Figure 6.11 will remind you that the gap in the pelvis through which the baby will pass is wider in one direction than another. Thus, after delivery of the head, the baby *rotates* so that its shoulders may be delivered more comfortably. This rotation is shown in Figure 6.13. In most cases, the head appears facing backwards, towards its mother's anus. As soon as the head is free, it will turn to one side so that the baby is now facing a thigh. This allows delivery of the shoulders, one at a time.

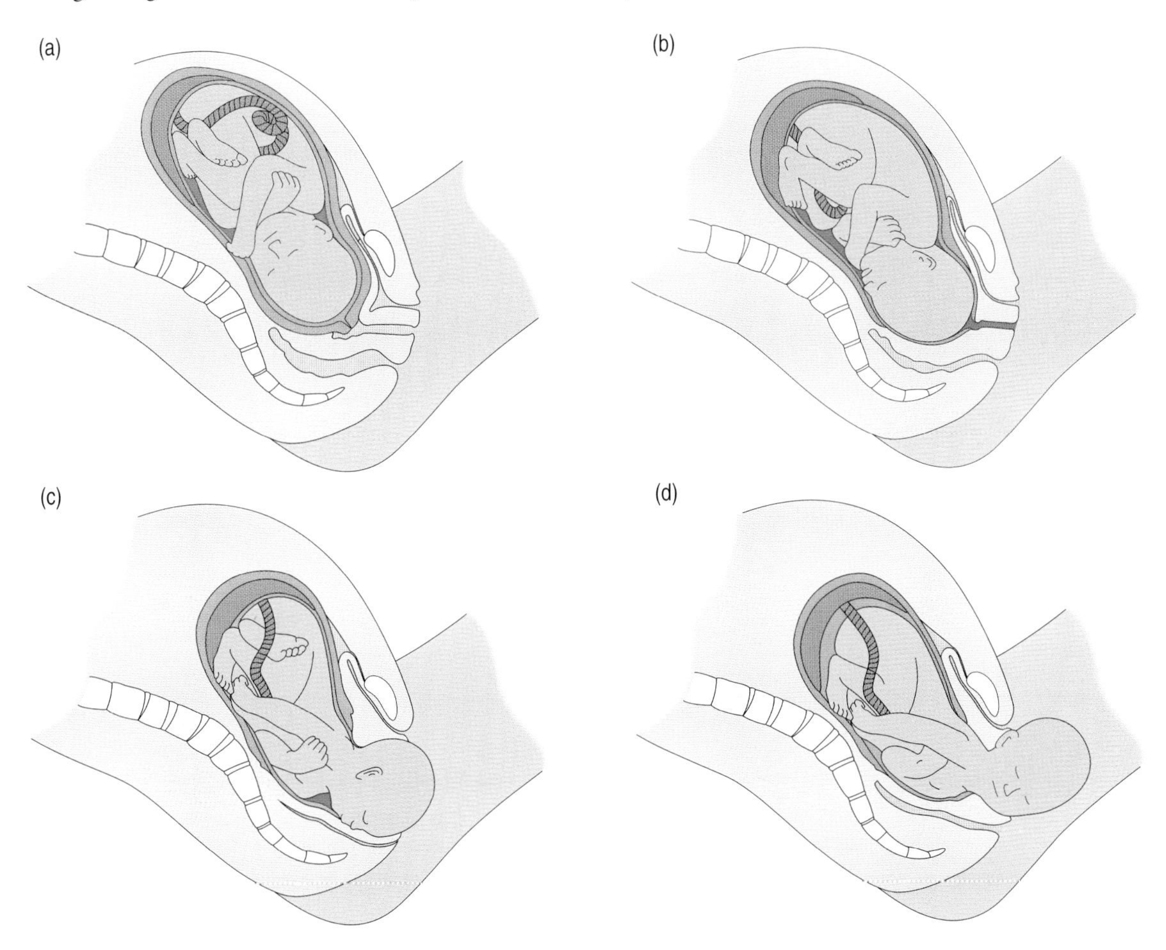

Figure 6.13 Position of baby during birth.

Even a well-dilated cervix offers a good deal of resistance to the passage of the baby, and another mechanism exists to help things along. The skull is made of several bony plates which are fused together to afford good protection for the brain. In the new-born, however, the fusion between the plates has not yet occurred, so there is a limited ability for the skull bones to slide across each other, making the size of the head a little smaller, and easing its passage through the birth canal. Once the head is out, only a few more contractions are needed to expel the rest of the baby. The appearance of the whole baby marks the end of the second stage of labour.

Because of the blood pressure changes that occur at birth (described in Section 6.8), blood flow through the umbilical cord stops soon after birth. The time at which this occurs can be seen in the new-born because the cord will stop pulsing as the blood stops being forced along it. It is important not to prevent the blood from flowing while the cord is still pulsing, as there is evidence to suggest that valuable nutrients are still passing through. After blood flow has stopped, however, the cord can safely be tied off and cut, at last separating the baby from its mother.

At some point during labour the membranes surrounding the baby will probably have ruptured, releasing the amniotic fluid which can then help to lubricate the birth canal. In these cases, the baby is born free of the membranes, but still attached to the placenta by the umbilical cord. Sometimes the membranes do not rupture, and here the baby is born still in its sac. It cannot draw its first breath until the membranes have been ruptured by the mother or her attendants. In the UK and elsewhere, the membranes are artificially ruptured as soon as they become visible during delivery. In any event, the remains of the membranes and the now redundant placenta must also be expelled from the mother's body, and this part is the third stage of labour (Figure 6.12). The strong uterine contractions which expelled the baby will also have begun to detach the placenta from the uterine wall, and usually only a few more contractions are needed to expel it altogether. From the mother's point of view, these hardly count as contractions at all – after her previous efforts – but they are extremely important as, if the placenta is not expelled, it will start to degenerate within the uterus, giving rise to infection. Since the placenta was firmly embedded in the wall of the uterus, and its whole raison d'être was to facilitate a blood supply, you would probably expect the third stage of labour to be accompanied by copious bleeding as the placenta is torn free. But, in fact, the uterine contractions have an important role in stopping bleeding by rapidly constricting the blood vessels, and the amount of blood lost at this stage is usually not too great. This is fortunate, because if blood flow were to continue unabated there would be a strong possibility of the mother bleeding to death. So great is the risk of this that many obstetricians and midwives routinely give the mother an injection of a drug called Syntometrine, which constricts blood vessels and prevents uterine bleeding.

Summary of Section 6.12

1 The baby's head must be engaged before labour can proceed.

2 The first stage of labour stretches the cervix around the baby's head until the opening is about 10 cm in diameter. The baby's head can then pass through.

3 The second stage of labour expels the baby. The head comes first, facing backwards, then the baby rotates so that one shoulder at a time can be delivered. This process is helped by the baby's skull bones sliding across each other to make the head slightly smaller.

4 The third stage of labour is the expulsion of the placenta and membranes. There is a risk of excessive bleeding at this stage.

6.13 A healthy baby

The pressure which the baby experiences during birth is generally enough to squeeze the fluid from its lungs. If enough surfactant has been produced the baby will find it easy to inflate its lungs with its first breath.

❑ What is the role of the surfactant?

■ It reduces the surface tension that holds the inside surfaces of the alveoli together.

Thus, the first action which a baby must be able to carry out is breathing.

❑ What do you think the second thing might be?

■ The baby must be able to feed.

As you saw above, the baby is primed with a supply of glycogen and fats which will tide it over until its first meal. So the baby does not have to eat *immediately*, although the first feed should not be delayed for too long. In many cases, the mother cuddles and feeds the baby straight after birth. This supplies not only nutrients, but also a comforting, and in particular, a *warm* environment for the baby. It is very important to keep a new-born baby warm.

❑ Why might the baby be cold?

■ It has emerged from an environment of 37 °C to one which, at best, is probably only around 20 °C. Also, it is dripping wet, so will feel the cold more keenly – as the liquid evaporates from the baby's surface, it will remove heat.

In births that are attended by medical practitioners, including midwives, i.e. most births in Britain and the USA, the baby will first be examined before being given to its mother. The sex of the baby will be assigned by the appearance of its genitals (see Book 4 for more details), it will be weighed and its length measured. The baby will also be subjected to a series of tests, collectively called the Apgar test, which are designed to assess the baby's level of awareness and mental and physical development. Apgar tests are repeated at intervals after birth. The physical stress associated with being born, and the sudden drop in temperature, sometimes mean that the baby does not score very highly on tests straight after its birth. However, when the tests are repeated five minutes after birth, most babies score highly. Those who do not can be given the appropriate treatment.

What is the Apgar test? The parameters measured are ability to breathe, heart rate, colour, response to touch, and general muscle tone.

The ability to breathe is fairly easily assessed, as the baby may well be crying, and needs to be able to breathe to produce sound. Even quiet babies can be quickly seen to be breathing.

Heart rate can be measured easily, either by taking a pulse, which can actually be quite tricky, as new-born babies are so small, or by listening directly to the heart by means of a stethoscope. A new-born heart beats quite quickly, and 120–140 beats per minute is considered normal. This is considerably higher than the heart rate of a resting adult, which is normally around 70 beats per minute.

The colour of the baby refers not to the colour of its skin, which obviously varies a great deal, but to the colour of non-pigmented areas, such as the inside of the mouth and the eyelids. In a healthy baby these areas will be pink as a result of oxygen-carrying blood flowing through blood vessels near the surface. If the colour is purplish, this indicates that the blood is not carrying enough oxygen and the baby's tissues are deprived of oxygen. The consequences of this can be very serious. Even healthy babies often appear to have blue fingers and toes, but this is likely to be due to cold rather than to bad circulation. Brick red membranes can suggest carbon monoxide poisoning.

Babies are normally born with certain reflexes in place. The most widely known of these is the so-called *rooting reflex*, which is a turning of the head towards a touch on the cheek. This enables the baby to latch on to a nipple for a feed, so is obviously an important reflex for the baby to develop if it is to survive. Other reflexes, such as a grabbing motion when the palm of the hand is stroked, or curling of the toes when the sole of the foot is stroked, may also be assessed.

Muscle tone, shown by a resistance to forced movement, is difficult to assess objectively. Babies are not known for their muscular strength, and all new-borns will appear to be fairly 'floppy'. However, experience allows the detection of particularly floppy babies, who offer little or no resistance to their limbs being moved, and so have poorly developed neuromuscular coordination.

Each of these five components of the Apgar test is given a score of 0 to 2, with 2 indicating a good response, and 0 no response at all. The baby's total score is used to assess its maturity and health. Scores between 7 and 10 are usual; even babies who score lower than this at one minute after birth can generally achieve a higher score by five minutes after birth.

The Apgar test seems remarkably simple, and can be carried out in just a few moments, yet it has not been surpassed as an indicator of a new-born baby's health.

❑ Think back to the main body systems described in Sections 6.5–6.9. Which of them are assessed by the Apgar test?

■ Lungs: breathing; gut: not tested; nervous and musculo-skeletal systems: reflexes in place; circulation: colour and heart rate; immune system: not tested.

So the Apgar test is by no means exhaustive, but checks merely the baby's capacity for immediate survival. Later a more detailed examination will be carried out; but for now, we have safely delivered a healthy baby (see Figure 6.14), and this is the ideal outcome of 38 weeks of development, growth and maturation, representing significant effort by both the baby and its mother.

RAJ SHARMA

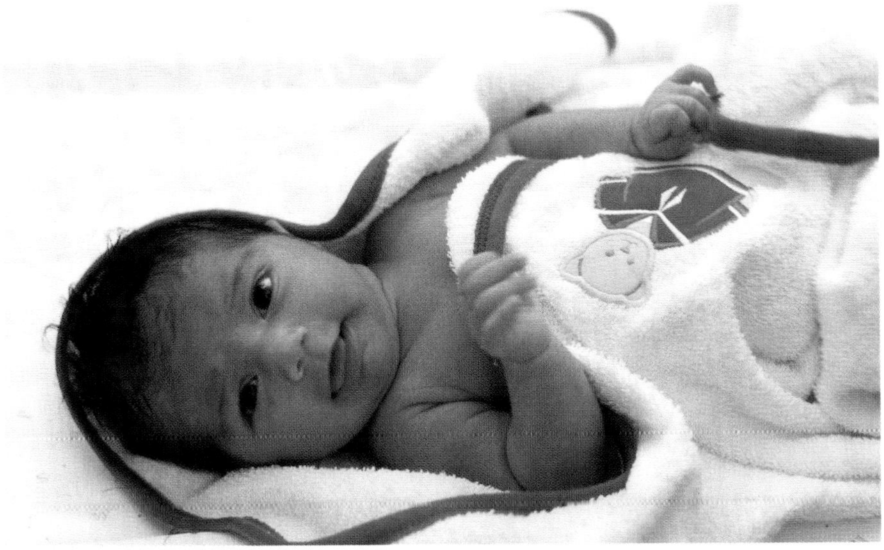

Figure 6.14 A healthy baby.

Summary of Section 6.13

1 New-born babies are subjected to the Apgar test. This checks their ability to breathe, their heart rate, their colour, their response to touch and their muscle tone.

2 A new-born baby needs to be fed quite soon after birth, and needs to be kept warm and dry.

Objectives for Chapter 6

After completing this chapter, you should be able to:

6.1 Define and use, or recognize definitions and applications of, each of the terms printed in **bold** in the text.

6.2 Describe using diagrams how a fetus increases in size during gestation. (*Question 6.1*)

6.3 Show that maternal nutrition influences fetal health. (*Questions 6.2 and 6.3*)

6.4 List some of the factors controlling fetal growth. (*Question 6.3*)

6.5 Show that fetal maturation is a dynamic process, with different systems maturing at different times. (*Questions 6.3 and 6.4*)

6.6 Describe the maturation of the principal body systems needed before birth. (*Question 6.4*)

6.7 List some of the factors believed to influence the onset of labour. (*Question 6.5*)

6.8 Describe the three stages of labour.

Questions for Chapter 6

Question 6.1 (*Objective 6.2*)

The graph shown in Figure 6.15 describes the increase in head circumference of a fetus. Using Figure 6.2b, which shows data for the 'average' fetus, suggest whether the fetus from Figure 6.15 is growing too quickly, too slowly or at just the right rate.

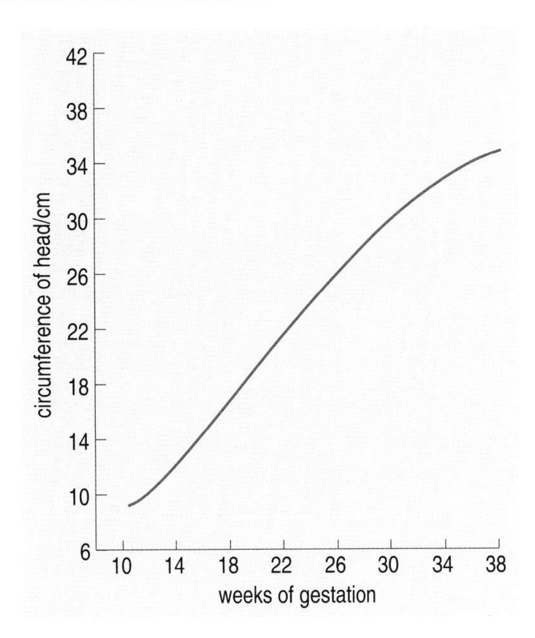

Figure 6.15 Graph showing the increase of the head circumference of a fetus with time.

Question 6.2 (*Objective 6.3*)

Predict the effect on birth weight of the babies of women who had a daily intake of (a) 1 000 kcal, (b) 2 500 kcal, or (c) 4 000 kcal. Assume that all three diets were balanced in terms of essential nutrients.

Question 6.3 (*Objectives 6.3–6.5*)

'The mother is completely responsible for the growth patterns of the fetus.' State, giving reasons, whether or not you agree with this statement.

Question 6.4 (*Objectives 6.5 and 6.6*)

Describe two ways in which the development of the gut and the musculo-skeletal systems are linked.

Question 6.5 (*Objective 6.7*)

What characteristics might the trigger of labour have?

ANSWERS TO QUESTIONS

Chapter 1

Question 1.1

A person in good health in their 90s has lived significantly longer than the average human lifespan, even in industrialized nations (where life expectancy is in the mid-70s). A possible biological contribution to longevity with good health is the genes a person inherits: certain genes may be associated with reduced susceptibility to degenerative diseases such as heart disease, cancers and strokes, or to increased resistance to infection; other genes may enable the person's body functions to occur with greater efficiency; for example, in terms of extracting nutrients from food or eliminating potentially toxic waste products. These biological advantages may interact with psychological characteristics to promote health: for example, intelligence and reasoning may have enabled the person to acquire education about factors that affect health, such as diet, smoking, exercise, etc. and to act on that knowledge. The same psychological characteristics are also likely to affect the person's sociological world; for example, by increasing the prospects of employment and the ability to afford the kinds of housing, nutrition, etc. which protect health. To complete the circle, we could add that the biological characteristic of resistance to illness also enhances the person's ability to attend school and hold down a job, thereby influencing his or her psychological development and sociological characteristics.

Question 1.2

The main strength of reductionist approaches to research is that they allow investigations to proceed with great attention to detail on a narrow, and hence manageable, front. The intensity of focus on well defined areas of study has produced the vast wealth of scientific knowledge of the human body and human psychology that informs us today. The weaknesses are twofold. First, the attention to details can be at the expense of attempts to study the whole; influences on a complex phenomenon that fall within the field of study of another discipline are ignored and interactions between different influences may remain invisible. Second, advocates of the reductionist philosophy have tended to discredit explanations for complex phenomena which have been generated by other fields of study (for example, some influential biologists have represented sociology as an inferior discipline), thereby damaging the reputation of other disciplines in the competition for research funds and students.

The main strength of holistic approaches to research is that they seek to understand all the interacting influences on a complex phenomenon. Knowledge generated by one discipline may inform and enhance the investigations carried out in another field of study, and reveal the interactions between them. The weaknesses are that interdisciplinary studies are intrinsically difficult and complex enterprises; they may not get very far and can produce superficial accounts because so few researchers are fluent in the concepts and methods of more than one discipline. An additional problem is that the recent popularity of holism as a philosophy may be elevating it to the status of a dogma which cannot be criticised – in effect, it may have become the 'new reductionism'!

Question 1.3

The twins should show identical features of growth and development, despite their different environments, since – to a biological determinist – every aspect of their biology is determined by their identical genes. (Studies of twins in this situation have revealed that certain aspects of growth and development, such as height, do indeed seem to be determined largely by genetic inheritance (nature), whereas other characteristics, such as weight, are influenced very significantly by the environment (nurture) in which the child is brought up.)

Question 1.4

The pads of fat are an adaptive characteristic in the Arctic environment, which confers a survival advantage on those individuals who have them. Evolution proceeds by natural selection in which individuals with the most adaptive characteristics in a given environment are 'selected'. These individuals have greater fitness than others, which means they have a greater chance of surviving long enough to reproduce and raise a larger number of offspring relative to others who do not have the fat pads. They pass on their genes to their offspring, who in turn survive to reproduce because they too have the survival advantage conferred by the fat pads. Over many generations, the number of individuals in the population who have the fat pads increases, until everyone has them.

Chapter 2

Question 2.1

Williamson's and Pearse's definition is underpinned by a positive concept of health. It contains ideas about growth, health enhancement and self-realization. It has affinities with the WHO's (1984) comment that health is about satisfying needs and realizing aspirations within an environment.

Question 2.2

You could respond by telling her to 'pull herself together' and make a 'health choice', but having read this chapter you should be more aware of how difficult it is to make changes and how people's choices are heavily restricted by their environment and social situation. For example, your neighbour is probably worried about herself, her husband and her child's illness and probably has money problems too. You could help by reassuring her that she is not to blame and giving her support if she tries the diet again.

Question 2.3

You could reply that researchers can draw on long runs of roughly comparable national data on death rates and causes of death so they can highlight the relative fortunes of the different occupational classes over time. Three major reports, The Black Report (1980), The Health Divide (1987) and Variations in Health (1995) all concluded that there were unacceptably large variations in health and attributed these to differential exposure to physical and social environments.

Question 2.4

Based on your reading of Section 2.3.2, there are three possible explanations. First, male workers in occupational class V may simply be more irresponsible and careless in their behaviour. Second, male workers in social class V do unskilled work which may be dangerous and dirty, perhaps with long hours and poor working conditions and it is this type of environment which creates the high level of accidents. Third, economic and peer pressures may influence workers to take risks and the dangerous and dirty working environment determines that such risk-taking results in a high accident rate. In this case, the environment influences both the behaviour of the workers and the outcomes of that behaviour.

Question 2.5

It is correct to say that epidemiology cannot identify specific causes of conditions such as cardiovascular disease. However, by demonstrating statistically significant correlations, as between hypertension and placental to birth weight ratio in a group of individuals, such studies can lead to the formulation of hypotheses about specific causes of disease which can then be studied by more direct means, leading to ways of improving health.

Question 2.6

You could confirm that heavy smoking does tend to reduce the baby's birth weight, but that it does so by reducing the amount of oxygen and nutrients that the baby receives from the mother. You could point out that this could have long-term consequences for the person's future health, as suggested by epidemiological studies. You might then explain how this could happen through the effects on the blood vessels of the body: reduced oxygen and nutrients for the fetus result in preferential blood flow to the brain, so the body experiences reduced blood pressure and the vessels may then fail to develop normal elasticity, which could be irreversible. Finally, to reassure her, you could add that there are many built-in protective factors for the fetus, and that reasonable care is all that is required for a successful childbirth.

Chapter 3

Question 3.1

The nitrogen atom has a valency of 3, and so forms bonds to three hydrogen atoms (Figure 3.54, left). In this way, both the N and the H atoms achieve a filled outer electronic shell, as shown in Figure 3.54, right. Nitrogen is much more electronegative than hydrogen, so the N—H bonds are polar (with an excess of negative charge on the N atom and an excess of positive charge on the H atom). Not surprisingly, NH_3 is very soluble in water, a polar solvent.

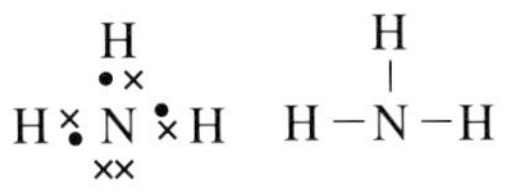

```
    H          H
   •×          |
H ×• N •× H  H−N−H
   ××
```

Figure 3.54 Structural formula (right) and outer electronic structure (left) of a molecule of ammonia, NH_3.

Question 3.2

Table 3.6 lists the main differences in the structure and function of insulin and glycogen. (Don't worry if your list does not include all these points – you will learn more about insulin and other hormones in Book 2.)

Table 3.6 For answer to Question 3.2.

Insulin	Glycogen
protein	carbohydrate polymer (polysaccharide)
made up mainly of the elements C, H, O and N	made up of C, H and O
composed of a variety of amino acid units in a precise sequence	composed of monosaccharide units, all of which are glucose
a fixed size	chains can vary in length
functions as a hormone	functions as an energy store
soluble molecule that enters the bloodstream	insoluble molecule which remains intracellular until broken down to soluble glucose
produced in very small quantities	can be stored in considerable quantities

Question 3.3

The enzyme is a regulatory one and functions in a catabolic (energy-producing) pathway. Catabolism of substrate leads to increased levels of ATP, which binds to the allosteric site thereby reducing the activity of the enzyme. When ATP levels drop, the inhibition is alleviated and the enzyme becomes active again. In this way the level of ATP, the cell's energy currency, can be matched to requirements.

Question 3.4

(a) Mitochondria are the power-houses of the cell and new-born babies can therefore generate a lot of heat by the fat 'burning' activity of the numerous mitochondria. (In fact, brown-fat mitochondria are specially adapted to producing heat in preference to generating energy as ATP.)

(b) The Golgi sacs process – in this case, add sugar chains to – the proteins destined for export as mucus which are synthesized by ribosomes on the ER. Vesicles containing the finished product (the mucus glycoprotein) bud off from the Golgi sacs and move up the cell to fuse with the cell membrane and so release their contents into the lumen of the intestine.

(c) The number and size of the granules will increase after a meal as blood glucose is stored as glycogen and will diminish between meals as the glycogen is converted back into glucose which goes into the bloodstream to be taken up and used by the cells of the body (Figure 3.10).

Question 3.5

If replication were *conservative*, then after one round of cell division (one generation) in the ^{14}N medium there would be heavy (HH) and light (LL) DNA but no DNA of intermediate density (HL). A generation later there would be the same components in the mixture but with an increased proportion of the LL product (Figure 3.55a), the HH DNA being progressively diluted out through successive generations of growth in the ^{14}N medium.

If replication were *semi-conservative*, then after one round of cell division in the ^{14}N medium all the DNA would be of intermediate density (HL). The second division would yield an equal mixture of light (LL) and intermediate density (HL) DNA (Figure 3.55b).

What Meselson and Stahl actually did was to separate the DNA extracted from the bacteria, on the basis of its density in a gradient of caesium chloride (a salt) concentration: heavy DNA 'floats' in the high-salt, dense part of the gradient nearer the bottom, light DNA forms a layer in the low-salt, less dense region nearer the top, and the DNA of intermediate density settles at a level between the two. By analysing the extracted DNA in this way, they were able to show that replication must be semi-conservative.

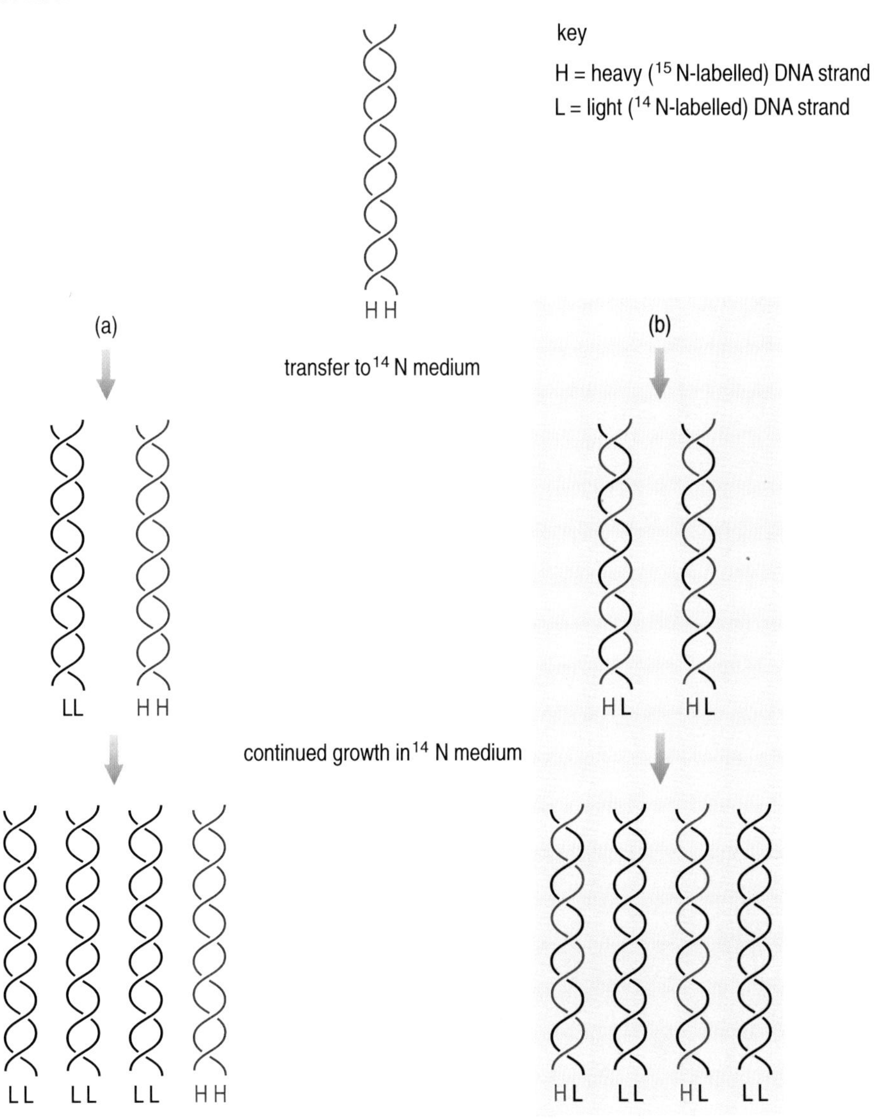

Figure 3.55 Changes from heavy to light DNA predicted through two generations in ^{14}N medium: (a) for conservative replication; (b) for semi conservative replication.

Question 3.6

Figure 3.56 is a correctly labelled version of Figure 3.54.

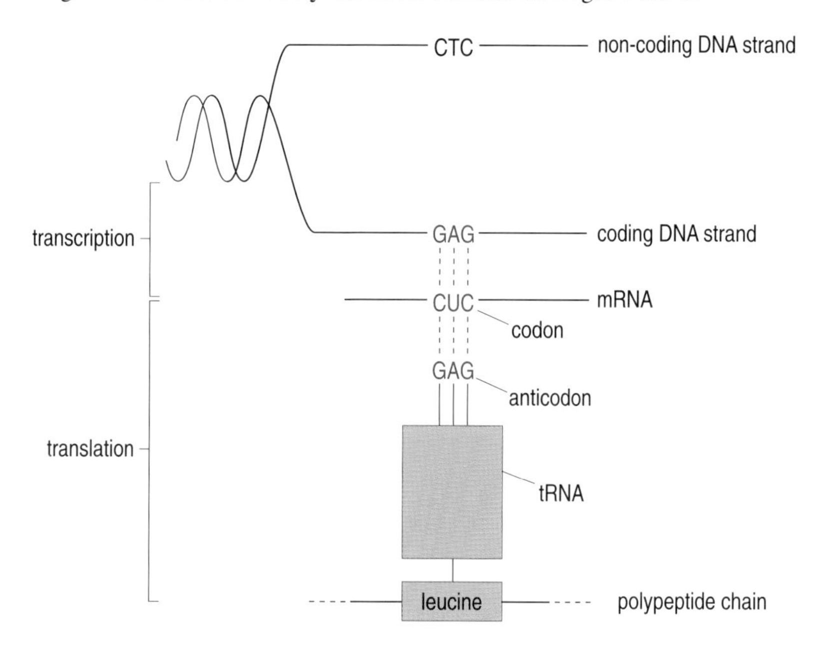

Figure 3.56 For answer to Question 3.6.

Question 3.7

The skin consists of two layers, an outer epidermis and an inner dermis. The epidermis is a strong and compact epithelium, consisting of several layers of cells arranged in hexagonal units called epidermal proliferative units (EPUs). At their base is a basement membrane attached to which are growing and dividing cells. Some of these migrate upward, becoming flat and differentiating into tough, water-impermeable cells filled with keratin and joined together tightly by junctions. Cells lost from the surface are replaced by differentiating cells moving up through the EPUs, thus maintaining the structure.

The dermis is made primarily of loose connective tissue consisting of an extracellular matrix with many collagen and elastin fibres running through it, and fibroblasts which maintain it. The dermis also contains hairs, sebaceous glands, nerve endings, blood vessels and lymphatics, all of which can regenerate after damage, provided the damage is not too severe.

Question 3.8

After a cell has duplicated all its cellular components, including its DNA and the associated proteins (histones) that make up the chromosomes, the nuclear envelope disappears and the chromosomes gather at the centre of the cell. A mitotic apparatus forms with two poles from which fibres radiate and attach to one of each pair of duplicated chromosomes. One of each type of chromosome then moves, pulled by the contracting fibres, to each pole and a new nuclear envelope forms around each complete set of chromosomes. Finally, the cytoplasm between the nuclei contracts, resulting in separation into two identical progeny cells.

Question 3.9

Persistent tenderness and redness of the skin indicates that the inflammation and early repair phases are being prolonged. This is when blood vessels are enlarged, releasing cells that clear up debris at the site of damage, and fibroblasts lay down new connective tissue. A cause of the delay in healing is likely to be the stress experienced by the carer, which reduces such factors as the level of interleukin in white blood cells, resulting in decreased fibroblast activity and slower connective tissue formation.

You might advise your friend to:

- arrange some relief from the care-giving until the wound heals properly;
- eat a healthier diet so that nutrition is improved;
- seek the assistance of a healer.

Chapter 4

Question 4.1

(a) This woman is probably ovulating around days 12–16 of her cycle. At this time the cervical mucus is of low viscosity. This would allow easier entry by sperm, and fertilization might occur if sperm were present when an egg was ovulated.

(b) Progestogen increases the viscosity of cervical mucus. Therefore if it were taken every day, the drop in viscosity would not occur (or at least not to the same extent). Thus it might prove impossible for sperm to pass through the cervix, and fertilization would not occur. Furthermore, the persistent high level of progestogen would prevent the wall of the uterus from becoming receptive to an implanting embryo.

Question 4.2

(a) In females, meiosis starts before birth, then stops until puberty, and from then on until the menopause a few divisions are completed each month, resulting in the production of (usually) just one egg. In males, all the stages of meiosis occur after puberty, and the process then carries on continuously during adulthood, resulting in a constant and copious supply of sperm (millions per day).

(b) In females, of the four products of meiosis only one will become an egg. In males, all of the meiotic products can become sperm.

Question 4.3

The first four divisions are cleavage divisions, taking place without net cell growth. The first three (1 to 2, 2 to 4, and 4 to 8 cells) all involve symmetrical divisions across the centre of the cells, yielding daughter cells of equal sizes. But the fourth, 8- to 16-cell, division is asymmetric, yielding two populations of cells, one made up of large cells (which will give rise to the trophoblast), and the other of smaller cells (which will give rise to the inner cell mass).

Question 4.4

By differential adhesion. The sponge cell membranes carry molecules which can recognize and stick to some surface molecules but not others. Thus the inner cells will stick tightly to each other, and less tightly to the outer cells. These can also stick to each other, but more loosely still. Over time the tightly stuck cells will find themselves in the centre of the mass, and the more loosely stuck cells will surround them.

Question 4.5

Yes – the two populations are X-bearing sperm and Y-bearing sperm. This relates to the sex chromosome they receive at meiosis. There are equal numbers of X- and Y-bearing sperm, so equal chances of getting male or female embryos. The X chromosome is larger and therefore heavier than the Y chromosome, so it was suggested that this might be a basis for separating the populations of sperm, for use in selecting the sex of an embryo in IVF or GIFT. But the random swimming paths taken by sperm largely negate any effects the weight difference might have.

Question 4.6

It used to be thought that the soul did not enter the fetus until its movements could be felt (quickening). Before this, the fetus was not considered to be alive, so could be aborted without sin. But in 1869 Pope Pius VI decreed that the soul entered the embryo at fertilization. This meant that the embryo was alive from this moment onwards, so abortion was seen to be taking a life.

Chapter 5

Question 5.1

Cell division and change of cell shape and size between the inner and outer cells of the conceptus lead to differential gene expression and so different cell surface properties and thus differential adhesion, with the result that the inner cell mass becomes separate from the outer layer (the trophoblast). Cell growth, cell division and change of cell shape produces the amniotic cavity in the ectodermal layer of the inner cell mass; growth, division and migration of cells in first the endoderm and then the ectoderm results in formation of the yolk sac and the chorionic cavity.

Question 5.2

Ectoderm cells along the midline of the germ disc move into the space between ectoderm and endoderm, forming the mesoderm. A central rod, the notochord (precursor of the backbone) forms by strong adhesion of mesodermal cells, while somites form alongside this rod as bead-like structures. The neural tube arises by upward bulging of the growing ectoderm to produce folds which fuse together to make a tube, starting in the middle and proceeding towards both ends (anterior and posterior). Genetic changes that affect the division, adhesion or movement of cells can disturb these processes, as can certain dietary deficiencies such as inadequate folic acid intake.

Question 5.3

The placenta forms from cells of the trophoblast as implantation occurs. Trophoblast cells divide and grow into the wall of the uterus, which responds by releasing nutrients from its disintegrating cells and filling the spaces with blood, which go on to develop into the maternal contribution to the placenta. The trophoblast also develops a system of blood vessels, which is part of the embryonic circulatory system, and separated from the maternal blood supply. The placenta provides nourishment for the embryo and fetus, removes its wastes, prevents access by pathogens and harmful chemicals and provides protection from immunological attack by the mother. It also allows some maternal antibodies to reach the embryo and fetus which give some immunity to pathogens.

Question 5.4

The circulatory system depends for its formation and function on the response of cells to the blood as a fluid. Small pools of blood which form in the yolk sac develop into a network of elastic tubes which become the blood vessels, which naturally expand and contract as blood flows through them. In response to pressure, the cells of the blood vessels produce more elastin and collagen, thereby becoming more elastic. The heart goes on to develop from one of these tubes into a specialized organ of blood accumulation and rhythmic contraction and relaxation, the whole system of heart and vessels working together to circulate blood throughout the embryo, to and from the placenta.

Question 5.5

If the supply of nutrients from the mother is inadequate, the placenta is induced to grow larger, and the fetal blood is diverted preferentially to the brain rather than the body. This causes a fall in the blood pressure in the vessels of the body, leading to reduced elasticity of their walls, which could persist throughout the life of the individual, resulting in proneness to hypertension.

Question 5.6

Every embryonic structure forms at a characteristic time in development, which is when the embryo is most sensitive to disturbances caused by the expression of defective genes or by environmental influences. The example of thalidomide illustrates this: the limbs begin to form in the middle of the fourth week of gestation (24–26 days) and have their basic structure in place by the beginning of the sixth week (36 days), which corresponds to the sensitive period of exposure to thalidomide (Figure 5.27).

Question 5.7

No. Twins with separate placentas can arise from either two separate zygotes or from the separation of one zygote into two cells after the first cleavage (Figure 5.28a).

Chapter 6

Question 6.1

Figure 6.15 shows that the head circumference of the fetus is increasing, but initially not as much for its age as the 'average' fetus. However, the data for this fetus still fall within the lower 95% line in Figure 6.2b. Later on in development the rate of growth of the head increases and the curve goes above the line in Figure 6.2b, but keeps within the upper 95% line. The growth of the fetus is therefore proceeding normally.

Question 6.2

(a) An intake of 1 000 kcal per day is not enough for an adult woman, never mind one who is pregnant. Although the mother's body will preferentially give resources to the fetus, probably at the expense of breaking down her own body tissues to do so, it is unlikely that she will be able to maintain the pregnancy without a larger energy intake. There is a high probability of miscarriage, and, if the fetus does survive, it will probably be very small, and possibly deformed.

(b) This energy intake is about right for a (smallish) pregnant woman. She should have no problem in nourishing her fetus, and the baby's birth weight is likely to be normal.

(c) This energy intake is excessive, unless the woman is very large and is physically very active. She is likely to put on more weight than would be expected in pregnancy. However, there are unlikely to be any problems for the fetus in getting enough nourishment, unless the mother's excessive weight gain causes medical problems for her which indirectly have an effect on the fetus. The baby is likely to be of normal size at birth.

Question 6.3

Although, of course, the mother plays a very important role in the growth of the fetus, the growth *patterns* are also influenced by the fetus itself, and, most importantly, by the placenta, which produces many growth factors and hormones.

Question 6.4

Two ways in which the development of the gut and the musculo-skeletal systems are linked are:

(a) in the ability to suck, when tongue, palate, and peristaltic movements must all be coordinated;

(b) in the production of blood and immune system cells, which are made first in the yolk sac, then by the fetal liver, and finally by the bone marrow, once the bones are long enough to contain it.

Question 6.5

The trigger of labour must have one of the following characteristics:

(a) It must be produced for some time before the onset of labour, then must reach a threshold which triggers the process.

(b) It must be produced in a burst just before the onset of labour.

(c) It must have been produced throughout pregnancy as an inhibitor of labour, then be destroyed or inactivated just before the onset of labour.

ACKNOWLEDGEMENTS

We are grateful to Jean MacQueen who prepared the index for this book.

Grateful acknowledgement is made to the following sources for permission to reproduce material in this book:

Figures

Figure 1.3: Courtesy of Griffith, J., from Griffith, J., Huberman, J. A. and Kornberg, A. 1971, 'Electron microscopy of DNA-polymerase bound to DNA', *Journal of Molecular Biology*, **55**, pp. 209–214, Plate 1b; *Figure 2.1*: Source: *Social Trends* 1994, crown copyright 1994, reproduced by permission of the Controller of HMSO and the Central Statistical Office; *Figures 2.2, 2.3:* Excerpted from *High Level Wellness*, copyright © 1986 by Donald B. Ardell with permission from Ten Speed Press, P. O. Box 7123, Berkeley, CA 94707; *Figure 2.5*: Townsend, P., Davidson, N. and Whitehead, M. 1988, *Inequalities in Health: The Black Report and The Health Divide*, Penguin Books Ltd; *Figure 3.1*: Courtesy of Graham Hills; *Figures 3.26, 3.27(b), (c), (d)*: *Molecular Biology of the Cell*, New York, Garland Publishing 1983, Figure 4.6, p. 147; *Figure 3.32*: Copyright © Dorothy F. Bainton; *Figures 3.43(a), 3.44, 3.38(c)*: Alberts, B. *et al.*, 1989, *Molecular Biology of the Cell*, 2nd edn, p. 216, Garland Publishing Inc.; *Figure 3.38(d)*: Driscoll, Youngquist and Baldeschwieler, Caltec/Science Photo Library; *Figures 3.46(a), (b), (c):* Copyright © Dr Michael Stewart, Open University; *Figure 3.52*: Kiecolt-Glaser, J. K., Marucha, P. T., Malarkey, W. B., Mercado, A. M. and Glaser, R. 1995, 'Slowing of wound healing by psychological stress', *The Lancet*, **346**, 4 November 1995, p. 1195, Lancet Ltd; *Figure 4.1:* Strickberger, M. W. 1985, *Genetics*, Figure 1.1, Blackie and Son Ltd; *Figure 4.2*: Adapted from McEvedy, C. and Jones, R. 1978, *Atlas of World Population History*, Penguin Books Ltd; *Figure 4.4(b)*: Copyright © British Museum; *Figures 4.14, 4.16:* Copyright © R. G. Edwards; *Figures 5.1, 5.9, 5.17:* Smith, C. P. W. and Williams, P. L. 1984, *Basic Human Embryology*, 3rd edn, Churchill Livingstone; *Figures 5.2, 5.3, 5.4, 5.5, 5.7, 5.8, 5.10, 5.11, 5.12, 5.13, 5.14, 5.16, 5.19, 5.22, 5.28*: Larsen, W. J. 1993, *Human Embryology*, Churchill Livingstone, New York; *Figures 5.20, 5.21, 5.23, 5.24, 5.25, 5.26*: Gilbert, S. G. 1989, *Pictorial Human Embryology*, University of Washington Press; *Figure 5.27*: Gilbert, F. S. 1994, *Developmental Biology*, 4th edn, p. 639, Sinauer Associates, Inc.

Tables

Table 2.1(a), (b): Cox, B. D. *et al.*, 1987, *The Health and Lifestyle Survey: preliminary report*, Tables 3.1 and 3.2, The Health Promotion Trust; *Table 2.3*: Jones, L. J. 1994, *Working Together for Better Health*, pp. 18–19, The Health of the Nation, Department of Health; *Table 2.4*: Excerpted from *High Level Wellness*, copyright © 1986 by Donald B. Ardell with permission from Ten Speed Press, P. O. Box 7123, Berkeley, CA 94707; *Table 2.6*: Barker, D. J. P. 1992, 'Foetal and infant origins of adult disease', *British Medical Journal*, **304**(6820), p. 178, 18 January 1992, BMJ Publishing; *Table 4.1*: *Population*, Population Concern.

Front cover image: © 1995 Comstock Inc. and © Pictor International; *Back cover image*: © Superstock Ltd, 'Fetal ultrasound at 3.5 months'.

INDEX

Note: Entries in **bold** are key terms. Page numbers in *italics* refer to figures and tables.